Solid State Chemistry: An Introduction

Solid State Chemistry:
An Introduction

Edited by
Aden Crowley

Larsen & Keller
www.larsen-keller.com

Solid State Chemistry: An Introduction
Edited by Aden Crowley
ISBN: 978-1-63549-265-1 (Hardback)

☰ Larsen & Keller

Published by Larsen and Keller Education,
5 Penn Plaza,
19th Floor,
New York, NY 10001, USA

Cataloging-in-Publication Data

Solid state chemistry : an introduction / edited by Aden Crowley.
    p. cm.
Includes bibliographical references and index.
ISBN 978-1-63549-265-1
1. Solid state chemistry. 2. Chemistry, Physical and theoretical. I. Crowley, Aden.
QD478 .S65 2017
541.042 1--dc23

The publisher's policy is to use permanent paper from mills that operate a sustainable forestry policy. Furthermore, the publisher ensures that the text paper and cover boards used have met acceptable environmental accreditation standards.

Printed and bound in the United States of America.

For more information regarding Larsen and Keller Education and its products, please visit the publisher's website www.larsen-keller.com

Table of Contents

Preface

Solid-state chemistry refers to that branch of chemistry which deals with the multidisciplinary study of the properties, synthesis, and structure of various solid-state materials. It is closely linked with other branches of chemistry like crystallography, metallurgy, solid-state physics, ceramics, materials science, thermodynamics, and electronics, etc. This book explores all the important aspects of solid-state chemistry in the present day scenario. The topics covered in this extensive text deal with the core subjects of the subject. Fundamental approaches, evaluations and methodologies have been included in it. This textbook is a complete source of knowledge on the present status of this important field.

Given below is the chapter wise description of the book:

Chapter 1- Solid-state chemistry studies structures and properties of solid materials. This subject has similarities with mineralogy, crystallography, ceramics and metallurgy. This chapter will provide an integrated understanding of solid-state chemistry.

Chapter 2- The key concepts of solid-state chemistry are nanocomposites, hybrid materials, liquid crystals, radioactive decays etc. Nanocomposites are solid materials which comprise of less than 100 nanometers. The text strategically encompasses and incorporates the major components and key concepts of solid-state chemistry, providing a complete understanding.

Chapter 3- Solids can be characterized as per the nature of bonding within their atomic components. These bonds can be distinguished into four types which are covalent bonding, ionic bonding, metallic bonding and weak inter molecular bonding. This text is an overview of the subject matter incorporating all the major aspects of the bonds found in solids.

Chapter 4- A crystal is a solid material. The components of a crystal are always arranged in a systematic structure. Crystal, crystal structure, crystal twinning, atomic packing factor and crystal structures in the periodic table are some of the topics elucidated in this section. The chapter serves as a source to understand the basic concept of crystals.

Chapter 5- X-ray crystallography is a method that is used to determine the atomic and molecular structure of a crystal. The methods that are used for characterizing solid apart from x-ray crystallography are neutron diffraction, electron microscope, X-ray absorption spectroscopy, powder diffraction etc. This section discusses the methods of characterizing solids in a critical manner providing key analysis to the subject matter.

Chapter 6- Salt is an ionic compound; there are different varieties of salts. The chapter also focuses on bresle method, acid salts, alkali salts, soap and alum. This section has been carefully written to provide an easy understanding of all the significant aspects of salt and its various applications.

Chapter 7- Electronic band structures are a spectrum of energy that determines the range of modification that electrons undergo when using quantum energy. In order to completely understand electronic band structure it is necessary to understand the process related to it. Some of the topics included in this chapter are band gap, Fermi level, tight binding and nearly free electron model.

At the end, I would like to thank all those who dedicated their time and efforts for the successful completion of this book. I also wish to convey my gratitude towards my friends and family who supported me at every step.

Editor

Introduction to Solid State Chemistry

Solid-state chemistry studies structures and properties of solid materials. This subject has similarities with mineralogy, crystallography, ceramics and metallurgy. This chapter will provide an integrated understanding of solid-state chemistry.

Solid-state Chemistry

Solid-state chemistry, also sometimes referred to as materials chemistry, is the study of the synthesis, structure, and properties of solid phase materials, particularly, but not necessarily exclusively of, non-molecular solids. It therefore has a strong overlap with solid-state physics, mineralogy, crystallography, ceramics, metallurgy, thermodynamics, materials science and electronics with a focus on the synthesis of novel materials and their characterization.

History

Because of its direct relevance to products of commerce, solid state inorganic chemistry has been strongly driven by technology. Progress in the field has often been fueled by the demands of industry, well ahead of purely academic curiosity. Applications discovered in the 20th century include zeolite and platinum-based catalysts for petroleum processing in the 1950s, high-purity silicon as a core component of microelectronic devices in the 1960s, and "high temperature" superconductivity in the 1980s. The invention of X-ray crystallography in the early 1900s by William Lawrence Bragg enabled further innovation. Our understanding of how reactions proceed at the atomic level in the solid state was advanced considerably by Carl Wagner's work on oxidation rate theory, counter diffusion of ions, and defect chemistry. Because of this, he has sometimes been referred to as the *father of solid state chemistry*.

Synthetic Methods

Given the diversity of solid state compounds, an equally diverse array of methods are used for their preparation. For organic materials, such as charge transfer salts, the methods operate near room temperature and are often similar to the techniques of organic synthesis. Redox reactions are sometimes conducted by electrocrystallisation, as illustrated by the preparation of the Bechgaard salts from tetrathiafulvalene.

Oven Techniques

For thermally robust materials, high temperature methods are often employed. For example, bulk solids are prepared using tube furnaces, which allow reactions to be conducted up to ca. 1100 °C. Special equipment e.g. ovens consisting of a tantalum tube through which an electric current is passed can be used for even higher temperatures up to 2000 °C. Such high temperatures are at times required to induce diffusion of the reactants, but this depends strongly on the system studied. Some solid state reactions already proceed at temperatures as low as 100 °C.

Melt Methods

One method often employed is to melt the reactants together and then later anneal the solidified melt. If volatile reactants are involved the reactants are often put in an ampoule that is evacuated -often while keeping the reactant mixture cold e.g. by keeping the bottom of the ampoule in liquid nitrogen- and then sealed. The sealed ampoule is then put in an oven and given a certain heat treatment.

Solution Methods

It is possible to use solvents to prepare solids by precipitation or by evaporation. At times the solvent is used hydrothermally, i.e. under pressure at temperatures higher than the normal boiling point. A variation on this theme is the use of flux methods, where a salt of relatively low melting point is added to the mixture to act as a high temperature solvent in which the desired reaction can take place.

Gas Reactions

Many solids react vigorously with reactive gas species like chlorine, iodine, oxygen etc. Others form adducts with other gases, e.g. CO or ethylene. Such reactions are often carried out in a tube that is open ended on both sides and through which the gas is passed. A variation of this is to let the reaction take place inside a measuring device such as a TGA. In that case stoichiometric information can be obtained during the reaction, which helps identify the products.

A special case of a gas reaction is a chemical transport reaction. These are often carried out in a sealed ampoule to which a small amount of a transport agent, e.g. iodine is added. The ampoule is then placed in a zone oven. This is essentially two tube ovens attached to each other which allows a temperature gradient to be imposed. Such a method can be used to obtain the product in the form of single crystals suitable for structure determination by X-ray diffraction.

Chemical vapour deposition is a high temperature method that is widely employed for the preparation of coatings and semiconductors from molecular precursors.

Air and Moisture Sensitive Materials

Many solids are hygroscopic and/or oxygen sensitive. Many halides e.g. are very 'thirsty' and can only be studied in their anhydrous form if they are handled in a glove box filled with dry (and/or oxygen-free) gas, usually nitrogen.

Characterization

New Phases, Phase Diagrams, Structures

The synthetic methodology and the characterization of the product often go hand in hand in the sense that not one but a series of reaction mixtures are prepared and subjected to heat treatment. The stoichiometry is typically *varied* in a systematic way to find which stoichiometries will lead to new solid compounds or to solid solutions between known ones. A prime method to characterize the reaction products is powder diffraction, because many solid state reactions will produce poly-cristalline ingots or powders. Powder diffraction will facilitate the identification of known phases in the mixture. If a pattern is found that is not known in the diffraction data libraries an attempt can be made to index the pattern, i.e. to identify the symmetry and the size of the unit cell. (If the product is not crystalline the characterization is typically much more difficult.)

Once the unit cell of a new phase is known, the next step is to establish the stoichiometry of the phase. This can be done in a number of ways. Sometimes the composition of the original mixture will give a clue, if one finds only one product -a single powder pattern- or if one was trying to make a phase of a certain composition by analogy to known materials but this is rare. Often considerable effort in refining the synthetic methodology is required to obtain a pure sample of the new material. If it is possible to separate the product from the rest of the reaction mixture elemental analysis can be used. Another way involves SEM and the generation of characteristic X-rays in the electron beam. The easiest way to solve the structure is by using single crystal X-ray diffraction.

The latter often requires *revisiting* and refining the preparative procedures and that is linked to the question which phases are stable at what composition and what stoichiometry. In other words, what does the phase diagram looks like. An important tool in establishing this is thermal analysis techniques like DSC or DTA and increasingly also, thanks to the advent of synchrotrons temperature-dependent powder diffraction. Increased knowledge of the phase relations often leads to further refinement in synthetic procedures in an iterative way. New phases are thus characterized by their melting points and their stoichiometric domains. The latter is important for the many solids that are non-stoichiometric compounds. The cell parameters obtained from XRD are particularly helpful to characterize the homogeneity ranges of the latter.

Key Concepts of Solid State Chemistry

The key concepts of solid-state chemistry are nanocomposites, hybrid materials, liquid crystals, radioactive decays etc. Nanocomposites are solid materials which comprise of less than 100 nanometers. The text strategically encompasses and incorporates the major components and key concepts of solid-state chemistry, providing a complete understanding.

Nanocomposite

Nanocomposite is a multiphase solid material where one of the phases has one, two or three dimensions of less than 100 nanometers (nm), or structures having nano-scale repeat distances between the different phases that make up the material. In the broadest sense this definition can include porous media, colloids, gels and copolymers, but is more usually taken to mean the solid combination of a bulk matrix and nano-dimensional phase(s) differing in properties due to dissimilarities in structure and chemistry. The mechanical, electrical, thermal, optical, electrochemical, catalytic properties of the nanocomposite will differ markedly from that of the component materials. Size limits for these effects have been proposed, <5 nm for catalytic activity, <20 nm for making a hard magnetic material soft, <50 nm for refractive index changes, and <100 nm for achieving superparamagnetism, mechanical strengthening or restricting matrix dislocation movement.

Nanocomposites are found in nature, for example in the structure of the abalone shell and bone. The use of nanoparticle-rich materials long predates the understanding of the physical and chemical nature of these materials. Jose-Yacaman *et al.* investigated the origin of the depth of colour and the resistance to acids and bio-corrosion of Maya blue paint, attributing it to a nanoparticle mechanism. From the mid-1950s nanoscale organo-clays have been used to control flow of polymer solutions (e.g. as paint viscosifiers) or the constitution of gels (e.g. as a thickening substance in cosmetics, keeping the preparations in homogeneous form). By the 1970s polymer/clay composites were the topic of textbooks, although the term "nanocomposites" was not in common use.

In mechanical terms, nanocomposites differ from conventional composite materials due to the exceptionally high surface to volume ratio of the reinforcing phase and/or its exceptionally high aspect ratio. The reinforcing material can be made up of particles (e.g. minerals), sheets (e.g. exfoliated clay stacks) or fibres (e.g. carbon nanotubes or electrospun fibres). The area of the interface between the matrix and reinforcement phase(s) is typically an order of magnitude greater than for conventional composite materials. The matrix material properties are significantly affected in the vicinity of the reinforcement. Ajayan *et al.* note that with polymer nanocomposites, properties related to local chemistry, degree of thermoset cure, polymer chain mobility, polymer chain conformation, degree of polymer chain ordering or crystallinity can all vary significantly and continuously from the interface with the reinforcement into the bulk of the matrix.

This large amount of reinforcement surface area means that a relatively small amount of nanoscale reinforcement can have an observable effect on the macroscale properties of the composite. For example, adding carbon nanotubes improves the electrical and thermal conductivity. Other kinds of nanoparticulates may result in enhanced optical properties, dielectric properties, heat resistance or mechanical properties such as stiffness, strength and resistance to wear and damage. In general, the nano reinforcement is dispersed into the matrix during processing. The percentage by weight (called *mass fraction*) of the nanoparticulates introduced can remain very low (on the order of 0.5% to 5%) due to the low filler percolation threshold, especially for the most commonly used non-spherical, high aspect ratio fillers (e.g. nanometer-thin platelets, such as clays, or nanometer-diameter cylinders, such as carbon nanotubes). The orientation and arrangement of asymmetric nanoparticles, thermal property mismatch at the interface, interface density per unit volume of nanocomposite, and polydispersity of nanoparticles significantly affect the effective thermal conductivity of nanocomposites.

Ceramic-matrix Nanocomposites

In this group of composites the main part of the volume is occupied by a ceramic, i.e. a chemical compound from the group of oxides, nitrides, borides, silicides etc.. In most cases, ceramic-matrix nanocomposites encompass a metal as the second component. Ideally both components, the metallic one and the ceramic one, are finely dispersed in each other in order to elicit the particular nanoscopic properties. Nanocomposite from these combinations were demonstrated in improving their optical, electrical and magnetic properties as well as tribological, corrosion-resistance and other protective properties.

The binary phase diagram of the mixture should be considered in designing ceramic-metal nanocomposites and measures have to be taken to avoid a chemical reaction between both components. The last point mainly is of importance for the metallic component that may easily react with the ceramic and thereby lose its metallic character. This is not an easily obeyed constraint, because the preparation of the ceramic component generally requires high process temperatures. The most safe measure thus is to carefully choose immiscible metal and ceramic phases. A good example for such a combination is represented by the ceramic-metal composite of TiO_2 and Cu, the mixtures of which were found immiscible over large areas in the Gibbs' triangle of Cu-O-Ti.

The concept of ceramic-matrix nanocomposites was also applied to thin films that are solid layers of a few nm to some tens of µm thickness deposited upon an underlying substrate and that play an important role in the functionalization of technical surfaces. Gas flow sputtering by the hollow cathode technique turned out as a rather effective technique for the preparation of nanocomposite layers. The process operates as a vacuum-based deposition technique and is associated with high deposition rates up to some µm/s and the growth of nanoparticles in the gas phase. Nanocomposite layers in the ceramics range of composition were prepared from TiO_2 and Cu by the hollow cathode technique that showed a high mechanical hardness, small coefficients of friction and a high resistance to corrosion.

Metal-matrix Nanocomposites

Metal matrix nanocomposites can also be defined as reinforced metal matrix composites. This type of composites can be classified as continuous and non-continuous reinforced materials. One of the more important nanocomposites is Carbon nanotube metal matrix composites, which is an emerging new material that is being developed to take advantage of the high tensile strength and electrical conductivity of carbon nanotube materials. Critical to the realization of CNT-MMC possessing optimal properties in these areas are the development of synthetic techniques that are (a) economically producible, (b) provide for a homogeneous dispersion of nanotubes in the metallic matrix, and (c) lead to strong interfacial adhesion between the metallic matrix and the carbon nanotubes. In addition to carbon nanotube metal matrix composites, boron nitride reinforced metal matrix composites and carbon nitride metal matrix composites are the new research areas on metal matrix nanocomposites.

A recent study, comparing the mechanical properties (Young's modulus, compressive yield strength, flexural modulus and flexural yield strength) of single- and multi-walled reinforced polymeric (polypropylene fumarate—PPF) nanocomposites to tungsten disulfide nanotubes reinforced PPF nanocomposites suggest that tungsten disulfide nanotubes reinforced PPF nanocomposites possess significantly higher mechanical properties and tungsten disulfide nanotubes are better reinforcing agents than carbon nanotubes. Increases in the mechanical properties can be attributed to a uniform dispersion of inorganic nanotubes in the polymer matrix (compared to carbon nanotubes that exist as micron sized aggregates) and increased crosslinking density of the polymer in the presence of tungsten disulfide nanotubes (increase in crosslinking density leads to an increase in the mechanical properties). These results suggest that inorganic nanomaterials, in general, may be better reinforcing agents compared to carbon nanotubes.

Another kind of nanocomposite is the energetic nanocomposite, generally as a hybrid sol–gel with a silica base, which, when combined with metal oxides and nano-scale aluminum powder, can form *superthermite* materials.

Polymer-matrix Nanocomposites

In the simplest case, appropriately adding nanoparticulates to a polymer matrix can enhance its performance, often dramatically, by simply capitalizing on the nature and properties of the nanoscale filler (these materials are better described by the term *nanofilled polymer composites*). This strategy is particularly effective in yielding high performance composites, when good dispersion of the filler is achieved and the properties of the nanoscale filler are substantially different or better than those of the matrix.

Nanoparticles such as graphene, carbon nanotubes, molybdenum disulfide and tungsten disulfide are being used as reinforcing agents to fabricate mechanically strong biodegradable polymeric nanocomposites for bone tissue engineering applications. The addition of these nanoparticles in the polymer matrix at low concentrations (~0.2 weight %) cause significant improvements in the compressive and flexural mechanical properties of polymeric nanocomposites. Potentially, these nanocomposites may be used as a novel, mechanically strong, light weight composite as bone implants. The results suggest that mechanical reinforcement is dependent on the nanostructure morphology, defects, dispersion of nanomaterials in the polymer matrix, and the cross-linking density of the polymer. In general, two-dimensional nanostructures can reinforce the polymer better than one-dimensional nanostructures, and inorganic nanomaterials are better reinforcing agents than carbon based nanomaterials. In addition to mechanical properties, polymer nanocomposites based on carbon nanotubes or graphene have been used to enhance a wide range of properties, giving rise to functional materials for a wide range of high added value applications in fields such as energy conversion and storage, sensing and biomedical tissue engineering. For example, multi-walled carbon nanotubes based polymer nanocomposites have been used for the enhancement of the electrical conductivity.

Nanoscale dispersion of filler or controlled nanostructures in the composite can introduce new physical properties and novel behaviors that are absent in the unfilled matrices. This effectively changes the nature of the original matrix (such composite materials can be better described by the term *genuine nanocomposites or hybrids*). Some examples of such new properties are fire resistance or flame retardancy, and accelerated biodegradability.

A range of polymeric nanocomposites are used for biomedical applications such as tissue engineering, drug delivery, cellular therapies. Due to unique interactions between polymer and nanoparticles, a range of property combinations can be engineered to mimic native tissue structure and properties. A range of natural and synthetic polymers are used to design polymeric nanocomposites for biomedical applications including starch, cellulose, alginate, chitosan, collagen, gelatin, and fibrin, poly(vinyl alcohol) (PVA), poly(ethylene glycol) (PEG), poly(caprolactone) (PCL), poly(lactic-co-glycolic acid) (PLGA), and poly(glycerol sebacate) (PGS). A range of nanoparticles including ceramic, polymeric, metal oxide and carbon-based nanomaterials are incorporated within polymeric network to obtain desired property combinations.

Hybrid Material

Hybrid materials are composites consisting of two constituents at the nanometer or molecular level. Commonly one of these compounds is inorganic and the other one organic in nature. Thus, they differ from traditional composites where the constituents are at the macroscopic (micrometer to millimeter) level. Mixing at the microscopic scale leads to a more homogeneous material that either show characteristics in between the two original phases or even new properties.

Introduction

Hybrid Materials in Nature

Many natural materials consist of inorganic and organic building blocks distributed on the nanoscale. In most cases the inorganic part provides mechanical strength and an overall structure to the natural objects while the organic part delivers bonding between the inorganic building blocks and/or the soft tissue. Typical examples of such materials are bone, or nacre.

Development of Hybrid Materials

The first hybrid materials were the paints made from inorganic and organic components that were used thousands of years ago. Rubber is an example of the use of inorganic materials as fillers for organic polymers. The sol–gel process developed in the 1930s was one of the major driving forces what has become the broad field of inorganic–organic hybrid materials.

Classification

Hybrid materials can be classified based on the possible interactions connecting the inorganic and organic species. *Class I* hybrid materials are those that show weak interactions between the two phases, such as van der Waals, hydrogen bonding or weak electrostatic interactions. *Class II* hybrid materials are those that show strong chemical interactions between the components such as covalent bonds.

Structural properties can also be used to distinguish between various hybrid materials. An organic moiety containing a functional group that allows the attachment to an inorganic network, e.g. a trialkoxysilane group, can act as a *network modifier* because in the final structure the inorganic network is only modified by the organic group. Phenyltrialkoxysilanes are an example for such compounds; they modify the silica network in the sol–gel process via the reaction of the trialkoxysilane group without supplying additional functional groups intended to undergo further chemical reactions to the material formed. If a reactive functional group is incorporated the system is called a *network functionalizer*. The situation is different if two or three of such anchor groups modify an organic segment; this leads to materials in which the inorganic group is afterwards an integral part of the hybrid network. The latter type of system is known as *network builder*

Blends are formed if no strong chemical interactions exist between the inorganic and organic building blocks. One example for such a material is the combination of inorganic clusters or particles with organic polymers lacking a strong (e.g. covalent) interaction between the components. In this case a material is formed that consists for example of an organic polymer with entrapped discrete inorganic moieties in which, depending on the functionalities of the components, for example weak crosslinking occurs by the entrapped inorganic units through physical interactions or the inorganic components are entrapped in a crosslinked polymer matrix. If an inorganic and an organic network interpenetrate each other without strong chemical interactions, so called interpenetrating networks (IPNs) are formed, which is for example the case if a sol–gel material is formed in presence of an organic polymer or vice versa. Both materials described belong to class I hybrids. Class II hybrids are formed when the discrete inorganic building blocks, e.g. clusters, are covalently bonded to the organic polymers or inorganic and organic polymers are covalently connected with each other.

Distinction between Nanocomposites and Hybrid Materials

The term nanocomposite is used if the combination of organic and inorganic structural units yield a material with composite properties. That is to say that the original properties of the separate organic and inorganic components are still present in the composite and are unchanged by mixing these materials. However, if a new property emerges from the intimate mixture, then the material becomes a hybrid. A macroscopic example is the mule, which is more suited for hard work than either of its parents, the horse and the donkey. The size of the individual components and the nature of their interaction (covalent, electrostatic, etc.) do not enter into the definition of a hybrid material.

Advantages of Hybrid Materials Over Traditional Composites

- Inorganic clusters or nanoparticles with specific optical, electronic or magnetic properties can be incorporated in organic polymer matrices.

- Contrary to pure solid state inorganic materials that often require a high temperature treatment for their processing, hybrid materials show a more polymer-like handling, either because of their large organic content or because of the formation of crosslinked inorganic networks from small molecular precursors just like in polymerization reactions.

- Light scattering in homogeneous hybrid material can be avoided and therefore optical transparency of the resulting hybrid materials and nanocomposites can be achieved.

Synthesis

Two different approaches can be used for the formation of hybrid materials: Either well-defined preformed building blocks are applied that react with each other to form the final hybrid material in which the precursors still at least partially keep their original integrity or one or both structural units are formed from the precursors that are transformed into a new (network) structure. It is important that the interface between the inorganic and organic materials which has to be tailored to overcome serious problems in the preparation of hybrid materials. Different building blocks and approaches can be used for their preparation and these have to be adapted to bridge the differences of inorganic and organic materials.

Building Block Approach

Building blocks at least partially keep their molecular integrity throughout the material formation, which means that structural units that are present in these sources for materials formation can also be found in the final material. At the same time typical properties of these building blocks usually survive the matrix formation, which is not the case if material precursors are transferred into novel materials. Representative examples of such well-defined building blocks are modified inorganic clusters or nanoparticles with attached reactive organic groups.

Cluster compounds often consist of at least one functional group that allows an interaction with an organic matrix, for example by copolymerization. Depending on the number of groups that can interact, these building blocks are able to modify an organic matrix (one functional group) or form partially or fully crosslinked materials (more than one group). For instance, two reactive groups can lead to the formation of chain structures. If the building blocks contain at least three reactive groups they can be used without additional molecules for the formation of a crosslinked material.

Beside the molecular building blocks mentioned, nanosized building blocks, such as particles or nanorods, can also be used to form nanocomposites. The building block approach has one large advantage compared with the in situ formation of the inorganic or organic entities: because at least one structural unit (the building block) is well-defined and usually does not undergo significant structural changes during the matrix formation, better structure–property predictions are possible. Furthermore, the building blocks can be designed in such a way to give the best performance in the materials' formation, for example good solubility of inorganic compounds in organic monomers by surface groups showing a similar polarity as the monomers.

In recent years many building blocks have been synthesized and used for the preparation of hybrid materials. Chemists can design these compounds on a molecular scale with highly sophisticated methods and the resulting systems are used for the formation of functional hybrid materials. Many future applications, in particular in nanotechnology, focus on a bottom-up approach in which complex structures are hierarchically formed by these small building blocks. This idea is also one of the driving forces of the building block approach in hybrid materials.

In Situ Formation of the Components

The in situ formation of the hybrid materials is based on the chemical transformation of the precursors used throughout materials' preparation. Typically this is the case if organic polymers are formed but also if the sol–gel process is applied to produce the inorganic component. In these cases well-defined discrete molecules are transformed to multidimensional structures, which often show totally different properties from the original precursors. Generally simple, commercially available molecules are applied and the internal structure of the final material is determined by the composition of these precursors but also by the reaction conditions. Therefore control over the latter is a crucial step in this process. Changing one parameter can often lead to two very different materials. If, for example, the inorganic species is a silica derivative formed by the sol–gel process, the change from base to acid catalysis makes a large difference because base catalysis leads to a more particle-like microstructure while acid catalysis leads to a polymer-like microstructure. Hence, the final performance of the derived materials is strongly dependent on their processing and its optimization.

In Situ Formation of Inorganic Materials

Many of the classical inorganic solid state materials are formed using solid precursors and high temperature processes, which are often not compatible with the presence of organic groups because they are decomposed at elevated temperatures. Hence, these high temperature processes are not suitable for the in situ formation of hybrid materials. Reactions that are employed should have more the character of classical covalent bond formation in solutions. One of the most prominent processes which fulfill these demands is the sol–gel process. However, such rather low temperature processes often do not lead to the thermodynamically most stable structure but to kinetic products, which has some implications for the structures obtained. For example low temperature derived inorganic materials are often amorphous or crystallinity is only observed on a very small length scale, i.e. the nanometer range. An example of the latter is the formation of metal nanoparticles in organic or inorganic matrices by reduction of metal salts or organometallic precursors.

Some methods of in situ formation of inorganic materials are:

- Sol-gel process

- Nonhydrolytic sol–gel process

- Sol–gel reactions of non-silicates

Formation of Organic Polymers in Presence of Preformed Inorganic Materials

If the organic polymerization occurs in the presence of an inorganic material to form the hybrid material one has to distinguish between several possibilities to overcome the incompatibility of the two species. The inorganic material can either have no surface functionalization but the bare material surface; it can be modified with nonreactive organic groups (e.g. alkyl chains); or it can contain reactive surface groups such as polymerizable functionalities. Depending on these prerequisites the material can be pretreated, for example a pure inorganic surface can be treated with surfactants or silane coupling agents to make it compatible with the organic monomers, or functional monomers can be added that react with the surface of the inorganic material. If the inorganic component has nonreactive organic groups attached to its surface and it can be dissolved in a monomer which is subsequently polymerized, the resulting material after the organic polymerization, is a blend. In this case the inorganic component interact only weakly or not at all with the organic polymer; hence, a class I material is formed. Homogeneous materials are only obtained in this case if agglomeration of the inorganic components in the organic environment is prevented. This can be achieved if the interactions between the inorganic components and the monomers are better or at least the same as between the inorganic components. However, if no strong chemical interactions are formed, the long-term stability of a once homogeneous material is questionable because of diffusion effects in the resulting hybrid material. The stronger the respective interaction between the components, the more stable is the final material. The strongest interaction is achieved if class II materials are formed, for example with covalent interactions.

Hybrid Materials by Simultaneous Formation of Both Components

Simultaneous formation of the inorganic and organic polymers can result in the most homogeneous type of interpenetrating networks. Usually the precursors for the sol–gel process are mixed with monomers for the organic polymerization and both processes are carried out at the same time with or without solvent. Applying this method, three processes are competing with each other:

(a) the kinetics of the hydrolysis and condensation forming the inorganic phase,

(b) the kinetics of the polymerization of the organic phase, and

(c) the thermodynamics of the phase separation between the two phases.

By tailoring the kinetics of the two polymerizations in such a way that they occur simultaneously and rapidly enough, phase separation is avoided or minimized. Additional parameters such as attractive interactions between the two moieties, as described above can also be used to avoid phase separation.

One problem that also arises from the simultaneous formation of both networks is the sensitivity of many organic polymerization processes for sol–gel conditions or the composition of the materials formed. Ionic polymerizations, for example, often interact with the precursors or intermediates formed in the sol–gel process. Therefore, they are not usually applied in these reactions.

Madelung Constant

The Madelung constant is used in determining the electrostatic potential of a single ion in a crystal by approximating the ions by point charges. It is named after Erwin Madelung, a German physicist.

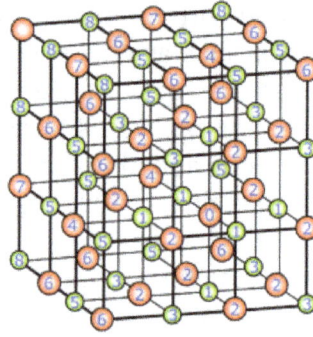

The Madelung constant being calculated for the NaCl ion labeled o in the expanding spheres method. Each number designates the order in which it is summed. Note that in this case, the sum is divergent, but there are methods for summing it which give a converging series.

Because the anions and cations in an ionic solid are attracting each other by virtue of their opposing charges, separating the ions requires a certain amount of energy. This energy must be given to the system in order to break the anion-cation bonds. The energy required to break these bonds for one mole of an ionic solid under standard conditions is the lattice energy.

Formal Expression

The Madelung constant shall allow for the calculation of the electric potential V_i of all ions of the lattice felt by the ion at position r_i

$$V_i = \frac{e}{4\pi\epsilon_0} \sum_{j\neq i} \frac{z_j}{r_{ij}}$$

where $r_{ij} = |r_i - r_j|$ is the distance between the ith and the jth ion. In addition,

z_j = number of charges of the jth ion

$e = 1.6022\times10^{-19}$ C

$4\pi\epsilon_0 = 1.112\times10^{-10}$ C^2/(J m).

If the distances r_{ij} are normalized to the nearest neighbor distance r_0 the potential may be written

$$V_i = \frac{e}{4\pi\epsilon_0 r_0} \sum_j \frac{z_j r_0}{r_{ij}} = \frac{e}{4\pi\epsilon_0 r_0} M_i$$

with M_i being the (dimensionless) Madelung constant of the ith ion

$$M_i = \sum_j \frac{z_j}{r_{ij}/r_0}.$$

The electrostatic energy of the ion at site r_i then is the product of its charge with the potential acting at its site

$$E_{el,i} = z_i e V_i = \frac{e^2}{4\pi\epsilon_0 r_0} z_i M_i.$$

There occur as many Madelung constants M_i in a crystal structure as ions occupy different lattice sites. For example, for the ionic crystal NaCl, there arise two Madelung constants – one for Na and another for Cl. Since both ions, however, occupy lattice sites of the same symmetry they both are of the same magnitude and differ only by sign. The electrical charge of the Na$^+$ and Cl$^-$ ion are assumed to be one-fold positive and negative, respectively, $z_{Na} = 1$ and $z_{Cl} = -1$. The nearest neighbour distance amounts to half the lattice parameter of the cubic unit cell $r_0 = a / 2$ and the Madelung constants become

$$M_{\text{Na}} = -M_{\text{Cl}} = \sum_{j,k,\ell=-\infty}^{\infty}{}' \frac{(-1)^{j+k+\ell}}{(j^2 + k^2 + \ell^2)^{1/2}}.$$

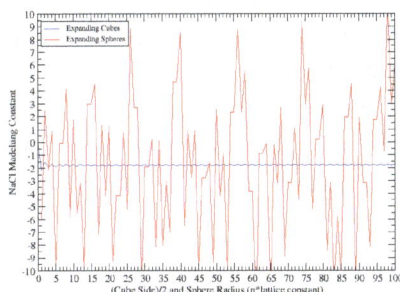

Madelung Constant for Expanding Spheres vs Expanding Cubes

This graph demonstrates the non-convergence of the expanding spheres method for calculating the Madelung constant for NaCl as compared to the expanding cubes method, which is convergent.

The prime indicates that the term $j = k = \ell = 0$ is to be left out. Since this sum is conditionally convergent it is not suitable as definition of Madelung's constant unless the order of summation is also specified. There are two "obvious" methods of summing this series, by expanding cubes or expanding spheres. The latter, though devoid of a meaningful physical interpretation (there are no spherical crystals) is rather popular because of its simplicity. Thus, the following expansion is often found in the literature:

$$M = -6 + 12/\sqrt{2} - 8/\sqrt{3} + 6/2 - 24/\sqrt{5} + \cdots = -1.74756\ldots.$$

However, this is wrong as this series diverges as was shown by Emersleben in 1951. The summation over expanding cubes converges to the correct value. An unambiguous mathematical definition is given by Borwein, Borwein and Taylor by means of analytic continuation of an absolutely convergent series.

There are many practical methods for calculating Madelung's constant using either direct summation (for example, the Evjen method) or integral transforms, which are used in the Ewald method.

Examples of Madelung Constants		
Ion in crystalline compound	M **(based on r_0)**	$\bar{M}$ **(based on w))**
Cl$^-$ and Na$^+$ in rocksalt NaCl	±1.748	±3.495
S^{2-} and Zn^{2+} in sphalerite ZnS	±1.638	±3.783
S$^-$ in pyrite FeS$_2$	...	1.957
Fe^{2+} in pyrite FeS$_2$	...	-7.458

Generalization

It is assumed for the calculation of Madelung constants that an ion's charge density may be approximated by a point charge. This is allowed, if the electron distribution of the ion is spherically symmetric. In particular cases, however, when the ions reside on lattice site of certain crystallographic point groups, the inclusion of higher order moments, i.e. multipole moments of the charge density might be required. It is shown by electrostatics that the interaction between two point charges only accounts for the first term of a general Taylor series describing the interaction between two charge distributions of arbitrary shape. Accordingly, the Madelung constant only represents the monopole-monopole term.

The electrostatic interaction model of ions in solids has thus been extended to a point multipole concept that also includes higher multipole moments like dipoles, quadrupoles etc. These concepts require the determination of higher order Madelung constants or so-called electrostatic lattice constants. In their case, instead of the nearest neighbor distance r_0 another standard length like the cube root of the unit cell volume $w = \sqrt[3]{V}$ is appropriately used for purposes of normalization. For instance, the Madelung constant then reads

$$\overline{M}_i = \sum_j \frac{z_j}{r_{ij} / w}.$$

The proper calculation of electrostatic lattice constants has to consider the crystallographic point groups of ionic lattice sites; for instance, dipole moments may only arise on polar lattice sites, i. e. exhibiting a C_1, C_{1h}, C_n or C_{nv} site symmetry (n = 2, 3, 4 or 6). These second order Madelung constants turned out to have significant effects on the lattice energy and other physical properties of heteropolar crystals.

Application to Organic Salts

The Madelung Constant is also a useful quantity in describing the lattice energy of organic salts. Izgorodina and coworkers have described a generalised method (called the EUGEN method) of calculating the Madelung constant for any crystal structure.

Quasicrystal

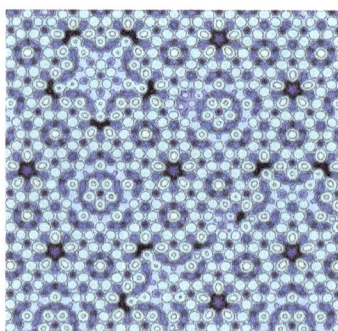

Potential energy surface for silver depositing on an aluminium-palladium-manganese (Al-Pd-Mn) quasicrystal surface. Similar to Fig. 6 in Ref.

A quasiperiodic crystal, or quasicrystal, is a structure that is ordered but not periodic. A quasicrystalline pattern can continuously fill all available space, but it lacks translational symmetry. While crystals, according to the classical crystallographic restriction theorem, can possess only two, three, four, and six-fold rotational symmetries, the Bragg diffraction pattern of quasicrystals shows sharp peaks with other symmetry orders, for instance five-fold.

Aperiodic tilings were discovered by mathematicians in the early 1960s, and, some twenty years later, they were found to apply to the study of quasicrystals. The discovery of these aperiodic forms in nature has produced a paradigm shift in the fields of crystallography. Quasicrystals had been investigated and observed earlier, but, until the 1980s, they were disregarded in favor of the prevailing views about the atomic structure of matter. In 2009, after a dedicated search, a mineralogical finding, icosahedrite, offered evidence for the existence of natural quasicrystals.

Roughly, an ordering is non-periodic if it lacks translational symmetry, which means that a shifted copy will never match exactly with its original. The more precise mathematical definition is that there is never translational symmetry in more than $n - 1$ linearly independent directions, where n is the dimension of the space filled, e.g., the three-dimensional tiling displayed in a quasicrystal may have translational symmetry in two dimensions. The ability to diffract comes from the existence of an indefinitely large number of elements with a regular spacing, a property loosely described as long-range order. Experimentally, the aperiodicity is revealed in the unusual symmetry of the diffraction pattern, that is, symmetry of orders other than two, three, four, or six. In 1982 materials scientist Dan Shechtman observed that certain aluminium-manganese alloys produced the unusual diffractograms which today are seen as revelatory of quasicrystal structures. Due to fear of the scientific community's reaction, it took him two years to publish the results for which he was awarded the Nobel Prize in Chemistry in 2011.

History

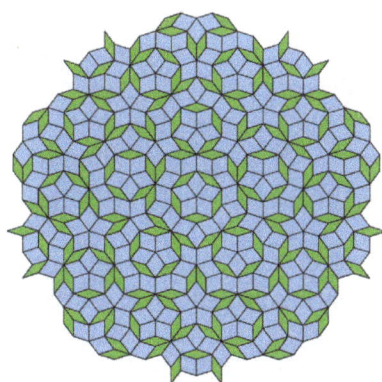

A Penrose tiling

In 1961, Hao Wang asked whether determining if a set of tiles admits a tiling of the plane is an algorithmically unsolvable problem or not. He conjectured that it is solvable, relying on the hypothesis that every set of tiles that can tile the plane can do it *periodically* (hence, it would suffice to try to tile bigger and bigger patterns until obtaining one that tiles periodically). Nevertheless, two years later, his student Robert Berger constructed a set of some 20,000 square tiles (now called Wang tiles) that can tile the plane but not in a periodic fashion. As further aperiodic sets of tiles were discovered, sets with fewer and fewer shapes were found. In 1976 Roger Penrose discovered a set of just two tiles, now referred to as Penrose tiles, that produced only non-periodic tilings of the plane. These tilings displayed instances of fivefold symmetry. One year later Alan Mackay showed experimentally that the diffraction pattern from the Penrose tiling had a two-dimensional Fourier transform consisting of sharp 'delta' peaks arranged in a fivefold symmetric pattern. Around the same time Robert Ammann created a set of aperiodic tiles that produced eightfold symmetry.

Mathematically, quasicrystals have been shown to be derivable from a general method that treats them as projections of a higher-dimensional lattice. Just as circles, ellipses, and hyperbolic curves in the plane can be obtained as sections from a three-dimensional double cone, so too various (aperiodic or periodic) arrangements in two and three dimensions can be obtained from postulated hyperlattices with four or more dimensions. Icosahedral quasicrystals in three dimensions were projected from a six-dimensional hypercubic lattice by Peter Kramer and Roberto Neri in 1984. The tiling is formed by two tiles with rhombohedral shape.

Shechtman first observed ten-fold electron diffraction patterns in 1982, as described in his notebook. The observation was made during a routine investigation, by electron microscopy, of a rapidly cooled alloy of aluminium and manganese prepared at the US National Bureau of Standards (later NIST).

In the summer of the same year Shechtman visited Ilan Blech and related his observation to him. Blech responded that such diffractions had been seen before. Around that time, Shechtman also related his finding to John Cahn of NIST who did not offer any explanation and challenged him to solve the observation. Shechtman quoted Cahn as saying: "Danny, this material is telling us something and I challenge you to find out what it is".

The observation of the ten-fold diffraction pattern lay unexplained for two years until the spring of 1984, when Blech asked Shechtman to show him his results again. A quick study of Shechtman's results showed that the common explanation for a ten-fold symmetrical diffraction pattern, the existence of twins, was ruled out by his experiments. Since periodicity and twins were ruled out, Blech, unaware of the two-dimensional tiling work, was looking for another possibility: a completely new structure containing cells connected to each other by defined angles and distances but without translational periodicity. Blech decided to use a computer simulation to calculate the diffraction intensity from a cluster of such a material without long-range translational order but still not random. He termed this new structure multiple polyhedral.

The idea of a new structure was the necessary paradigm shift to break the impasse. The "Eureka moment" came when the computer simulation showed sharp ten-fold diffraction patterns, similar to the observed ones, emanating from the three-dimensional structure devoid of periodicity. The multiple polyhedral structure was termed later by many researchers as icosahedral glass but in effect it embraces *any arrangement of polyhedra connected with definite angles and distances* (this general definition includes tiling, for example).

Shechtman accepted Blech's discovery of a new type of material and it gave him the courage to publish his experimental observation. Shechtman and Blech jointly wrote a paper entitled "The Microstructure of Rapidly Solidified Al_6Mn" and sent it for publication around June 1984 to the *Journal of Applied Physics* (JAP). The JAP editor promptly rejected the paper as being better fit for a metallurgical readership. As a result, the same paper was re-submitted for publication to the *Metallurgical Transactions A*, where it was accepted. Although not noted in the body of the published text, the published paper was slightly revised prior to publication.

Meanwhile, on seeing the draft of the Shechtman-Blech paper in the summer of 1984, John Cahn suggested that Shechtman's experimental results merit a fast publication in a more appropriate scientific journal. Shechtman agreed and, in hindsight, called this fast publication "a winning move". This paper, published in the *Physical Review Letters* (PRL), repeated Shechtman's observation and used the same illustrations as the original Shechtman-Blech paper in the *Metallurgical Transactions A*. The PRL paper, the first to appear in print, caused considerable excitement in the scientific community.

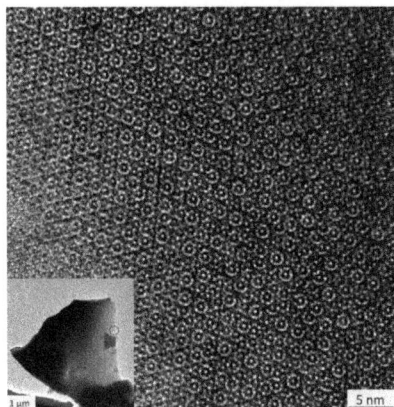

Next year Ishimasa *et al.* reported twelvefold symmetry in Ni-Cr particles. Soon, eightfold diffraction patterns were recorded in V-Ni-Si and Cr-Ni-Si alloys. Over the years, hundreds of quasicrystals with various compositions and different symmetries have been discovered. The first quasicrystalline materials were thermodynamically unstable—when heated, they formed regular crystals. However, in 1987, the first of many stable quasicrystals were discovered, making it possible to produce large samples for study and opening the door to potential applications. In 2009, following a 10-year systematic search, scientists reported the first natural quasicrystal, a mineral found in the Khatyrka River in eastern Russia. This natural quasicrystal exhibits high crystalline quality, equalling the best artificial examples. The natural quasicrystal phase, with a composition of $Al_{63}Cu_{24}Fe_{13}$, was named icosahedrite and it was approved by the International Mineralogical Association in 2010. Furthermore, analysis indicates it may be meteoritic in origin, possibly delivered from a carbonaceous chondrite asteroid.

Atomic image of a micron-sized grain of the natural $Al_{71}Ni_{24}Fe_5$ quasicrystal (shown in the inset) from a Khatyrka meteorite. The corresponding diffraction patterns reveal a ten-fold symmetry.

A further study of Khatyrka meteorites revealed micron-sized grains of another natural quasicrystal, which has a ten-fold symmetry and a chemical formula of $Al_{71}Ni_{24}Fe_5$. This quasicrystal is stable in a narrow temperature range, from 1120 to 1200 K at ambient pressure, which suggests that natural quasicrystals are formed by rapid quenching of a meteorite heated during an impact-induced shock.

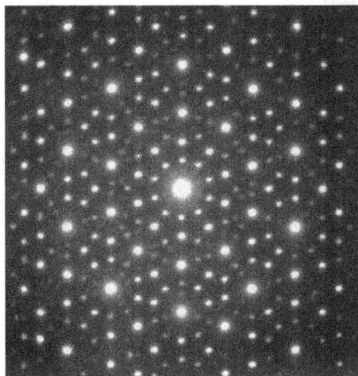

Electron diffraction pattern of an icosahedral Ho-Mg-Zn quasicrystal

In 1972 de Wolf and van Aalst reported that the diffraction pattern produced by a crystal of sodium carbonate cannot be labeled with three indices but needed one more, which implied that the underlying structure had four dimensions in reciprocal space. Other puzzling cases have been reported, but until the concept of quasicrystal came to be established, they were explained away or denied. However, at the end of the 1980s the idea became acceptable, and in 1992 the International Union of Crystallography altered its definition of a crystal, broadening it as a result of Shechtman's findings, reducing it to the ability to produce a clear-cut diffraction pattern and acknowledging the possibility of the ordering to be either periodic or aperiodic. Now, the symmetries compatible with translations are defined as "crystallographic", leaving room for other "non-crystallographic" symmetries. Therefore, aperiodic or quasiperiodic structures can be divided into two main classes: those with crystallographic point-group symmetry, to which the incommensurately modulated structures and composite structures belong, and those with non-crystallographic point-group symmetry, to which quasicrystal structures belong.

Originally, the new form of matter was dubbed "Shechtmanite". The term "quasicrystal" was first used in print by Steinhardt and Levine shortly after Shechtman's paper was published. The adjective *quasicrystalline* had already been in use, but now it came to be applied to any pattern with unusual symmetry. 'Quasiperiodical' structures were claimed to be observed in some decorative tilings devised by medieval Islamic architects. For example, Girih tiles in a medieval Islamic mosque in Isfahan, Iran, are arranged in a two-dimensional quasicrystalline pattern. These claims have, however, been under some debate.

Shechtman was awarded the Nobel Prize in Chemistry in 2011 for his work on quasicrystals. "His discovery of quasicrystals revealed a new principle for packing of atoms and molecules," stated the Nobel Committee and pointed that "this led to a paradigm shift within chemistry."

Mathematics

A penteract (5-cube) pattern using 5D orthographic projection to 2D using Petrie polygon basis vectors overlaid on the diffractogram from an Icosahedral Ho-Mg-Zn quasicrystal

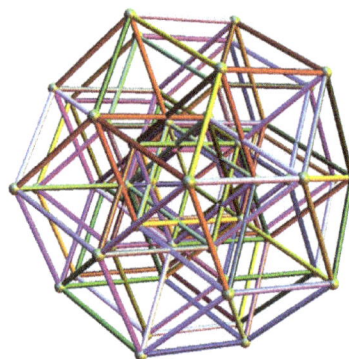

A hexeract (6-cube) pattern using 6D orthographic projection to a 3D Perspective (visual) object (the Rhombic triacontahedron) using the Golden ratio in the basis vectors. This is used to understand the aperiodic Icosahedral structure of Quasicrystals.

There are several ways to mathematically define quasicrystalline patterns. One definition, the "cut and project" construction, is based on the work of Harald Bohr (mathematician brother of Niels Bohr). The concept of an almost periodic function (also called a quasiperiodic function) was studied by Bohr, including work of Bohl and Escanglon. He introduced the notion of a superspace. Bohr showed that quasiperiodic functions arise as restrictions of high-dimensional periodic functions to an irrational slice (an intersection with one or more hyperplanes), and discussed their Fourier point spectrum. These functions are not exactly periodic, but they are arbitrarily close in some sense, as well as being a projection of an exactly periodic function.

In order that the quasicrystal itself be aperiodic, this slice must avoid any lattice plane of the higher-dimensional lattice. De Bruijn showed that Penrose tilings can be viewed as two-dimensional slices of five-dimensional hypercubic structures. Equivalently, the Fourier transform of such a quasicrystal is nonzero only at a dense set of points spanned by integer multiples of a finite set of basis vectors (the projections of the primitive reciprocal lattice vectors of the higher-dimensional lattice). The intuitive considerations obtained from simple model aperiodic tilings are formally expressed in the concepts of Meyer and Delone sets. The mathematical counterpart of physical diffraction is the Fourier transform and the qualitative description of a diffraction picture as 'clear cut' or 'sharp' means that singularities are present in the Fourier spectrum. There are different methods to construct model quasicrystals. These are the same methods that produce aperiodic tilings with the additional constraint for the diffractive property. Thus, for a substitution tiling the eigenvalues of the substitution matrix should be Pisot numbers. The aperiodic structures obtained by the cut-and-project method are made diffractive by choosing a suitable orientation for the construction; this is a geometric approach that has also a great appeal for physicists.

Classical theory of crystals reduces crystals to point lattices where each point is the center of mass of one of the identical units of the crystal. The structure of crystals can be analyzed by defining an associated group. Quasicrystals, on the other hand, are composed of more than one type of unit, so, instead of lattices, quasilattices must be used. Instead of groups, groupoids, the mathematical generalization of groups in category theory, is the appropriate tool for studying quasicrystals.

Using mathematics for construction and analysis of quasicrystal structures is a difficult task for most experimentalists. Computer modeling, based on the existing theories of quasicrystals, however, greatly facilitated this task. Advanced programs have been developed allowing one to construct, visualize and analyze quasicrystal structures and their diffraction patterns.

Interacting spins were also analyzed in quasicrystals: AKLT Model and 8-vertex model were solved in quasicrystals analytically

Study of quasicrystals may shed light on the most basic notions related to quantum critical point observed in heavy fermion metals. Experimental measurements on the gold-aluminium-ytterbium quasicrystal have revealed a quantum critical point defining the divergence of the magnetic susceptibility as temperature tends to zero. It is suggested that the electronic system of some quasicrystals is located at quantum critical point without tuning, while quasicrystals exhibit the typical scaling behaviour of their thermodynamic properties and belong to the famous family of heavy-fermion metals.

Materials Science

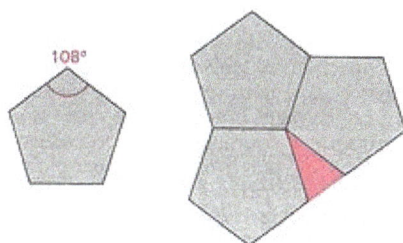

Tiling of a plane by regular pentagons is impossible but can be realized on a sphere in the form of pentagonal dodecahedron.

A Ho-Mg-Zn icosahedral quasicrystal formed as a pentagonal dodecahedron, the dual of the icosahedron. Unlike the similar pyritohedron shape of some cubic-system crystals such as pyrite, the quasicrystal has faces that are true regular pentagons

TiMn quasicrystal approximant lattice.

Since the original discovery by Dan Shechtman, hundreds of quasicrystals have been reported and confirmed. Undoubtedly, the quasicrystals are no longer a unique form of solid; they exist universally in many metallic alloys and some polymers. Quasicrystals are found most often in aluminium alloys (Al-Li-Cu, Al-Mn-Si, Al-Ni-Co, Al-Pd-Mn, Al-Cu-Fe, Al-Cu-V, etc.), but numerous other compositions are also known (Cd-Yb, Ti-Zr-Ni, Zn-Mg-Ho, Zn-Mg-Sc, In-Ag-Yb, Pd-U-Si, etc.).

Two types of quasicrystals are known. The first type, polygonal (dihedral) quasicrystals, have an axis of 8, 10, or 12-fold local symmetry (octagonal, decagonal, or dodecagonal quasicrystals, respectively). They are periodic along this axis and quasiperiodic in planes normal to it. The second type, icosahedral quasicrystals, are aperiodic in all directions.

Quasicrystals fall into three groups of different thermal stability:

- Stable quasicrystals grown by slow cooling or casting with subsequent annealing,

- Metastable quasicrystals prepared by melt spinning, and

- Metastable quasicrystals formed by the crystallization of the amorphous phase.

Except for the Al−Li−Cu system, all the stable quasicrystals are almost free of defects and disorder, as evidenced by X-ray and electron diffraction revealing peak widths as sharp as those of perfect crystals such as Si. Diffraction patterns exhibit fivefold, threefold, and twofold symmetries, and reflections are arranged quasiperiodically in three dimensions.

The origin of the stabilization mechanism is different for the stable and metastable quasicrystals. Nevertheless, there is a common feature observed in most quasicrystal-forming liquid alloys or their undercooled liquids: a local icosahedral order. The icosahedral order is in equilibrium in the *liquid state* for the stable quasicrystals, whereas the icosahedral order prevails in the *undercooled liquid state* for the metastable quasicrystals.

A nanoscale icosahedral phase was formed in Zr-, Cu- and Hf-based bulk metallic glasses alloyed with noble metals.

Most quasicrystals have ceramic-like properties including high thermal and electrical resistance, hardness and brittleness, resistance to corrosion, and non-stick properties. Many metallic quasicrystalline substances are impractical for most applications due to their thermal instability; the Al-Cu-Fe ternary system and the Al-Cu-Fe-Cr and Al-Co-Fe-Cr quaternary systems, thermally stable up to 700 °C, are notable exceptions.

Applications

Quasicrystalline substances have potential applications in several forms. The tendency to brittleness is a problem that must be overcome. Recent studies show typically brittle quasicrystals can exhibit remarkable ductility of over 50% strains at room temperature and sub-micrometer scales (<500 nm).

Metallic quasicrystalline coatings can be applied by plasma-coating or magnetron sputtering. A problem that must be resolved is the tendency for cracking due to the materials' extreme brittleness. The cracking could be suppressed by reducing sample dimensions or coating thickness.

An application was the use of low-friction Al-Cu-Fe-Cr quasicrystals as a coating for frying pans. Food did not stick to it as much as to stainless steel making the pan moderately non-stick and easy to clean; heat transfer and durability were better than PTFE non-stick cookware and the pan was free from perfluorooctanoic acid (PFOA); the surface was very hard, claimed to be ten times harder than stainless steel, and not harmed by metal utensils or cleaning in a dishwasher; and the pan could withstand temperatures of 1,000 °C (1,800 °F) without harm. However, cooking with a lot of salt would etch the quasicrystalline coating used, and the pans were eventually withdrawn from production. Shechtman had one of these pans.

The Nobel citation said that quasicrystals, while brittle, could reinforce steel "like armor". When Shechtman was asked about potential applications of quasicrstals he said that a precipitation-hardened stainless steel is produced that is strengthened by small quasicrystalline particles. It does not corrode and is extremely strong, suitable for razor blades and surgery instruments. The small quasicrystalline particles impede the motion of dislocation in the material.

Quasicrystals were also being used to develop heat insulation, LEDs, diesel engines, and new materials that convert heat to electricity. Shechtman suggested new applications taking advantage of the low coefficient of friction and the hardness of some quasicrystalline materials, for example embedding particles in plastic to make strong, hard-wearing, low-friction plastic gears. The low heat conductivity of some quasicrystals makes them good for heat insulating coatings.

Other potential applications include selective solar absorbers for power conversion, broad-wavelength reflectors, and bone repair and prostheses applications where biocompatibility, low friction and corrosion resistance are required. Magnetron sputtering can be readily applied to other stable quasicrystalline alloys such as Al-Pd-Mn.

While saying that the discovery of icosahedrite, the first quasicrystal found in nature, was important, Shechtman saw no practical applications.

Liquid Crystal

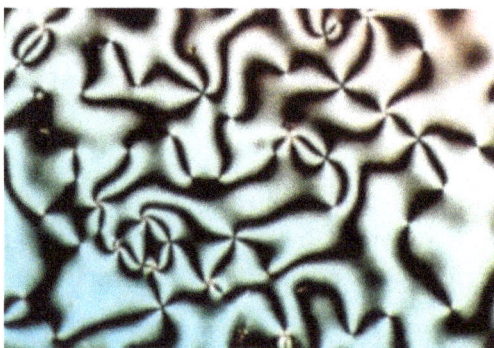

Schlieren texture of liquid crystal nematic phase

Liquid crystals (LCs) are matter in a state which has properties between those of conventional liquids and those of solid crystals. For instance, a liquid crystal may flow like a liquid, but its molecules may be oriented in a crystal-like way. There are many different types of liquid-crystal phases, which can be distinguished by their different optical properties (such as birefringence). When viewed under a microscope using a polarized light source, different liquid crystal phases will appear to have distinct textures. The contrasting areas in the textures correspond to domains where the liquid-crystal molecules are oriented in different directions. Within a domain, however, the molecules are well ordered. LC materials may not always be in a liquid-crystal phase (just as water may turn into ice or steam).

Liquid crystals can be divided into thermotropic, lyotropic and metallotropic phases. Thermotropic and lyotropic liquid crystals consist mostly of organic molecules, although a few minerals are also known. Thermotropic LCs exhibit a phase transition into the liquid-crystal phase as temperature is changed. Lyotropic LCs exhibit phase transitions as a function of both temperature and concentration of the liquid-crystal molecules in a solvent (typically water). Metallotropic LCs are composed of both organic and inorganic molecules; their liquid-crystal transition depends not only on temperature and concentration, but also on the inorganic-organic composition ratio.

Examples of liquid crystals can be found both in the natural world and in technological applications. Most contemporary electronic displays use liquid crystals. Lyotropic liquid-crystalline phases are abundant in living systems but can also be found in the mineral world. For example, many proteins and cell membranes are liquid crystals. Other well-known examples of liquid crystals are solutions of soap and various related detergents, as well as the tobacco mosaic virus, and some clays.

History

In 1888, Austrian botanical physiologist Friedrich Reinitzer, working at the Karl-Ferdinands-Universität, examined the physico-chemical properties of various derivatives of cholesterol which now belong to the class of materials known as cholesteric liquid crystals. Previously, other researchers had observed distinct color effects when cooling cholesterol derivatives just above the freezing point, but had not associated it with a new phenomenon. Reinitzer perceived that color changes in a derivative cholesteryl benzoate were not the most peculiar feature.

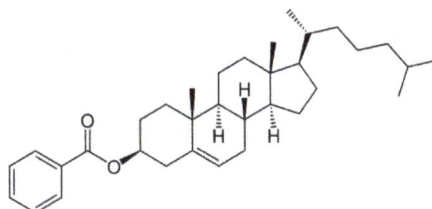

Chemical structure of cholesteryl benzoate molecule

He found that cholesteryl benzoate does not melt in the same manner as other compounds, but has two melting points. At 145.5 °C (293.9 °F) it melts into a cloudy liquid, and at 178.5 °C (353.3 °F) it melts again and the cloudy liquid becomes clear. The phenomenon is reversible. Seeking help from a physicist, on March 14, 1888, he wrote to Otto Lehmann, at that time a *Privatdozent* in Aachen. They exchanged letters and samples. Lehmann examined the intermediate cloudy fluid, and reported seeing crystallites. Reinitzer's Viennese colleague von Zepharovich also indicated that the intermediate "fluid" was crystalline. The exchange of letters with Lehmann ended on April 24, with many questions unanswered. Reinitzer presented his results, with credits to Lehmann and von Zepharovich, at a meeting of the Vienna Chemical Society on May 3, 1888.

By that time, Reinitzer had discovered and described three important features of cholesteric liquid crystals (the name coined by Otto Lehmann in 1904): the existence of two melting points, the reflection of circularly polarized light, and the ability to rotate the polarization direction of light.

After his accidental discovery, Reinitzer did not pursue studying liquid crystals further. The research was continued by Lehmann, who realized that he had encountered a new phenomenon and was in a position to investigate it: In his postdoctoral years he had acquired expertise in crystallography and microscopy. Lehmann started a systematic study, first of cholesteryl benzoate, and then of related compounds which exhibited the double-melting phenomenon. He was able to make observations in polarized light, and his microscope was equipped with a hot stage (sample holder equipped with a heater) enabling high temperature observations. The intermediate cloudy phase clearly sustained flow, but other features, particularly the signature under a microscope, convinced Lehmann that he was dealing with a solid. By the end of August 1889 he had published his results in the Zeitschrift für Physikalische Chemie.

Lehmann's work was continued and significantly expanded by the German chemist Daniel Vorländer, who from the beginning of 20th century until his retirement in 1935, had synthesized most of the liquid crystals known. However, liquid crystals were not popular among scientists and the material remained a pure scientific curiosity for about 80 years.

Otto Lehmann

After World War II work on the synthesis of liquid crystals was restarted at university research laboratories in Europe. George William Gray, a prominent researcher of liquid crystals, began investigating these materials in England in the late 1940s. His group synthesized many new materials that exhibited the liquid crystalline state and developed a better understanding of how to design molecules that exhibit the state. His book *Molecular Structure and the Properties of Liquid Crystals* became a guidebook on the subject. One of the first U.S. chemists to study liquid crystals was Glenn H. Brown, starting in 1953 at the University of Cincinnati and later at Kent State University. In 1965, he organized the first international conference on liquid crystals, in Kent, Ohio, with about 100 of the world's top liquid crystal scientists in attendance. This conference marked the beginning of a worldwide effort to perform research in this field, which soon led to the development of practical applications for these unique materials.

Liquid crystal materials became a focus of research in the development of flat panel electronic displays beginning in 1962 at RCA Laboratories. When physical chemist Richard Williams applied an electric field to a thin layer of a nematic liquid crystal at 125 °C, he observed the formation of a regular pattern that he called domains (now known as Williams Domains). This led his colleague George H. Heilmeier to perform research on a liquid crystal-based flat panel display to replace the cathode ray vacuum tube used in televisions. But the para-Azoxyanisole that Williams and Heilmeier used exhibits the nematic liquid crystal state only above 116 °C, which made it impractical to use in a commercial display product. A material that could be operated at room temperature was clearly needed.

In 1966, Joel E. Goldmacher and Joseph A. Castellano, research chemists in Heilmeier group at RCA, discovered that mixtures made exclusively of nematic compounds that differed only in the number of carbon atoms in the terminal side chains could yield room-temperature nematic liquid crystals. A ternary mixture of Schiff base compounds resulted in a material that had a nematic range of 22–105 °C. Operation at room temperature enabled the first practical display device to be made. The team then proceeded to prepare numerous mixtures of nematic compounds many of which had much lower melting points. This technique of mixing nematic compounds to obtain wide operating temperature range eventually became the industry standard and is used to this very day to tailor materials to meet specific applications.

In 1969, Hans Kelker succeeded in synthesizing a substance that had a nematic phase at room temperature, MBBA, which is one of the most popular subjects of liquid crystal research. The next step to commercialization of liquid crystal displays was the synthesis of further chemically stable substances (cyanobiphenyls) with low melting temperatures by George Gray. That work with Ken Harrison and the UK MOD (RRE Malvern), in 1973, led to design of new materials resulting in rapid adoption of small area LCDs within electronic products.

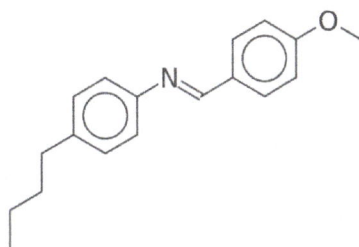

Chemical structure of N-(4-Methoxybenzylidene)-4-butylaniline (MBBA) molecule

These molecules are rod-shaped, some created in the lab and some appearing spontaneously in nature. Since then, two new types of LC molecules have been discovered, both man-made: disc-shaped (created by S. Chandrasekhar's group in India, 1977) and bowl-shaped (invented by Lui Lam in China, 1982, and synthesized in Europe three years later).

In 1991, when liquid crystal displays were already well established, Pierre-Gilles de Gennes working at the Université Paris-Sud received the Nobel Prize in physics "for discovering that methods developed for studying order phenomena in simple systems can be generalized to more complex forms of matter, in particular to liquid crystals and polymers".

Design of Liquid Crystalline Materials

A large number of chemical compounds are known to exhibit one or several liquid crystalline phases. Despite significant differences in chemical composition, these molecules have some common features in chemical and physical properties. There are three types of thermotropic liquid crystals: discotics, bowlics and rod-shaped molecules. Discotics are flat disc-like molecules consisting of a core of adjacent aromatic rings; the core in a bowlic is not flat but like a rice bowl (a three-dimensional object). This allows for two dimensional columnar ordering, for both discotics and bowlics. Rod-shaped molecules have an elongated, anisotropic geometry which allows for preferential alignment along one spatial direction.

- The molecular shape should be relatively thin, flat or bowl-like, especially within rigid molecular frameworks.

- The molecular length should be at least 1.3 nm, consistent with the presence of long alkyl group on many room-temperature liquid crystals.

- The structure should not be branched or angular, except for the bowlics.

- A low melting point is preferable in order to avoid metastable, monotropic liquid crystalline phases. Low-temperature mesomorphic behavior in general is technologically more useful, and alkyl terminal groups promote this.

An extended, structurally rigid, highly anisotropic shape seems to be the main criterion for liquid crystalline behavior, and as a result many liquid crystalline materials are based on benzene rings.

Liquid-crystal Phases

The various liquid-crystal phases (called mesophases) can be characterized by the type of ordering. One can distinguish positional order (whether molecules are arranged in any sort of ordered lattice) and orientational order (whether molecules are mostly pointing in the same direction), and moreover order can be either short-range (only between molecules close to each other) or long-range (extending to larger, sometimes macroscopic, dimensions). Most thermotropic LCs will have an isotropic phase at high temperature. That is that heating will eventually drive them into a conventional liquid phase characterized by random and isotropic molecular ordering (little to no long-range order), and fluid-like flow behavior. Under other conditions (for instance, lower temperature), a LC might inhabit one or more phases with significant anisotropic orientational structure and short-range orientational order while still having an ability to flow.

The ordering of liquid crystalline phases is extensive on the molecular scale. This order extends up to the entire domain size, which may be on the order of micrometers, but usually does not extend to the macroscopic scale as often occurs in classical crystalline solids. However some techniques, such as the use of boundaries or an applied electric field, can be used to enforce a single ordered domain in a macroscopic liquid crystal sample. The ordering in a liquid crystal might extend along only one dimension, with the material being essentially disordered in the other two directions.

Thermotropic Liquid Crystals

Thermotropic phases are those that occur in a certain temperature range. If the temperature rise is too high, thermal motion will destroy the delicate cooperative ordering of the LC phase, pushing the material into a conventional isotropic liquid phase. At too low temperature, most LC materials will form a conventional crystal. Many thermotropic LCs exhibit a variety of phases as temperature is changed. For instance, on heating a particular type of LC molecule (called mesogen) may exhibit various smectic phases followed by the nematic phase and finally the isotropic phase as temperature is increased. An example of a compound displaying thermotropic LC behavior is para-azoxyanisole.

Nematic Phase

Alignment in a nematic phase.

Phase transition between a nematic (left) and smectic A (right) phases observed between crossed polarizers. The black color corresponds to isotropic medium.

One of the most common LC phases is the nematic. This term originates from the thread-like topological defects observed in nematics, which are formally called 'disclinations'. Nematics also exhibit so-called "hedgehog" topological defects. In a nematic phase, the *calamitic* or rod-shaped organic molecules have no positional order, but they self-align to have long-range directional order with their long axes roughly parallel. Thus, the molecules are free to flow and their center of mass positions are randomly distributed as in a liquid, but still maintain their long-range directional order. Most nematics are uniaxial: they have one axis that is longer and preferred, with the other two being equivalent (can be approximated as cylinders or rods). However, some liquid crystals are biaxial nematics, meaning that in addition to orienting their long axis, they also orient along a secondary axis. Nematics have fluidity similar to that of ordinary (isotropic) liquids but they can be easily aligned by an external magnetic or electric field. Aligned nematics have the optical prop-erties of uniaxial crystals and this makes them extremely useful in liquid crystal displays (LCD).

Smectic Phases

Schematic of alignment in the smectic phases. The smectic A phase (left) has molecules organized into layers. In the smectic C phase (right), the molecules are tilted inside the layers.

The smectic phases, which are found at lower temperatures than the nematic, form well-defined layers that can slide over one another in a manner similar to that of soap. The word "smectic" originates from the Latin word "smecticus", meaning cleaning, or having soap-like properties. The smectics are thus positionally ordered along one direction. In the Smectic A phase, the molecules are oriented along the layer normal, while in the Smectic C phase they are tilted away from it. These phases are liquid-like within the layers. There are many different smectic phases, all charac-terized by different types and degrees of positional and orientational order.

Chiral Phases

Schematic of ordering in chiral liquid crystal phases. The chiral nematic phase (left), also called the cholesteric phase, and the smectic C* phase (right).

The chiral nematic phase exhibits chirality (handedness). This phase is often called the *cholesteric* phase because it was first observed for cholesterol derivatives. Only chiral molecules (i.e., those that have no internal planes of symmetry) can give rise to such a phase. This phase exhibits a twisting of the molecules perpendicular to the director, with the molecular axis parallel to the director. The finite twist angle between adjacent molecules is due to their asymmetric packing, which results in longer-range chiral order. In the smectic C* phase (an asterisk denotes a chiral phase), the molecules have positional ordering in a layered structure (as in the other smectic phases), with the molecules tilted by a finite angle with respect to the layer normal. The chirality induces a finite azimuthal twist from one layer to the next, producing a spiral twisting of the molecular axis along the layer normal.

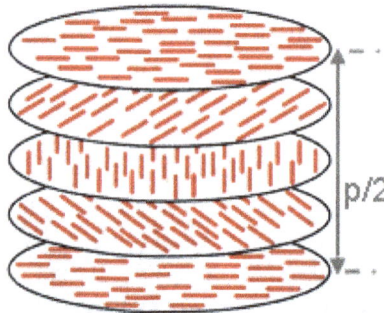

Chiral nematic phase; p refers to the chiral pitch

The *chiral pitch*, p, refers to the distance over which the LC molecules undergo a full 360° twist (but note that the structure of the chiral nematic phase repeats itself every half-pitch, since in this phase directors at 0° and ±180° are equivalent). The pitch, p, typically changes when the temperature is altered or when other molecules are added to the LC host (an achiral LC host material will form a chiral phase if doped with a chiral material), allowing the pitch of a given material to be tuned accordingly. In some liquid crystal systems, the pitch is of the same order as the wavelength of visible light. This causes these systems to exhibit unique optical properties, such as Bragg reflection and low-threshold laser emission, and these properties are exploited in a number of optical applications. For the case of Bragg reflection only the lowest-order reflection is allowed if the light is incident along the helical axis, whereas for oblique incidence higher-order reflections become permitted. Cholesteric liquid crystals also exhibit the unique property that they reflect circularly polarized light when it is incident along the helical axis and elliptically polarized if it comes in obliquely.

Blue phases are liquid crystal phases that appear in the temperature range between a chiral nematic phase and an isotropic liquid phase. Blue phases have a regular three-dimensional cubic structure of defects with lattice periods of several hundred nanometers, and thus they exhibit selective Bragg reflections in the wavelength range of visible light corresponding to the cubic lattice. It was theoretically predicted in 1981 that these phases can possess icosahedral symmetry similar to quasicrystals.

Although blue phases are of interest for fast light modulators or tunable photonic crystals, they exist in a very narrow temperature range, usually less than a few kelvin. Recently the stabilization of blue phases over a temperature range of more than 60 K including room temperature (260–326 K) has been demonstrated. Blue phases stabilized at room temperature allow electro-optical switching with response times of the order of 10^{-4} s.

In May 2008, the first Blue Phase Mode LCD panel had been developed.

Discotic Phases

Disk-shaped LC molecules can orient themselves in a layer-like fashion known as the discotic nematic phase. If the disks pack into stacks, the phase is called a discotic columnar. The columns themselves may be organized into rectangular or hexagonal arrays. Chiral discotic phases, similar to the chiral nematic phase, are also known.

Bowlic Phases

Bowl-shaped LC molecules, like in discotics, can form columnar phases. Other phases, such as nonpolar nematic, polar nematic, stringbean, donut and onion phases, have been predicted. Bowlic phases, except nonpolar nematic, are polar phases.

Lyotropic Liquid Crystals

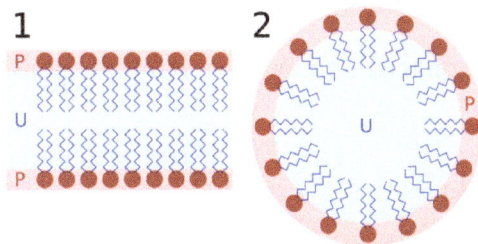

Structure of lyotropic liquid crystal. The red heads of surfactant molecules are in contact with water, whereas the tails are immersed in oil (blue): bilayer (left) and micelle (right).

A lyotropic liquid crystal consists of two or more components that exhibit liquid-crystalline properties in certain concentration ranges. In the lyotropic phases, solvent molecules fill the space around the compounds to provide fluidity to the system. In contrast to thermotropic liquid crystals, these lyotropics have another degree of freedom of concentration that enables them to induce a variety of different phases.

A compound that has two immiscible hydrophilic and hydrophobic parts within the same molecule is called an amphiphilic molecule. Many amphiphilic molecules show lyotropic liquid-crystalline phase sequences depending on the volume balances between the hydrophilic part and hydrophobic part. These structures are formed through the micro-phase segregation of two incompatible components on a nanometer scale. Soap is an everyday example of a lyotropic liquid crystal.

The content of water or other solvent molecules changes the self-assembled structures. At very low amphiphile concentration, the molecules will be dispersed randomly without any ordering. At slightly higher (but still low) concentration, amphiphilic molecules will spontaneously assemble into micelles or vesicles. This is done so as to 'hide' the hydrophobic tail of the amphiphile inside the micelle core, exposing a hydrophilic (water-soluble) surface to aqueous solution. These spherical objects do not order themselves in solution, however. At higher concentration, the assemblies will become ordered. A typical phase is a hexagonal columnar phase, where the amphiphiles form long cylinders (again with a hydrophilic surface) that arrange themselves into a roughly hexagonal lattice. This is called the middle soap phase. At still higher concentration, a lamellar phase (neat soap phase) may form, wherein extended sheets of amphiphiles are separated by thin layers of water. For some systems, a cubic (also called viscous isotropic) phase may exist between the hexagonal and lamellar phases, wherein spheres are formed that create a dense cubic lattice. These spheres may also be connected to one another, forming a bicontinuous cubic phase.

The objects created by amphiphiles are usually spherical (as in the case of micelles), but may also be disc-like (bicelles), rod-like, or biaxial (all three micelle axes are distinct). These anisotropic self-assembled nano-structures can then order themselves in much the same way as thermotropic liquid crystals do, forming large-scale versions of all the thermotropic phases (such as a nematic phase of rod-shaped micelles).

For some systems, at high concentrations, inverse phases are observed. That is, one may generate an inverse hexagonal columnar phase (columns of water encapsulated by amphiphiles) or an inverse micellar phase (a bulk liquid crystal sample with spherical water cavities).

A generic progression of phases, going from low to high amphiphile concentration, is:

- Discontinuous cubic phase (micellar cubic phase)

- Hexagonal phase (hexagonal columnar phase) (middle phase)

- Lamellar phase

- Bicontinuous cubic phase

- Reverse hexagonal columnar phase

- Inverse cubic phase (Inverse micellar phase)

Even within the same phases, their self-assembled structures are tunable by the concentration: for example, in lamellar phases, the layer distances increase with the solvent volume. Since lyotropic liquid crystals rely on a subtle balance of intermolecular interactions, it is more difficult to analyze their structures and properties than those of thermotropic liquid crystals.

Similar phases and characteristics can be observed in immiscible diblock copolymers.

Metallotropic Liquid Crystals

Liquid crystal phases can also be based on low-melting *inorganic* phases like $ZnCl_2$ that have a structure formed of linked tetrahedra and easily form glasses. The addition of long chain soap-like molecules leads to a series of new phases that show a variety of liquid crystalline behavior both as a function of the inorganic-organic composition ratio and of temperature. This class of materials has been named metallotropic.

Laboratory Analysis of Mesophases

Thermotropic mesophases are detected and characterized by two major methods, the original method was use of thermal optical microscopy, in which a small sample of the material was placed between two crossed polarizers; the sample was then heated and cooled. As the isotropic phase would not significantly affect the polarization of the light, it would appear very dark, whereas the crystal and liquid crystal phases will both polarize the light in a uniform way, leading to brightness and color gradients. This method allows for the characterization of the particular phase, as the different phases are defined by their particular order, which must be observed. The second method, differential scanning calorimetry (DSC), allows for more precise determination of phase transitions and transition enthalpies. In DSC, a small sample is heated in a way that generates a very precise change in temperature with respect to time. During phase transitions, the heat flow required to maintain this heating or cooling rate will change. These changes can be observed and attributed to various phase transitions, such as key liquid crystal transitions.

Lyotropic mesophases are analyzed in a similar fashion, though these experiments are somewhat more complex, as the concentration of mesogen is a key factor. These experiments are run at various concentrations of mesogen in order to analyze that impact.

Biological Liquid Crystals

Lyotropic liquid-crystalline phases are abundant in living systems, the study of which is referred to as lipid polymorphism. Accordingly, lyotropic liquid crystals attract particular attention in the field of biomimetic chemistry. In particular, biological membranes and cell membranes are a form of liquid crystal. Their constituent molecules (e.g. phospholipids) are perpendicular to the membrane surface, yet the membrane is flexible. These lipids vary in shape. The constituent molecules can inter-mingle easily, but tend not to leave the membrane due to the high energy requirement of this process. Lipid molecules can flip from one side of the membrane to the other, this process being catalyzed by flippases and floppases (depending on the direction of movement). These liquid crystal membrane phases can also host important proteins such as receptors freely "floating" inside, or partly outside, the membrane, e.g. CCT.

Many other biological structures exhibit liquid-crystal behavior. For instance, the concentrated protein solution that is extruded by a spider to generate silk is, in fact, a liquid crystal phase. The precise ordering of molecules in silk is critical to its renowned strength. DNA and many polypeptides can also form LC phases and this too forms an important part of current academic research.

Mineral Liquid Crystals

Examples of liquid crystals can also be found in the mineral world, most of them being lyotropics. The first discovered was Vanadium(V) oxide, by Zocher in 1925. Since then, few others have been discovered and studied in detail. The existence of a true nematic phase in the case of the smectic clays family was raised by Langmuir in 1938, but remained open for a very long time and was only solved recently. With the rapid development of nanosciences, and the synthesis of many new anisotropic nanoparticles, the number of such mineral liquid crystals is quickly increasing, with, for example, carbon nanotubes and graphene. A lamellar phase was even discovered, $H_3Sb_3P_2O_{14}$, which exhibits hyperswelling up to ~250 nm for the interlamellar distance.

Pattern Formation in Liquid Crystals

Anisotropy of liquid crystals is a property not observed in other fluids. This anisotropy makes flows of liquid crystals behave more differentially than those of ordinary fluids. For example, injection of a flux of a liquid crystal between two close parallel plates (viscous fingering) causes orientation of the molecules to couple with the flow, with the resulting emergence of dendritic patterns. This anisotropy is also manifested in the interfacial energy (surface tension) between different liquid crystal phases. This anisotropy determines the equilibrium shape at the coexistence temperature, and is so strong that usually facets appear. When temperature is changed one of the phases grows, forming different morphologies depending on the temperature change. Since growth is controlled by heat diffusion, anisotropy in thermal conductivity favors growth in specific directions, which has also an effect on the final shape.

Theoretical Treatment of Liquid Crystals

Microscopic theoretical treatment of fluid phases can become quite complicated, owing to the high material density, meaning that strong interactions, hard-core repulsions, and many-body correlations cannot be ignored. In the case of liquid crystals, anisotropy in all of these interactions further complicates analysis. There are a number of fairly simple theories, however, that can at least predict the general behavior of the phase transitions in liquid crystal systems.

Director

As we already saw above, the nematic liquid crystals are composed of rod-like molecules with the long axes of neighboring molecules aligned approximately to one another. To describe this anisotropic structure, a dimensionless unit vector n called the *director*, is introduced to represent the direction of preferred orientation of molecules in the neighborhood of any point. Because there is no physical polarity along the director axis, n and $-n$ are fully equivalent.

Order Parameter

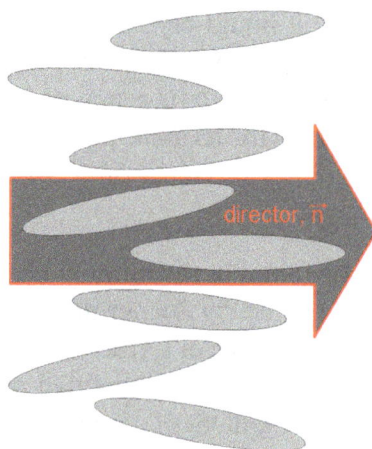

The *local nematic director*, which is also the *local optical axis*, is given by the spatial and temporal average of the long molecular axes

The description of liquid crystals involves an analysis of order. A second rank symmetric traceless tensor order parameter is used to describe the orientational order of a nematic liquid crystal, although a scalar order parameter is usually sufficient to describe uniaxial nematic liquid crystals. To make this quantitative, an orientational order parameter is usually defined based on the average of the second Legendre polynomial:

$$S = \langle P_2(\cos\theta) \rangle = \left\langle \frac{3\cos^2\theta - 1}{2} \right\rangle$$

where θ is the angle between the liquid-crystal molecular axis and the *local director* (which is the 'preferred direction' in a volume element of a liquid crystal sample, also representing its *local optical axis*). The brackets denote both a temporal and spatial average. This definition is convenient, since for a completely random and isotropic sample, S=0, whereas for a perfectly aligned sample S=1. For a typical liquid crystal sample, S is on the order of 0.3 to 0.8, and generally decreases as the temperature is raised. In particular, a sharp drop of the order parameter to 0 is observed when the system undergoes a phase transition from an LC phase into the isotropic phase. The order parameter can be measured experimentally in a number of ways; for instance, diamagnetism, birefringence, Raman scattering, NMR and EPR can be used to determine S.

The order of a liquid crystal could also be characterized by using other even Legendre polynomials (all the odd polynomials average to zero since the director can point in either of two antiparallel directions). These higher-order averages are more difficult to measure, but can yield additional information about molecular ordering.

A positional order parameter is also used to describe the ordering of a liquid crystal. It is characterized by the variation of the density of the center of mass of the liquid crystal molecules along a given vector. In the case of positional variation along the z-axis the density $\rho(z)$ is often given by:

$$\rho(\mathbf{r}) = \rho(z) = \rho_0 + \rho_1 \cos\left(q_s z - \phi\right) + \cdots$$

The complex positional order parameter is defined as $\psi(\mathbf{r}) = \rho_1(\mathbf{r})e^{i\phi(\mathbf{r})}$ and ρ_0 the average density. Typically only the first two terms are kept and higher order terms are ignored since most phases can be described adequately using sinusoidal functions. For a perfect nematic $\psi = 0$ and for a smectic phase ψ will take on complex values. The complex nature of this order parameter allows for many parallels between nematic to smectic phase transitions and conductor to superconductor transitions.

Onsager Hard-rod Model

A simple model which predicts lyotropic phase transitions is the hard-rod model proposed by Lars Onsager. This theory considers the volume excluded from the center-of-mass of one idealized cylinder as it approaches another. Specifically, if the cylinders are oriented parallel to one another, there is very little volume that is excluded from the center-of-mass of the approaching cylinder (it can come quite close to the other cylinder). If, however, the cylinders are at some angle to one another, then there is a large volume surrounding the cylinder which the approaching cylinder's center-of-mass cannot enter (due to the hard-rod repulsion between the two idealized objects). Thus, this angular arrangement sees a *decrease* in the net positional entropy of the approaching cylinder (there are fewer states available to it).

The fundamental insight here is that, whilst parallel arrangements of anisotropic objects lead to a decrease in orientational entropy, there is an increase in positional entropy. Thus in some case greater positional order will be entropically favorable. This theory thus predicts that a solution of rod-shaped objects will undergo a phase transition, at sufficient concentration, into a nematic phase. Although this model is conceptually helpful, its mathematical formulation makes several assumptions that limit its applicability to real systems.

Maier–Saupe Mean Field Theory

This statistical theory, proposed by Alfred Saupe and Wilhelm Maier, includes contributions from an attractive intermolecular potential from an induced dipole moment between adjacent liquid crystal molecules. The anisotropic attraction stabilizes parallel alignment of neighboring molecules, and the theory then considers a mean-field average of the interaction. Solved self-consistently, this theory predicts thermotropic nematic-isotropic phase transitions, consistent with experiment.

McMillan's Model

McMillan's model, proposed by William McMillan, is an extension of the Maier–Saupe mean field theory used to describe the phase transition of a liquid crystal from a nematic to a smectic A phase. It predicts that the phase transition can be either continuous or discontinuous depending on the strength of the short-range interaction between the molecules. As a result, it allows for a triple critical point where the nematic, isotropic, and smectic A phase meet. Although it predicts the existence of a triple critical point, it does not successfully predict its value. The model utilizes two order parameters that describe the orientational and positional order of the liquid crystal. The first is simply the average of the second Legendre polynomial and the second order parameter is given by:

$$\sigma = \left\langle \cos\left(\frac{2\pi z_i}{d}\right)\left(\frac{3}{2}\cos^2\theta_i - \frac{1}{2}\right)\right\rangle$$

The values z_i, θ_i, and d are the position of the molecule, the angle between the molecular axis and director, and the layer spacing. The postulated potential energy of a single molecule is given by:

$$U_i(\theta_i, z_i) = -U_0\left(S + \alpha\sigma\cos\left(\frac{2\pi z_i}{d}\right)\right)\left(\frac{3}{2}\cos^2\theta_i - \frac{1}{2}\right)$$

Here constant α quantifies the strength of the interaction between adjacent molecules. The potential is then used to derive the thermodynamic properties of the system assuming thermal equilibrium. It results in two self-consistency equations that must be solved numerically, the solutions of which are the three stable phases of the liquid crystal.

Elastic Continuum Theory

In this formalism, a liquid crystal material is treated as a continuum; molecular details are entirely ignored. Rather, this theory considers perturbations to a presumed oriented sample. The distortions of the liquid crystal are commonly described by the Frank free energy density. One can identify three types of distortions that could occur in an oriented sample: (1) twists of the material, where neighboring molecules are forced to be angled with respect to one another, rather than aligned; (2) splay of the material, where bending occurs perpendicular to the director; and (3) bend of the material, where the distortion is parallel to the director and molecular axis. All three of these types of distortions incur an energy penalty. They are distortions that are induced by the boundary conditions at domain walls or the enclosing container. The response of the material can then be decomposed into terms based on the elastic constants corresponding to the three types of distortions. Elastic continuum theory is a particularly powerful tool for modeling liquid crystal devices and lipid bilayers.

External Influences on Liquid Crystals

Scientists and engineers are able to use liquid crystals in a variety of applications because external perturbation can cause significant changes in the macroscopic properties of the liquid crystal system. Both electric and magnetic fields can be used to induce these changes. The magnitude of the fields, as well as the speed at which the molecules align are important characteristics industry deals with. Special surface treatments can be used in liquid crystal devices to force specific orientations of the director.

Electric and Magnetic Field Effects

The ability of the director to align along an external field is caused by the electric nature of the molecules. Permanent electric dipoles result when one end of a molecule has a net positive charge while the other end has a net negative charge. When an external electric field is applied to the liquid crystal, the dipole molecules tend to orient themselves along the direction of the field.

Even if a molecule does not form a permanent dipole, it can still be influenced by an electric field. In some cases, the field produces slight re-arrangement of electrons and protons in molecules such that an induced electric dipole results. While not as strong as permanent dipoles, orientation with the external field still occurs. The effects of magnetic fields on liquid crystal molecules are analogous to electric fields. Because magnetic fields are generated by moving electric charges, permanent magnetic dipoles are produced by electrons moving about atoms. When a magnetic field is applied, the molecules will tend to align with or against the field.

Surface Preparations

In the absence of an external field, the director of a liquid crystal is free to point in any direction. It is possible, however, to force the director to point in a specific direction by introducing an outside agent to the system. For example, when a thin polymer coating (usually a polyimide) is spread on a glass substrate and rubbed in a single direction with a cloth, it is observed that liquid crystal molecules in contact with that surface align with the rubbing direction. The currently accepted mechanism for this is believed to be an epitaxial growth of the liquid crystal layers on the partially aligned polymer chains in the near surface layers of the polyimide.

Fredericks Transition

The competition between orientation produced by surface anchoring and by electric field effects is often exploited in liquid crystal devices. Consider the case in which liquid crystal molecules are aligned parallel to the surface and an electric field is applied perpendicular to the cell. At first, as the electric field increases in magnitude, no change in alignment occurs. However at a threshold magnitude of electric field, deformation occurs. Deformation occurs where the director changes its orientation from one molecule to the next. The occurrence of such a change from an aligned to a deformed state is called a Fredericks transition and can also be produced by the application of a magnetic field of sufficient strength.

The Fredericks transition is fundamental to the operation of many liquid crystal displays because the director orientation (and thus the properties) can be controlled easily by the application of a field.

Effect of Chirality

As already described, chiral liquid-crystal molecules usually give rise to chiral mesophases. This means that the molecule must possess some form of asymmetry, usually a stereogenic center. An additional requirement is that the system not be racemic: a mixture of right- and left-handed molecules will cancel the chiral effect. Due to the cooperative nature of liquid crystal ordering, however, a small amount of chiral dopant in an otherwise achiral mesophase is often enough to select out one domain handedness, making the system overall chiral.

Chiral phases usually have a helical twisting of the molecules. If the pitch of this twist is on the order of the wavelength of visible light, then interesting optical interference effects can be observed. The chiral twisting that occurs in chiral LC phases also makes the system respond differently from right- and left-handed circularly polarized light. These materials can thus be used as polarization filters.

It is possible for chiral LC molecules to produce essentially achiral mesophases. For instance, in certain ranges of concentration and molecular weight, DNA will form an achiral line hexatic phase. An interesting recent observation is of the formation of chiral mesophases from achiral LC molecules. Specifically, bent-core molecules (sometimes called banana liquid crystals) have been shown to form liquid crystal phases that are chiral. In any particular sample, various domains will have opposite handedness, but within any given domain, strong chiral ordering will be present. The appearance mechanism of this macroscopic chirality is not yet entirely clear. It appears that the molecules stack in layers and orient themselves in a tilted fashion inside the layers. These liquid crystals phases may be ferroelectric or anti-ferroelectric, both of which are of interest for applications.

Chirality can also be incorporated into a phase by adding a chiral dopant, which may not form LCs itself. Twisted-nematic or super-twisted nematic mixtures often contain a small amount of such dopants.

Applications of Liquid Crystals

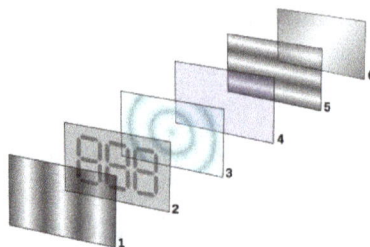

Structure of liquid crystal display: 1 – vertical polarization filter, 2,4 – glass with electrodes, 3 – liquid crystals, 5 – horizontal polarization filter, 6 – reflector

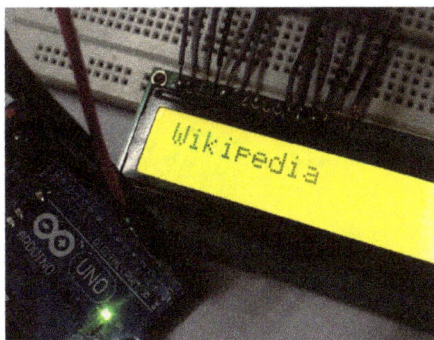

"Wikipedia" displayed on an LCD

Liquid crystals find wide use in liquid crystal displays, which rely on the optical properties of certain liquid crystalline substances in the presence or absence of an electric field. In a typical device, a liquid crystal layer (typically 4 μm thick) sits between two polarizers that are crossed (oriented at 90° to one another). The liquid crystal alignment is chosen so that its relaxed phase is a twisted one. This twisted phase reorients light that has passed through the first polarizer, allowing its transmission through the second polarizer (and reflected back to the observer if a reflector is provided). The device thus appears transparent. When an electric field is applied to the LC layer, the long molecular axes tend to align parallel to the electric field thus gradually untwisting in the center of the liquid crystal layer. In this state, the LC molecules do not reorient light, so the light polarized at the first polarizer is absorbed at the second polarizer, and the device loses transparency with increasing voltage. In this way, the electric field can be used to make a pixel switch between transparent or opaque on command. Color LCD systems use the same technique, with color filters used to generate red, green, and blue pixels. Chiral smectic liquid crystals are used in ferroelectric LCDs which are fast-switching binary light modulators. Similar principles can be used to make other liquid crystal based optical devices.

Liquid crystal tunable filters are used as electrooptical devices, e.g., in hyperspectral imaging.

Thermotropic chiral LCs whose pitch varies strongly with temperature can be used as crude liquid crystal thermometers, since the color of the material will change as the pitch is changed. Liquid crystal color transitions are used on many aquarium and pool thermometers as well as on thermometers for infants or baths. Other liquid crystal materials change color when stretched or stressed. Thus, liquid crystal sheets are often used in industry to look for hot spots, map heat flow, measure stress distribution patterns, and so on. Liquid crystal in fluid form is used to detect electrically generated hot spots for failure analysis in the semiconductor industry.

Liquid crystal lasers use a liquid crystal in the lasing medium as a distributed feedback mechanism instead of external mirrors. Emission at a photonic bandgap created by the periodic dielectric structure of the liquid crystal gives a low-threshold high-output device with stable monochromatic emission.

Polymer Dispersed Liquid Crystal (PDLC) sheets and rolls are available as adhesive backed Smart film which can be applied to windows and electrically switched between transparent and opaque to provide privacy.

Many common fluids, such as soapy water, are in fact liquid crystals. Soap forms a variety of LC phases depending on its concentration in water.

Bowlic columns could be used for fast switches.

Radioactive Decay

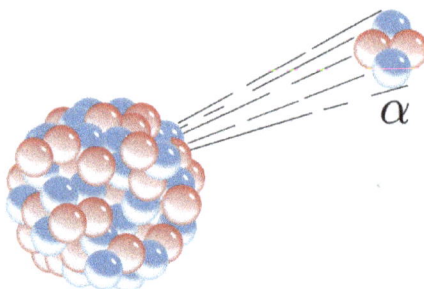

Alpha decay is one type of radioactive decay, in which an atomic nucleus emits an alpha particle, and thereby transforms (or "decays") into an atom with a mass number decreased by 4 and atomic number decreased by 2.

Radioactive decay (also known as nuclear decay or radioactivity) is the process by which the nucleus of an unstable atom loses energy by emitting radiation, including alpha particles, beta particles, gamma rays, and conversion electrons. A material that spontaneously emits such radiation is considered radioactive.

Radioactive decay is a stochastic (i.e. random) process at the level of single atoms, in that, according to quantum theory, it is impossible to predict when a particular atom will decay, regardless of how long the atom has existed. For a collection of atoms however, the collection's decay rate can be calculated from their measured decay constants or half-lives. This is the basis of radiometric dating. The half-lives of radioactive atoms have no known lower or upper limit, spanning a time range of over 55 orders of magnitude, from nearly instantaneous to far longer than the age of the universe. A radioactive source emits its decay products isotropically (all directions and without bias) in the absence of external influence.

There are many different types of radioactive decay. A decay, or loss of energy from the nucleus, results when an atom with an initial type of nucleus, called the *parent radionuclide* (or *parent radioisotope*), transforms into a *daughter nuclide*. The transformation produces an atom in a different state (a nucleus containing a different number of protons and neutrons). In some decays, the parent and the daughter nuclides are different chemical elements, and thus the decay process results in the creation of an atom of a different element. This is known as a nuclear transmutation.

The first decay processes to be discovered were alpha decay, beta decay, and gamma decay. Alpha decay occurs when the nucleus ejects an alpha particle (helium nucleus). This is the most common process of emitting nucleons, but in rarer types of decays, nuclei can eject protons, or in the case of cluster decay specific nuclei of other elements. Beta decay occurs when the nucleus emits an electron or positron and a neutrino, in a process that changes a proton to a neutron or the other way about. Highly excited neutron-rich nuclei, formed as the product of other types of decay, occasionally lose energy by way of neutron emission, resulting in a change from one isotope to another of the same element.. The nucleus may capture an orbiting electron, causing a proton to convert into a neutron in a process called electron capture. All of these processes result in a well-defined nuclear transmutation.

By contrast, there are radioactive decay processes that do not result in a nuclear transmutation. The energy of an excited nucleus may be emitted as a gamma ray in a process called gamma decay, or that energy may be lost when the nucleus interacts with an orbital electron causing its ejection from the atom, in a process called internal conversion.

Another type of radioactive decay results in products that are not defined, but appear in a range of "pieces" of the original nucleus. This decay, called spontaneous fission, happens when a large unstable nucleus spontaneously splits into two (and occasionally three) smaller daughter nuclei, and generally leads to the emission of gamma rays, neutrons, or other particles from those products.

For a summary table showing the number of stable and radioactive nuclides in each category, see radionuclide. There exist twenty-nine chemical elements on Earth that are radioactive. They are those that contain thirty-four radionuclides that date before the time of formation of the solar system, and are known as primordial nuclides. Well-known examples are uranium and thorium, but also included are naturally occurring long-lived radioisotopes such as potassium-40. Another fifty or so shorter-lived radionuclides, such as radium and radon, found on Earth, are the products of decay chains that began with the primordial nuclides, or are the product of ongoing cosmogenic processes, such as the production of carbon-14 from nitrogen-14 by cosmic rays. Radionuclides may also be produced artificially in particle accelerators or nuclear reactors, resulting in 650 of these with half-lives of over an hour, and several thousand more with even shorter half-lives. See this list of nuclides for a list of these, sorted by half life.

History of Discovery

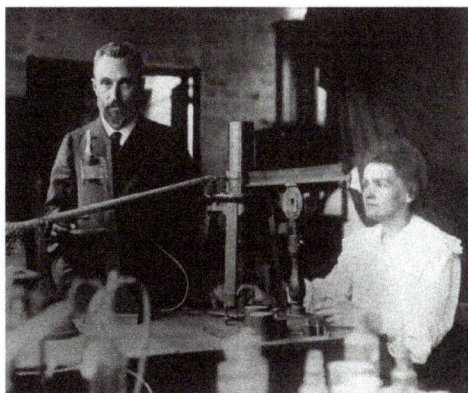

Pierre and Marie Curie in their Paris laboratory, before 1907

Radioactivity was discovered in 1896 by the French scientist Henri Becquerel, while working with phosphorescent materials. These materials glow in the dark after exposure to light, and he suspected that the glow produced in cathode ray tubes by X-rays might be associated with phosphorescence. He wrapped a photographic plate in black paper and placed various phosphorescent salts on it. All results were negative until he used uranium salts. The uranium salts caused a blackening of the plate in spite of the plate being wrapped in black paper. These radiations were given the name "Becquerel Rays".

It soon became clear that the blackening of the plate had nothing to do with phosphorescence, as the blackening was also produced by non-phosphorescent salts of uranium and metallic uranium. It became clear from these experiments that there was a form of invisible radiation that could pass through paper and was causing the plate to react as if exposed to light.

At first, it seemed as though the new radiation was similar to the then recently discovered X-rays. Further research by Becquerel, Ernest Rutherford, Paul Villard, Pierre Curie, Marie Curie, and others showed that this form of radioactivity was significantly more complicated. Rutherford was the first to realize that all such elements decay in accordance with the same mathematical exponential formula. Rutherford and his student Frederick Soddy were the first to realize that many decay processes resulted in the transmutation of one element to another. Subsequently, the radioactive displacement law of Fajans and Soddy was formulated to describe the products of alpha and beta decay.

The early researchers also discovered that many other chemical elements, besides uranium, have radioactive isotopes. A systematic search for the total radioactivity in uranium ores also guided Pierre and Marie Curie to isolate two new elements: polonium and radium. Except for the radioactivity of radium, the chemical similarity of radium to barium made these two elements difficult to distinguish.

Marie and Pierre Curie's study of radioactivity is an important factor in science and medicine. After their research on Becquerel's rays led them to the discovery on both radium and polonium, they coined the term "radioactivity." Their research on the penetrating rays in uranium and discovery of radium launched an era of using radium for treatment of cancer. Their exploration of radium could be seen as the first peaceful use of nuclear energy and the start of modern nuclear medicine.

Early Health Dangers

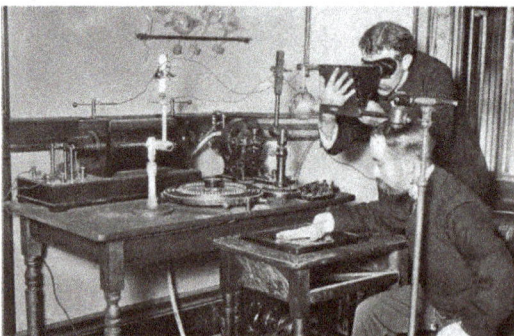

Taking an X-ray image with early Crookes tube apparatus in 1896. The Crookes tube is visible in the centre. The standing man is viewing his hand with a fluoroscope screen; this was a common way of setting up the tube. No precautions against radiation exposure are being taken; its hazards were not known at the time.

The dangers of ionizing radiation due to radioactivity and X-rays were not immediately recognized.

X-rays

The discovery of xrays by Wilhelm Röntgen in 1895 led to widespread experimentation by scientists, physicians, and inventors. Many people began recounting stories of burns, hair loss and worse in technical journals as early as 1896. In February of that year, Professor Daniel and Dr. Dudley of Vanderbilt University performed an experiment involving X-raying Dudley's head that resulted in his hair loss. A report by Dr. H.D. Hawks, of his suffering severe hand and chest burns in an X-ray demonstration, was the first of many other reports in *Electrical Review*.

Other experimenters including Elihu Thomson, and Nikola Tesla also reported burns. Thomson deliberately exposed a finger to an X-ray tube over a period of time and suffered pain, swelling, and blistering. Other effects, including ultraviolet rays and ozone were sometimes blamed for the damage, and many physicians still claimed that there were no effects from X-ray exposure at all.

Despite this, there were some early systematic hazard investigations, and as early as 1902 William Herbert Rollins wrote almost despairingly that his warnings about the dangers involved in careless use of X-rays was not being heeded, either by industry or by his colleagues. By this time Rollins had proved that X-rays could kill experimental animals, could cause a pregnant guinea pig to abort, and that they could kill a fetus. He also stressed that "animals vary in susceptibility to the external action of X-light" and warned that these differences be considered when patients were treated by means of X-rays.

Radioactive Substances

Radioactivity is characteristic of elements with large atomic number. Elements with at least one stable isotope are shown in light blue. Green shows elements whose most stable isotope has a half-life measured in millions of years. Yellow and orange are progressively less stable, with half-lives in thousands or hundreds of years, down toward one day. Red and purple show highly and extremely radioactive elements where the most stable isotopes exhibit half-lives measured on the order of one day and much less.

However, the biological effects of radiation due to radioactive substances were less easy to gauge. This gave the opportunity for many physicians and corporations to market radioactive substances as patent medicines. Examples were radium enema treatments, and radium-containing waters to be drunk as tonics. Marie Curie protested against this sort of treatment, warning that the effects of radiation on the human body were not well understood. Curie later died from aplastic anaemia, likely caused by exposure to ionizing radiation. By the 1930s, after a number of cases of bone necrosis and death of radium treatment enthusiasts, radium-containing medicinal products had been largely removed from the market (radioactive quackery).

Radiation Protection

Only a year after Röntgen's discovery of X rays, the American engineer Wolfram Fuchs (1896) gave what is probably the first protection advice, but it was not until 1925 that the first International Congress of Radiology (ICR) was held and considered establishing international protection standards. The effects of radiation on genes, including the effect of cancer risk, were recognized much later. In 1927, Hermann Joseph Muller published research showing genetic effects and, in 1946, was awarded the Nobel Prize in Physiology or Medicine for his findings.

The second ICR was held in Stockholm in 1928 and proposed the adoption of the rontgen unit, and the 'International X-ray and Radium Protection Committee' (IXRPC) was formed. Rolf Sievert was named Chairman, but a driving force was George Kaye of the British National Physical Laboratory. The committee met in 1931, 1934 and 1937.

After World War II the increased range and quantity of radioactive substances being handled as a result of military and civil nuclear programmes led to large groups of occupational workers and the public being potentially exposed to harmful levels of ionising radiation. This was considered at the first post-war ICR convened in London in 1950, when the present International Commission on Radiological Protection (ICRP) was born. Since then the ICRP has developed the present international system of radiation protection, covering all aspects of radiation hazard.

Units of Radioactivity

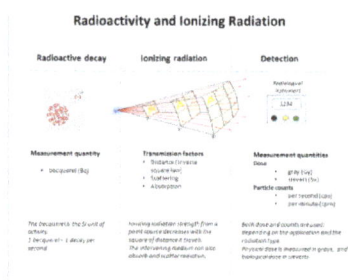

Graphic showing relationships between radioactivity and detected ionizing radiation

The International System of Units (SI) unit of radioactive activity is the becquerel (Bq), named in honour of the scientist Henri Becquerel. One Bq is defined as one transformation (or decay or disintegration) per second.

An older unit of radioactivity is the curie, Ci, which was originally defined as "the quantity or mass of radium emanation in equilibrium with one gram of radium (element)". Today, the curie is defined as 3.7×10^{10} disintegrations per second, so that 1 curie (Ci) = 3.7×10^{10} Bq. For radiological protection purposes, although the United States Nuclear Regulatory Commission permits the use of the unit curie alongside SI units, the European Union European units of measurement directives required that its use for "public health ... purposes" be phased out by 31 December 1985.

Types of Decay

Early researchers found that an electric or magnetic field could split radioactive emissions into three types of beams. The rays were given the names alpha, beta, and gamma, in order of their ability to penetrate matter. While alpha decay was observed only in heavier elements of atomic number 52 (tellurium) and greater, the other two types of decay were produced by all of the elements. Lead, atomic number 82, is the heaviest element to have any isotopes stable (to the limit of measurement) to radioactive decay. Radioactive decay is seen in all isotopes of all elements of atomic number 83 (bismuth) or greater. Bismuth, however, is only very slightly radioactive, with a half-life greater than the age of the universe; radioisotopes with extremely long half-lives are considered effectively stable for practical purposes.

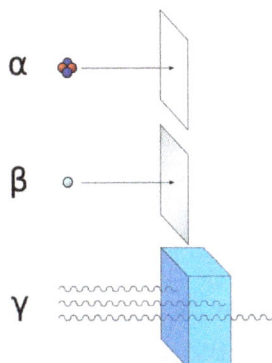

Alpha particles may be completely stopped by a sheet of paper, beta particles by aluminium shielding. Gamma rays can only be reduced by much more substantial mass, such as a very thick layer of lead.

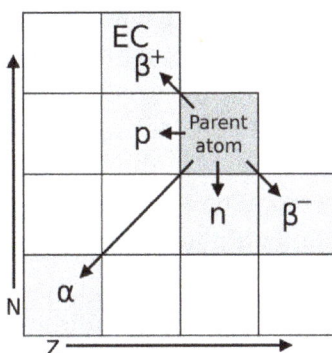

Transition diagram for decay modes of a radionuclide, with neutron number N and atomic number Z (shown are α, β±, p+, and n° emissions, EC denotes electron capture).

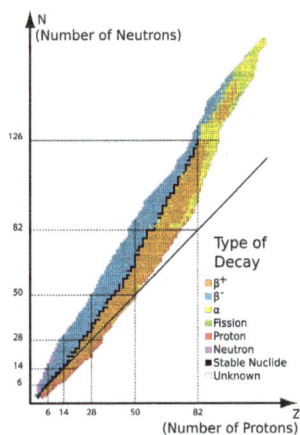

Types of radioactive decay related to N and Z numbers

In analysing the nature of the decay products, it was obvious from the direction of the electromagnetic forces applied to the radiations by external magnetic and electric fields that alpha particles carried a positive charge, beta particles carried a negative charge, and gamma rays were neutral. From the magnitude of deflection, it was clear that alpha particles were much more massive than beta particles. Passing alpha particles through a very thin glass window and trapping them in a discharge tube allowed researchers to study the emission spectrum of the captured particles, and ultimately proved that alpha particles are helium nuclei. Other experiments showed beta radiation, resulting from decay and cathode rays, were high-speed electrons. Likewise, gamma radiation and X-rays were found to be high-energy electromagnetic radiation.

The relationship between the types of decays also began to be examined: For example, gamma decay was almost always found to be associated with other types of decay, and occurred at about the same time, or afterwards. Gamma decay as a separate phenomenon, with its own half-life (now termed isomeric transition), was found in natural radioactivity to be a result of the gamma decay of excited metastable nuclear isomers, which were in turn created from other types of decay.

Although alpha, beta, and gamma radiations were most commonly found, other types of emission were eventually discovered. Shortly after the discovery of the positron in cosmic ray products, it was realized that the same process that operates in classical beta decay can also produce positrons (positron emission), along with neutrinos (classical beta decay produces antineutrinos). In a more common analogous process, called electron capture, some proton-rich nuclides were found to capture their own atomic electrons instead of emitting positrons, and subsequently these nuclides emit only a neutrino and a gamma ray from the excited nucleus (and often also Auger electrons and characteristic X-rays, as a result of the re-ordering of electrons to fill the place of the missing captured electron). These types of decay involve the nuclear capture of electrons or emission of electrons or positrons, and thus acts to move a nucleus toward the ratio of neutrons to protons that has the least energy for a given total number of nucleons. This consequently produces a more stable (lower energy) nucleus.

(A theoretical process of positron capture, analogous to electron capture, is possible in antimatter atoms, but has not been observed, as complex antimatter atoms beyond antihelium are not experimentally available. Such a decay would require antimatter atoms at least as complex as beryllium-7, which is the lightest known isotope of normal matter to undergo decay by electron capture.)

Shortly after the discovery of the neutron in 1932, Enrico Fermi realized that certain rare beta-decay reactions immediately yield neutrons as a decay particle (neutron emission). Isolated proton emission was eventually observed in some elements. It was also found that some heavy elements may undergo spontaneous fission into products that vary in composition. In a phenomenon called cluster decay, specific combinations of neutrons and protons other than alpha particles (helium nuclei) were found to be spontaneously emitted from atoms.

Other types of radioactive decay were found to emit previously-seen particles, but via different mechanisms. An example is internal conversion, which results in an initial electron emission, and then often further characteristic X-rays and Auger electrons emissions, although the internal conversion process involves neither beta nor gamma decay. A neutrino is not emitted, and none of the electron(s) and photon(s) emitted originate in the nucleus, even though the energy to emit all of them does originate there. Internal conversion decay, like isomeric transition gamma decay and neutron emission, involves the release of energy by an excited nuclide, without the transmutation of one element into another.

Rare events that involve a combination of two beta-decay type events happening simultaneously are known. Any decay process that does not violate the conservation of energy or momentum laws (and perhaps other particle conservation laws) is permitted to happen, although not all have been detected. An interesting example discussed in a final section, is bound state beta decay of rhenium-187. In this process, beta electron-decay of the parent nuclide is not accompanied by beta electron emission, because the beta particle has been captured into the K-shell of the emitting atom. An antineutrino is emitted, as in all negative beta decays.

Radionuclides can undergo a number of different reactions. These are summarized in the following table. A nucleus with mass number A and atomic number Z is represented as (A, Z). The column "Daughter nucleus" indicates the difference between the new nucleus and the original nucleus. Thus, $(A - 1, Z)$ means that the mass number is one less than before, but the atomic number is the same as before.

If energy circumstances are favorable, a given radionuclide may undergo many competing types of decay, with some atoms decaying by one route, and others decaying by another. An example is copper-64, which has 29 protons, and 35 neutrons, which decays with a half-life of about 12.7 hours. This isotope has one unpaired proton and one unpaired neutron, so either the proton or the neutron can decay to the opposite particle. This particular nuclide (though not all nuclides in this situation) is almost equally likely to decay through positron emission (18%), or through electron capture (43%), as it does through electron emission (39%). The excited energy states resulting from these decays which fail to end in a ground energy state, also produce later internal conversion and gamma decay in almost 0.5% of the time.

More common in heavy nuclides is competition between alpha and beta decay. The daughter nuclides will then normally decay through beta or alpha, respectively, to end up in the same place.

Mode of decay	Participating particles	Daughter nucleus
Decays with emission of nucleons:		
Alpha decay	An alpha particle ($A = 4$, $Z = 2$) emitted from nucleus	$(A - 4, Z - 2)$
Proton emission	A proton ejected from nucleus	$(A - 1, Z - 1)$
Neutron emission	A neutron ejected from nucleus	$(A - 1, Z)$
Double proton emission	Two protons ejected from nucleus simultaneously	$(A - 2, Z - 2)$
Spontaneous fission	Nucleus disintegrates into two or more smaller nuclei and other particles	—
Cluster decay	Nucleus emits a specific type of smaller nucleus (A_1, Z_1) which is larger than an alpha particle	$(A - A_1, Z - Z_1) + (A_1, Z_1)$
Different modes of beta decay:		
β^- decay	A nucleus emits an electron and an electron antineutrino	$(A, Z + 1)$
Positron emission (β^+ decay)	A nucleus emits a positron and an electron neutrino	$(A, Z - 1)$

Electron capture	A nucleus captures an orbiting electron and emits a neutrino; the daughter nucleus is left in an excited unstable state	$(A, Z - 1)$
Bound state beta decay	A free neutron or nucleus beta decays to electron and antineutrino, but the electron is not emitted, as it is captured into an empty K-shell; the daughter nucleus is left in an excited and unstable state. This process is a minority of free neutron decays (0.0004%) due to the low energy of hydrogen ionization, and is suppressed except in ionized atoms that have K-shell vacancies.	$(A, Z + 1)$
Double beta decay	A nucleus emits two electrons and two antineutrinos	$(A, Z + 2)$
Double electron capture	A nucleus absorbs two orbital electrons and emits two neutrinos – the daughter nucleus is left in an excited and unstable state	$(A, Z - 2)$
Electron capture with positron emission	A nucleus absorbs one orbital electron, emits one positron and two neutrinos	$(A, Z - 2)$
Double positron emission	A nucleus emits two positrons and two neutrinos	$(A, Z - 2)$
Transitions between states of the same nucleus:		
Isomeric transition	Excited nucleus releases a high-energy photon (gamma ray)	(A, Z)
Internal conversion	Excited nucleus transfers energy to an orbital electron, which is subsequently ejected from the atom	(A, Z)

Radioactive decay results in a reduction of summed rest mass, once the released energy (the *disintegration energy*) has escaped in some way. Although decay energy is sometimes defined as associated with the difference between the mass of the parent nuclide products and the mass of the decay products, this is true only of rest mass measurements, where some energy has been removed from the product system. This is true because the decay energy must always carry mass with it, wherever it appears according to the formula $E = mc^2$. The decay energy is initially released as the energy of emitted photons plus the kinetic energy of massive emitted particles (that is, particles that have rest mass). If these particles come to thermal equilib-rium with their surroundings and photons are absorbed, then the decay energy is transformed to thermal energy, which retains its mass.

Decay energy therefore remains associated with a certain measure of mass of the decay system, called invariant mass, which does not change during the decay, even though the energy of decay is distributed among decay particles. The energy of photons, the kinetic energy of emitted particles, and, later, the thermal energy of the surrounding matter, all contribute to the invariant mass of the system. Thus, while the sum of the rest masses of the particles is not conserved in radioactive decay, the *system* mass and system invariant mass (and also the system total energy) is conserved throughout any decay process. This is a restatement of the equivalent laws of conservation of energy and conservation of mass.

Radioactive Decay Rates

The *decay rate*, or *activity*, of a radioactive substance is characterized by:

Constant Quantities:

- The *half-life*—$t_{1/2}$, is the time taken for the activity of a given amount of a radioactive sub-stance to decay to half of its initial value.

- The *decay constant*— λ, "lambda" the inverse of the mean lifetime, sometimes referred to as simply *decay rate*.

- The *mean lifetime*— τ, "tau" the average lifetime (1/e life) of a radioactive particle before decay.

Although these are constants, they are associated with the statistical behavior of populations of atoms. In consequence, predictions using these constants are less accurate for minuscule samples of atoms.

In principle a half-life, a third-life, or even a $(1/\sqrt{2})$-life, can be used in exactly the same way as half-life; but the mean life and half-life $t_{1/2}$ have been adopted as standard times associated with exponential decay.

Time-variable Quantities:

- *Total activity*— A, is the number of decays per unit time of a radioactive sample.

- *Number of particles*—N, is the total number of particles in the sample.

- *Specific activity*—S_A, number of decays per unit time per amount of substance of the sample at time set to zero (t = 0). "Amount of substance" can be the mass, volume or moles of the initial sample.

These are related as follows:

$$t_{1/2} = \frac{\ln(2)}{\lambda} = \tau \ln(2)$$

$$A = -\frac{dN}{dt} = \lambda N$$

$$S_A a_0 = -\frac{dN}{dt}\Big|_{t=0} = \lambda N_0$$

where N_0 is the initial amount of active substance — substance that has the same percentage of unstable particles as when the substance was formed.

Mathematics of Radioactive Decay

Universal Law of Radioactive Decay

Radioactivity is one very frequently given example of exponential decay. The law describes the statistical behaviour of a large number of nuclides, rather than individual atoms. In the following formalism, the number of nuclides or the nuclide population N, is of course a discrete variable (a natural number)—but for any physical sample N is so large that it can be treated as a continuous variable. Differential calculus is needed to set up differential equations for the modelling the behaviour of the nuclear decay.

The mathematics of radioactive decay depend on a key assumption that a nucleus of a radionuclide has no "memory" or way of translating its history into its present behavior. A nucleus does not "age" with the passage of time. Thus, the probability of its breaking down does not increase with time, but stays constant no matter how long the nucleus has existed. This constant probability may vary greatly between different types of nuclei, leading to the many different observed decay rates. However, whatever the probability is, it does not change. This is in marked contrast to complex objects which do show aging, such as automobiles and humans. These systems do have a chance of breakdown per unit of time, that increases from the moment they begin their existence.

One-decay Process

Consider the case of a nuclide A that decays into another B by some process $A \rightarrow B$ (emission of other particles, like electron neutrinos ν e and electrons e^- as in beta decay, are irrelevant in what follows). The decay of an unstable nucleus is entirely random and it is impossible to predict when a particular atom will decay. However, it is equally likely to decay at any instant in time. Therefore, given a sample of a particular radioisotope, the number of decay events $-dN$ expected to occur in a small interval of time dt is proportional to the number of atoms present N, that is

$$-\frac{dN}{dt} \propto N.$$

Particular radionuclides decay at different rates, so each has its own decay constant λ. The expected decay $-dN/N$ is proportional to an increment of time, dt:

$$\boxed{-\frac{dN}{N} = \lambda dt}$$

The negative sign indicates that N decreases as time increases, as the decay events follow one after another. The solution to this first-order differential equation is the function:

$$N(t) = N_0 e^{-\lambda t} = N_0 e^{-t/\tau},$$

where N_0 is the value of N at time $t = 0$.

We have for all time t:

$$N_A + N_B = N_{total} = N_{A0},$$

where N_{total} is the constant number of particles throughout the decay process, which is equal to the initial number of A nuclides since this is the initial substance.

If the number of non-decayed A nuclei is:

$$N_A = N_{A0} e^{-\lambda t}$$

then the number of nuclei of B, i.e. the number of decayed A nuclei, is

$$N_B = N_{A0} - N_A = N_{A0} - N_{A0}e^{-\lambda t} = N_{A0}\left(1 - e^{-\lambda t}\right).$$

The number of decays observed over a given interval obeys Poisson statistics. If the average number of decays is $<N>$, the probability of a given number of decays N is

$$P(N) = \frac{N^N \exp(-N)}{N!}.$$

Chain-decay Processes

Chain of two Decays

Now consider the case of a chain of two decays: one nuclide A decaying into another B by one process, then B decaying into another C by a second process, i.e. $A \rightarrow B \rightarrow C$. The previous equation cannot be applied to the decay chain, but can be generalized as follows. Since A decays into B, *then* B decays into C, the activity of A adds to the total number of B nuclides in the present sample, *before* those B nuclides decay and reduce the number of nuclides leading to the later sample. In other words, the number of second generation nuclei B increases as a result of the first generation nuclei decay of A, and decreases as a result of its own decay into the third generation nuclei C. The sum of these two terms gives the law for a decay chain for two nuclides:

$$\frac{dN_B}{dt} = -\lambda_B N_B + \lambda_A N_A.$$

The rate of change of N_B, that is dN_B/dt, is related to the changes in the amounts of A and B, N_B can increase as B is produced from A and decrease as B produces C.

Re-writing using the previous results:

$$\boxed{\frac{dN_B}{dt} = -\lambda_B N_B + \lambda_A N_{A0}e^{-\lambda_A t}}$$

The subscripts simply refer to the respective nuclides, i.e. N_A is the number of nuclides of type A, N_{A0} is the initial number of nuclides of type A, λ_A is the decay constant for A - and similarly for nuclide B. Solving this equation for N_B gives:

$$N_B = \frac{N_{A0}\lambda_A}{\lambda_B - \lambda_A}\left(e^{-\lambda_A t} - e^{-\lambda_B t}\right).$$

In the case where B is a stable nuclide ($\lambda_B = 0$), this equation reduces to the previous solution:

$$\lim_{\lambda_B \to 0}\left[\frac{N_{A0}\lambda_A}{\lambda_B - \lambda_A}\left(e^{-\lambda_A t} - e^{-\lambda_B t}\right)\right] = \frac{N_{A0}\lambda_A}{0 - \lambda_A}\left(e^{-\lambda_A t} - 1\right) = N_{A0}\left(1 - e^{-\lambda_A t}\right),$$

as shown above for one decay. The solution can be found by the integration factor method, where the integrating factor is $e^{\lambda_B t}$. This case is perhaps the most useful, since it can derive both the one-decay equation (above) and the equation for multi-decay chains (below) more directly.

Chain of Any Number of Decays

For the general case of any number of consecutive decays in a decay chain, i.e. $A_1 \rightarrow A_2 \cdots \rightarrow A_i \cdots \rightarrow A_D$, where D is the number of decays and i is a dummy index ($i = _{1, 2, 3, \ldots D}$), each nuclide population can be found in terms of the previous population. In this case $N_2 = 0$, $N_3 = 0,\ldots, N_D = 0$. Using the above result in a recursive form:

$$\frac{\mathrm{d}N_j}{\mathrm{d}t} = -\lambda_j N_j + \lambda_{j-1} N_{(j-1)0} e^{-\lambda_{j-1} t}.$$

The general solution to the recursive problem is given by *Bateman's equations*:

> **Bateman's equations**
>
> $$N_D = \frac{N_1(0)}{\lambda_D} \sum_{i=1}^{D} \lambda_i c_i e^{-\lambda_i t}$$
>
> $$c_i = \prod_{j=1, i \neq j}^{D} \frac{\lambda_j}{\lambda_j - \lambda_i}$$

Alternative Decay Modes

In all of the above examples, the initial nuclide decays into only one product. Consider the case of one initial nuclide that can decay into either of two products, that is $A \rightarrow B$ and $A \rightarrow C$ in parallel. For example, in a sample of potassium-40, 89.3% of the nuclei decay to calcium-40 and 10.7% to argon-40. We have for all time t:

$$N = N_A + N_B + N_C$$

which is constant, since the total number of nuclides remains constant. Differentiating with respect to time:

$$ed \frac{\mathrm{d}N_A}{\mathrm{d}t} = -\left(\frac{\mathrm{d}N_B}{\mathrm{d}t} + \frac{\mathrm{d}N_C}{\mathrm{d}t} \right) - \lambda N_A = -N_A \left(\lambda_B + \lambda_C \right)$$

defining the *total decay constant* λ in terms of the sum of *partial decay constants* λ_B and λ_C:

$$\lambda = \lambda_B + \lambda_C.$$

Notice that

$$\frac{\mathrm{d}N_A}{\mathrm{d}t} < 0, \frac{\mathrm{d}N_B}{\mathrm{d}t} > 0, \frac{\mathrm{d}N_C}{\mathrm{d}t} > 0$$

Solving this equation for N_A:

$$N_A = N_{A0} e^{-\lambda t}.$$

where N_{A0} is the initial number of nuclide A. When measuring the production of one nuclide, one can only observe the total decay constant λ. The decay constants λ_B and λ_C determine the probability for the decay to result in products B or C as follows:

$$N_B = \frac{\lambda_B}{\lambda} N_{A0} \left(1 - e^{-\lambda t}\right),$$

$$N_C = \frac{\lambda_C}{\lambda} N_{A0} \left(1 - e^{-\lambda t}\right).$$

because the fraction λ_B/λ of nuclei decay into B while the fraction λ_C/λ of nuclei decay into C.

Corollaries of the Decay Laws

The above equations can also be written using quantities related to the number of nuclide particles N in a sample;

- The activity: $A = \lambda N$.

- The amount of substance: $n = N/L$.

- The mass: $M = A_r n = A_r N/L$.

where $L = 6.022 \times 10^{23}$ is Avogadro's constant, A_r is the relative atomic mass number, and the amount of the substance is in moles.

Decay Timing: Definitions and Relations

Time Constant and Mean-life

For the one-decay solution $A \rightarrow B$:

$$N = N_0 e^{-\lambda t} = N_0 e^{-t/\tau},$$

the equation indicates that the decay constant λ has units of t^{-1}, and can thus also be represented as $1/\tau$, where τ is a characteristic time of the process called the *time constant*.

In a radioactive decay process, this time constant is also the mean lifetime for decaying atoms. Each atom "lives" for a finite amount of time before it decays, and it may be shown that this mean lifetime is the arithmetic mean of all the atoms' lifetimes, and that it is τ, which again is related to the decay constant as follows:

$$\tau = \frac{1}{\lambda}.$$

This form is also true for two-decay processes simultaneously $A \rightarrow B + C$, inserting the equivalent values of decay constants (as given above)

$$\lambda = \lambda_B + \lambda_C$$

into the decay solution leads to:

$$\frac{1}{\tau} = \lambda = \lambda_B + \lambda_C = \frac{1}{\tau_B} + \frac{1}{\tau_C}$$

Half-life

A more commonly used parameter is the half-life. Given a sample of a particular radionuclide, the half-life is the time taken for half the radionuclide's atoms to decay. For the case of one-decay nuclear reactions:

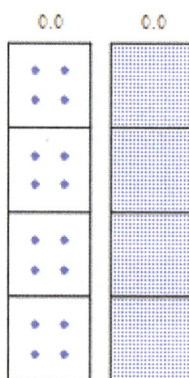

Simulation of many identical atoms undergoing radioactive decay, starting with either 4 atoms (left) or 400 (right). The number at the top indicates how many half-lives have elapsed.

$$N = N_0 e^{-\lambda t} = N_0 e^{-t/\tau},$$

the half-life is related to the decay constant as follows: set $N = N_0/2$ and $t = T_{1/2}$ to obtain

$$t_{1/2} = \frac{\ln 2}{\lambda} = \tau \ln 2.$$

This relationship between the half-life and the decay constant shows that highly radioactive substances are quickly spent, while those that radiate weakly endure longer. Half-lives of known radionuclides vary widely, from more than 10^{19} years, such as for the very nearly stable nuclide ^{209}Bi, to 10^{-23} seconds for highly unstable ones.

The factor of ln(2) in the above relations results from the fact that concept of "half-life" is merely a way of selecting a different base other than the natural base e for the lifetime expression. The time constant τ is the e -1 -life, the time until only $1/e$ remains, about 36.8%, rather than the 50% in the half-life of a radionuclide. Thus, τ is longer than $t_{1/2}$. The following equation can be shown to be valid:

$$N(t) = N_0 e^{-t/\tau} = N_0 2^{-t/t_{1/2}}.$$

Since radioactive decay is exponential with a constant probability, each process could as easily be described with a different constant time period that (for example) gave its "(1/3)-life" (how long until only 1/3 is left) or "(1/10)-life" (a time period until only 10% is left), and so on. Thus, the choice of τ and $t_{1/2}$ for marker-times, are only for convenience, and from convention. They reflect a fundamental principle only in so much as they show that the *same proportion* of a given radioactive substance will decay, during any time-period that one chooses.

Mathematically, the nth life for the above situation would be found in the same way as above—by setting $N = N_o/n$, $t = T_{1/n}$ and substituting into the decay solution to obtain

$$t_{1/n} = \frac{\ln n}{\lambda} = \tau \ln n.$$

Example

A sample of ^{14}C has a half-life of 5,730 years and a decay rate of 14 disintegration per minute (dpm) per gram of natural carbon.

If an artifact is found to have radioactivity of 4 dpm per gram of its present C, we can find the approximate age of the object using the above equation:

$$N = N_0 e^{-t/\tau},$$

$$\text{where: } \frac{N}{N_0} = 4/14 \approx 0.286,$$

$$\tau = \frac{T_{1/2}}{\ln 2} \approx 8267 \text{ years},$$

$$t = -\tau \ln \frac{N}{N_0} \approx 10356 \text{ years}.$$

Changing Decay Rates

The radioactive decay modes of electron capture and internal conversion are known to be slightly sensitive to chemical and environmental effects that change the electronic structure of the atom, which in turn affects the presence of 1s and 2s electrons that participate in the decay process. A small number of mostly light nuclides are affected. For example, chemical bonds can affect the rate of electron capture to a small degree (in general, less than 1%) depending on the proximity of electrons to the nucleus. In ^7Be, a difference of 0.9% has been observed between half-lives in metallic and insulating environments. This relatively large effect is because beryllium is a small atom whose valence electrons are in 2s atomic orbitals, which are subject to electron capture in ^7Be because (like all s atomic orbitals in all atoms) they naturally penetrate into the nucleus.

In 1992, Jung et al. of the Darmstadt Heavy-Ion Research group observed an accelerated β decay of $^{163}Dy^{66+}$. Although neutral ^{163}Dy is a stable isotope, the fully ionized $^{163}Dy^{66+}$ undergoes β decay into the K and L shells with a half-life of 47 days.

Rhenium-187 is another spectacular example. [187]Re normally beta decays to [187]Os with a half-life of 41.6×10^9 years, but studies using fully ionised [187]Re atoms (bare nuclei) have found that this can decrease to only 33 years. This is attributed to "bound-state β^- decay" of the fully ionised atom – the electron is emitted into the "K-shell" (1s atomic orbital), which cannot occur for neutral atoms in which all low-lying bound states are occupied.

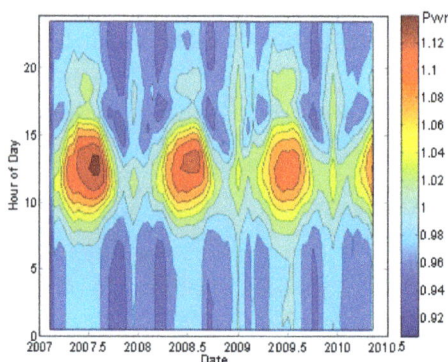

Decay rate of radon-222 as a function of date and time of day. The color-bar gives the power of the observed signal and represents ~4% seasonal decay rate variation.

A number of experiments have found that decay rates of other modes of artificial and naturally occurring radioisotopes are, to a high degree of precision, unaffected by external conditions such as temperature, pressure, the chemical environment, and electric, magnetic, or gravitational fields. Comparison of laboratory experiments over the last century, studies of the Oklo natural nuclear reactor (which exemplified the effects of thermal neutrons on nuclear decay), and astrophysical observations of the luminosity decays of distant supernovae (which occurred far away so the light has taken a great deal of time to reach us), for example, strongly indicate that unperturbed decay rates have been constant (at least to within the limitations of small experimental errors) as a function of time as well.

Recent results suggest the possibility that decay rates might have a weak dependence on environmental factors. It has been suggested that measurements of decay rates of silicon-32, manganese-54, and radium-226 exhibit small seasonal variations (of the order of 0.1%), while the decay of radon-222 exhibits large 4% peak-to-peak seasonal variations, proposed to be related to either solar flare activity or distance from the Sun. However, such measurements are highly susceptible to systematic errors, and a subsequent paper has found no evidence for such correlations in seven other isotopes ([22]Na, [44]Ti, [108]Ag, [121]Sn, [133]Ba, [241]Am, [238]Pu), and sets upper limits on the size of any such effects.

Theoretical Basis of Decay Phenomena

The neutrons and protons that constitute nuclei, as well as other particles that approach close enough to them, are governed by several interactions. The strong nuclear force, not observed at the familiar macroscopic scale, is the most powerful force over subatomic distances. The electrostatic force is almost always significant, and, in the case of beta decay, the weak nuclear force is also involved.

The interplay of these forces produces a number of different phenomena in which energy may be released by rearrangement of particles in the nucleus, or else the change of one type of particle into others. These rearrangements and transformations may be hindered energetically, so that they do not occur immediately. In certain cases, random quantum vacuum fluctuations are theorized to promote relaxation to a lower energy state (the "decay") in a phenomenon known as quantum tunneling. Radioactive decay half-life of nuclides has been measured over timescales of 55 orders of magnitude, from 2.3×10^{-23} seconds (for hydrogen-7) to 6.9×10^{31} seconds (for tellurium-128). The limits of these timescales are set by the sensitivity of instrumentation only, and there are no known natural limits to how brief or long a decay half life for radioactive decay of a radionuclide may be.

The decay process, like all hindered energy transformations, may be analogized by a snowfield on a mountain. While friction between the ice crystals may be supporting the snow's weight, the system is inherently unstable with regard to a state of lower potential energy. A disturbance would thus facilitate the path to a state of greater entropy: The system will move towards the ground state, producing heat, and the total energy will be distributable over a larger number of quantum states. Thus, an avalanche results. The *total* energy does not change in this process, but, because of the second law of thermodynamics, avalanches have only been observed in one direction and that is toward the "ground state" — the state with the largest number of ways in which the available energy could be distributed.

Such a collapse (a *decay event*) requires a specific activation energy. For a snow avalanche, this energy comes as a disturbance from outside the system, although such disturbances can be arbitrarily small. In the case of an excited atomic nucleus, the arbitrarily small disturbance comes from quantum vacuum fluctuations. A radioactive nucleus (or any excited system in quantum mechanics) is unstable, and can, thus, *spontaneously* stabilize to a less-excited system. The resulting transformation alters the structure of the nucleus and results in the emission of either a photon or a high-velocity particle that has mass (such as an electron, alpha particle, or other type).

Occurrence and Applications

According to the Big Bang theory, stable isotopes of the lightest five elements (H, He, and traces of Li, Be, and B) were produced very shortly after the emergence of the universe, in a process called Big Bang nucleosynthesis. These lightest stable nuclides (including deuterium) survive to today, but any radioactive isotopes of the light elements produced in the Big Bang (such as tritium) have long since decayed. Isotopes of elements heavier than boron were not produced at all in the Big Bang, and these first five elements do not have any long-lived radioisotopes. Thus, all radioactive nuclei are, therefore, relatively young with respect to the birth of the universe, having formed later in various other types of nucleosynthesis in stars (in particular, supernovae), and also during ongoing interactions between stable isotopes and energetic particles. For example, carbon-14, a radioactive nuclide with a half-life of only 5,730 years, is constantly produced in Earth's upper atmosphere due to interactions between cosmic rays and nitrogen.

Nuclides that are produced by radioactive decay are called radiogenic nuclides, whether they themselves are stable or not. There exist stable radiogenic nuclides that were formed from short-lived extinct radionuclides in the early solar system. The extra presence of these stable radiogenic nuclides (such as Xe-129 from primordial I-129) against the background of primordial stable nuclides can be inferred by various means.

Radioactive decay has been put to use in the technique of radioisotopic labeling, which is used to track the passage of a chemical substance through a complex system (such as a living organism). A sample of the substance is synthesized with a high concentration of unstable atoms. The presence of the substance in one or another part of the system is determined by detecting the locations of decay events.

On the premise that radioactive decay is truly random (rather than merely chaotic), it has been used in hardware random-number generators. Because the process is not thought to vary significantly in mechanism over time, it is also a valuable tool in estimating the absolute ages of certain materials. For geological materials, the radioisotopes and some of their decay products become trapped when a rock solidifies, and can then later be used (subject to many well-known qualifications) to estimate the date of the solidification. These include checking the results of several simultaneous processes and their products against each other, within the same sample. In a similar fashion, and also subject to qualification, the rate of formation of carbon-14 in various eras, the date of formation of organic matter within a certain period related to the isotope's half-life may be estimated, because the carbon-14 becomes trapped when the organic matter grows and incorporates the new carbon-14 from the air. Thereafter, the amount of carbon-14 in organic matter decreases according to decay processes that may also be independently cross-checked by other means (such as checking the carbon-14 in individual tree rings, for example).

Szilard–Chalmers Effect

The *Szilard–Chalmers effect* is defined as the breaking of a chemical bond between an atom and the molecule which the atom is part of, as a result of a nuclear reaction of the atom. The effect can be used to separate isotopes by chemical means. The discovery of this effect is due to L. Szilárd and T.A. Chalmers.

Origins of Radioactive Nuclides

Radioactive primordial nuclides found in the Earth are residues from ancient supernova explosions which occurred before the formation of the solar system. They are the fraction of radionuclides that survived from that time, through the formation of the primordial solar nebula, through planet accretion, and up to the present time. The naturally occurring short-lived radiogenic radionuclides found in today's rocks, are the daughters of those radioactive primordial nuclides. Another minor source of naturally occurring radioactive nuclides are cosmogenic nuclides, that are formed by cosmic ray bombardment of material in the Earth's atmosphere or crust. The decay of the radionuclides in rocks of the Earth's mantle and crust contribute significantly to Earth's internal heat budget.

Decay Chains and Multiple Modes

Gamma-ray energy spectrum of uranium ore (inset). Gamma-rays are emitted by decaying nuclides, and the gamma-ray energy can be used to characterize the decay (which nuclide is decaying to which). Here, using the gamma-ray spectrum, several nuclides that are typical of the decay chain of ^{238}U have been identified: ^{226}Ra, ^{214}Pb, ^{214}Bi.

The daughter nuclide of a decay event may also be unstable (radioactive). In this case, it too will decay, producing radiation. The resulting second daughter nuclide may also be radioactive. This can lead to a sequence of several decay events called a *decay chain*. Eventually, a stable nuclide is produced.

An example is the natural decay chain of ^{238}U:

- Uranium-238 decays, through alpha-emission, with a half-life of 4.5 billion years to thorium-234

- which decays, through beta-emission, with a half-life of 24 days to protactinium-234

- which decays, through beta-emission, with a half-life of 1.2 minutes to uranium-234

- which decays, through alpha-emission, with a half-life of 240 thousand years to thorium-230

- which decays, through alpha-emission, with a half-life of 77 thousand years to radium-226

- which decays, through alpha-emission, with a half-life of 1.6 thousand years to radon-222

- which decays, through alpha-emission, with a half-life of 3.8 days to polonium-218

- which decays, through alpha-emission, with a half-life of 3.1 minutes to lead-214

- which decays, through beta-emission, with a half-life of 27 minutes to bismuth-214

- which decays, through beta-emission, with a half-life of 20 minutes to polonium-214

- which decays, through alpha-emission, with a half-life of 160 microseconds to lead-210

- which decays, through beta-emission, with a half-life of 22 years to bismuth-210

- which decays, through beta-emission, with a half-life of 5 days to polonium-210

- which decays, through alpha-emission, with a half-life of 140 days to lead-206, which is a stable nuclide.

Some radionuclides may have several different paths of decay. For example, approximately 36% of bismuth-212 decays, through alpha-emission, to thallium-208 while approximately 64% of bismuth-212 decays, through beta-emission, to polonium-212. Both thallium-208 and polonium-212 are radioactive daughter products of bismuth-212, and both decay directly to stable lead-208.

References

- Suck, Jens-Boie; Schreiber, M.; Häussler, Peter (2002). Quasicrystals: An Introduction to Structure, Physical Properties and Applications. Springer Science & Business Media. pp. 1–. ISBN 978-3-540-64224-4

- Paterson, Alan L. T. (1999). Groupoids, inverse semigroups, and their operator algebras. Springer. p. 164. ISBN 0-8176-4051-7.

- Sluckin, T. J.; Dunmur, D. A. & Stegemeyer, H. (2004). Crystals That Flow – classic papers from the history of liquid crystals. London: Taylor & Francis. ISBN 0-415-25789-1.

- Castellano, Joseph A. (2005). Liquid Gold: The Story of Liquid Crystal Displays and the Creation of an Industry. World Scientific Publishing. ISBN 978-981-238-956-5.

- Stabin, Michael G. (2007). "3". Radiation Protection and Dosimetry: An Introduction to Health Physics. Springer. doi:10.1007/978-0-387-49983-3. ISBN 978-0-387-49982-6.

- Best, Lara; Rodrigues, George; Velker, Vikram (2013). "1.3". Radiation Oncology Primer and Review. Demos Medical Publishing. ISBN 978-1-62070-004-4.

- Loveland, W.; Morrissey, D.; Seaborg, G.T. (2006). Modern Nuclear Chemistry. Wiley-Interscience. p. 57. ISBN 0-471-11532-0.

- Shultis, John K.; Faw, Richard E. (2007). Fundamentals of Nuclear Science and Engineering (2nd ed.). CRC Press. p. 175. ISBN 978-1-4398-9408-8.

- L'Annunziata, Michael F. (2007). Radioactivity: Introduction and History. Amsterdam, Netherlands: Elsevier Science. p. 2. ISBN 9780080548883.

- Clayton, Donald D. (1983). Principles of Stellar Evolution and Nucleosynthesis (2nd ed.). University of Chicago Press. p. 75. ISBN 0-226-10953-4.

Bonding in Solids: An Integrated Study

Solids can be characterized as per the nature of bonding within their atomic components. These bonds can be distinguished into four types which are covalent bonding, ionic bonding, metallic bonding and weak inter molecular bonding. This text is an overview of the subject matter incorporating all the major aspects of the bonds found in solids.

Bonding in Solids

Solids can be classified according to the nature of the bonding between their atomic or molecular components. The traditional classification distinguishes four kinds of bonding:

- Covalent bonding, which forms network covalent solids (sometimes called simply "covalent solids")

- Ionic bonding, which forms ionic solids

- Metallic bonding, which forms metallic solids

- Weak inter molecular bonding, which forms molecular solids (sometimes anomalously called "covalent solids")

Typical members of these classes have distinctive electron distributions, thermodynamic, electronic, and mechanical properties. In particular, the binding energies of these interactions vary widely. Bonding in solids can be of mixed or intermediate kinds, however, hence not all solids have the typical properties of a particular class, and some can be described as intermediate forms.

Basic Classes of Solids

Network Covalent Solids

A network covalent solid consists of atoms held together by a network of covalent bonds (pairs of electrons shared between atoms of similar electronegativity), and hence can be regarded as a single, large molecule. The classic example is diamond; other examples include silicon, quartz and graphite.

Properties

- High strength

- High elastic modulus

- High melting point

- Brittle

Their strength, stiffness, and high melting points are consequences of the strength and stiffness of the covalent bonds that hold them together. They are also characteristically brittle because the directional nature of covalent bonds strongly resists the shearing motions associated with plastic flow, and are, in effect, broken when shear occurs. This property results in brittleness for reasons studied in the field of fracture mechanics. Network covalent solids vary from insulating to semi-conducting in their behavior, depending on the band gap of the material.

Ionic Solids

A standard ionic solid consists of atoms held together by ionic bonds, that is, by the electrostatic attraction of opposite charges (the result of transferring electrons from atoms with lower electro-negativity to atoms with higher electronegativity). Among the ionic solids are compounds formed by alkali and alkaline earth metals in combination with halogens; a classic example is table salt, sodium chloride.

Ionic solids are typically of intermediate strength and extremely brittle. Melting points are typically moderately high, but some combinations of molecular cations and anions yield an ionic liquid with a freezing point below room temperature. Vapor pressures in all instances are extraordinarily low; this is a consequence of the large energy required to move a bare charge (or charge pair) from an ionic medium into free space.

Metallic Solids

Metallic solids are held together by a high density of shared, delocalized electrons, resulting in metallic bonding. Classic examples are metals such as copper and aluminum, but some materials are metals in an electronic sense but have negligible metallic bonding in a mechanical or thermo-dynamic sense. Metallic solids have, by definition, no band gap at the Fermi level and hence are conducting.

Solids with purely metallic bonding are characteristically ductile and, in their pure forms, have low strength; melting points can be very low (*e.g.*, Mercury melts at 234 K (−39 °C)). These properties are consequences of the non-directional and non-polar nature of metallic bonding, which allows atoms (and planes of atoms in a crystal lattice) to move past one another without disrupting their bonding interactions. Metals can be strengthened by introducing crystal defects (for example, by alloying) that interfere with the motion of dislocations that mediate plastic deformation. Further, some transition metals exhibit directional bonding in addition to metallic bonding; this increases shear strength and reduces ductility, imparting some of the characteristics of a covalent solid (an intermediate case below).

Molecular Solids

A classic molecular solid consists of small, non-polar covalent molecules, and is held together by London dispersion forces (van der Waals forces); a classic example is paraffin wax. These forces are weak, resulting in pairwise interatomic binding energies on the order of 1/100 those of covalent, ionic, and metallic bonds. Binding energies tend to increase with increasing molecular size and polarity.

Solids that are composed of small, weakly bound molecules are mechanically weak and have low melting points; an extreme case is solid molecular hydrogen, which melts at 14 K (−259 °C). The non-directional nature of dispersion forces typically allows easy plastic deformation, as planes of molecules can slide over one another without seriously disrupting their attractive interactions. Molecular solids are typically insulators with large band gaps.

Solids of Intermediate Kinds

The four classes of solids permit six pairwise intermediate forms:

Ionic to Network Covalent

Covalent and ionic bonding form a continuum, with ionic character increasing with increasing difference in the electronegativity of the participating atoms. Covalent bonding corresponds to sharing of a pair of electrons between two atoms of essentially equal electronegativity (for example, C–C and C–H bonds in aliphatic hydrocarbons). As bonds become more polar, they become increasingly ionic in character. Metal oxides vary along the iono-covalent spectrum. The Si–O bonds in quartz, for example, are polar yet largely covalent, and are considered to be of mixed character.

Metallic to Network Covalent

What is in most respects a purely covalent structure can support metallic delocalization of electrons; metallic carbon nanotubes are one example. Transition metals and intermetallic compounds based on transition metals can exhibit mixed metallic and covalent bonding, resulting in high shear strength, low ductility, and elevated melting points; a classic example is tungsten.

Molecular to Network Covalent

Materials can be intermediate between molecular and network covalent solids either because of the intermediate organization of their covalent bonds, or because the bonds themselves are of an intermediate kind.

Intermediate Organization of Covalent Bonds:

Regarding the organization of covalent bonds, recall that classic molecular solids, as stated above, consist of small, non-polar covalent molecules. The example given, paraffin wax, is a member of a family of hydrocarbon molecules of differing chain lengths, with high-density polyethylene at the long-chain end of the series. High-density polyethylene can be a strong material: when the hydrocarbon chains are well aligned, the resulting fibers rival the strength of steel. The covalent bonds in this material form extended structures, but do not form a continuous network. With cross-linking, however, polymer networks can become continuous, and a series of materials spans the range from Cross-linked polyethylene, to rigid thermosetting resins, to hydrogen-rich amorphous solids, to vitreous carbon, diamond-like carbons, and ultimately to diamond itself. As this example shows, there can be no sharp boundary between molecular and network covalent solids.

Intermediate Kinds of Bonding:

A solid with extensive hydrogen bonding will be considered a molecular solid, yet strong hydrogen bonds can have a significant degree of covalent character. As noted above, covalent and ionic bonds form a continuum between shared and transferred electrons; covalent and weak bonds form a continuum between shared and unshared electrons. In addition, molecules can be polar, or have polar groups, and the resulting regions of positive and negative charge can interact to produce electrostatic bonding resembling that in ionic solids.

Molecular to Ionic

A large molecule with an ionized group is technically an ion, but its behavior may be largely the result of non-ionic interactions. For example, sodium stearate (the main constituent of traditional soaps) consists entirely of ions, yet it is a soft material quite unlike a typical ionic solid. There is a continuum between ionic solids and molecular solids with little ionic character in their bonding.

Metallic to Molecular

Metallic solids are bound by a high density of shared, delocalized electrons. Although weakly bound molecular components are incompatible with strong metallic bonding, low densities of shared, delocalized electrons can impart varying degrees of metallic bonding and conductivity overlaid on discrete, covalently bonded molecular units, especially in reduced-dimensional systems. Examples include charge transfer complexes.

Metallic to Ionic

The charged components that make up ionic solids cannot exist in the high-density sea of delocalized electrons characteristic of strong metallic bonding. Some molecular salts, however, feature both ionic bonding among molecules and substantial one-dimensional conductivity, indicating a degree of metallic bonding among structural components along the axis of conductivity. Examples include tetrathiafulvalene salts.

Covalent Bond

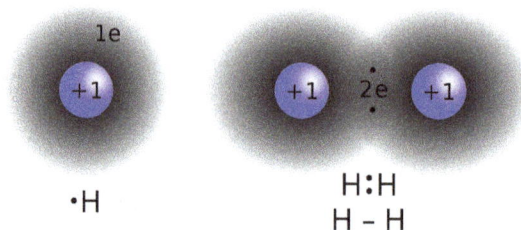

A covalent bond forming H_2 (right) where two hydrogen atoms share the two electrons

A covalent bond, also called a molecular bond, is a chemical bond that involves the sharing of electron pairs between atoms. These electron pairs are known as shared pairs or bonding pairs, and the stable balance of attractive and repulsive forces between atoms, when they share electrons, is known as covalent bonding. For many molecules, the sharing of electrons allows each atom to attain the equivalent of a full outer shell, corresponding to a stable electronic configuration.

Covalent bonding includes many kinds of interactions, including σ-bonding, π-bonding, metal-to-metal bonding, agostic interactions, bent bonds, and three-center two-electron bonds. The term *covalent bond* dates from 1939. The prefix co- means *jointly, associated in action, partnered to a lesser degree,* etc.; thus a "co-valent bond", in essence, means that the atoms share "valence", such as is discussed in valence bond theory.

In the molecule H 2, the hydrogen atoms share the two electrons via covalent bonding. Covalency is greatest between atoms of similar electronegativities. Thus, covalent bonding does not necessarily require that the two atoms be of the same elements, only that they be of comparable electronegativity. Covalent bonding that entails sharing of electrons over more than two atoms is said to be delocalized.

History

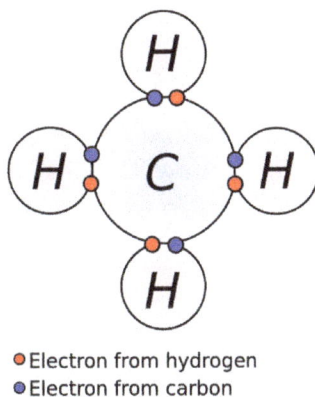

● Electron from hydrogen
● Electron from carbon

Early concepts in covalent bonding arose from this kind of image of the molecule of methane. Covalent bonding is implied in the Lewis structure by indicating electrons shared between atoms.

The term *covalence* in regard to bonding was first used in 1919 by Irving Langmuir in a *Journal of the American Chemical Society* article entitled "The Arrangement of Electrons in Atoms and Molecules". Langmuir wrote that "we shall denote by the term *covalence* the number of pairs of electrons that a given atom shares with its neighbors."

The idea of covalent bonding can be traced several years before 1919 to Gilbert N. Lewis, who in 1916 described the sharing of electron pairs between atoms. He introduced the *Lewis notation* or *electron dot notation* or *Lewis dot structure*, in which valence electrons (those in the outer shell) are represented as dots around the atomic symbols. Pairs of electrons located between atoms represent covalent bonds. Multiple pairs represent multiple bonds, such as double bonds and triple bonds. An alternative form of representation, not shown here, has bond-forming electron pairs represented as solid lines.

Lewis proposed that an atom forms enough covalent bonds to form a full (or closed) outer electron shell. In the methane diagram shown here, the carbon atom has a valence of four and is, therefore, surrounded by eight electrons (the octet rule), four from the carbon itself and four from the hydrogens bonded to it. Each hydrogen has a valence of one and is surrounded by two electrons (a duet rule) – its own one electron plus one from the carbon. The numbers of electrons correspond to full shells in the quantum theory of the atom; the outer shell of a carbon atom is the $n = 2$ shell, which can hold eight electrons, whereas the outer (and only) shell of a hydrogen atom is the $n = 1$ shell, which can hold only two.

While the idea of shared electron pairs provides an effective qualitative picture of covalent bonding, quantum mechanics is needed to understand the nature of these bonds and predict the structures and properties of simple molecules. Walter Heitler and Fritz London are credited with the first successful quantum mechanical explanation of a chemical bond (molecular hydrogen) in 1927. Their work was based on the valence bond model, which assumes that a chemical bond is formed when there is good overlap between the atomic orbitals of participating atoms.

Types of Covalent Bonds

Atomic orbitals (except for s orbitals) have specific directional properties leading to different types of covalent bonds. Sigma (σ) bonds are the strongest covalent bonds and are due to head-on overlapping of orbitals on two different atoms. A single bond is usually a σ bond. Pi (π) bonds are weaker and are due to lateral overlap between p (or d) orbitals. A double bond between two given atoms consists of one σ and one π bond, and a triple bond is one σ and two π bonds.

Covalent bonds are also affected by the electronegativity of the connected atoms which determines the chemical polarity of the bond. Two atoms with equal electronegativity will make nonpolar covalent bonds such as H–H. An unequal relationship creates a polar covalent bond such as with H–Cl.

Covalent Structures

There are several types of structures for covalent substances, including individual molecules, molecular structures, macromolecular structures and giant covalent structures. Individual molecules have strong bonds that hold the atoms together, but there are negligible forces of attraction between molecules. Such covalent substances are usually gases, for example, HCl, SO_2, CO_2, and CH_4. In molecular structures, there are weak forces of attraction. Such covalent substances are low-boiling-temperature liquids (such as ethanol), and low-melting-temperature solids (such as iodine and solid CO_2). Macromolecular structures have large numbers of atoms linked by covalent bonds in chains, including synthetic polymers such as polyethylene and nylon, and biopolymers such as proteins and starch. Network covalent structures (or giant covalent structures) contain large numbers of atoms linked in sheets (such as graphite), or 3-dimensional structures (such as diamond and quartz). These substances have high melting and boiling points, are frequently brittle, and tend to have high electrical resistivity. Elements that have high electronegativity, and the ability to form three or four electron pair bonds, often form such large macromolecular structures.

One- and Three-electron Bonds

Bonds with one or three electrons can be found in radical species, which have an odd number of electrons. The simplest example of a 1-electron bond is found in the dihydrogen cation, H_2^+. One-electron bonds often have about half the bond energy of a 2-electron bond, and are therefore called "half bonds". However, there are exceptions: in the case of dilithium, the bond is actually stronger for the 1-electron Li+ 2 than for the 2-electron Li_2. This exception can be explained in terms of hybridization and inner-shell effects.

2e bond (e.g., CH₄)

3e bond (e.g., NO)

Comparison of the electronic structure of the three-electron bond to the conventional covalent bond.

The simplest example of three-electron bonding can be found in the helium dimer cation, He+ 2. It is considered a "half bond" because it consists of only one shared electron (rather than two); in molecular orbital terms, the third electron is in an anti-bonding orbital which cancels out half of the bond formed by the other two electrons. Another example of a molecule containing a 3-electron bond, in addition to two 2-electron bonds, is nitric oxide, NO. The oxygen molecule, O_2 can also be regarded as having two 3-electron bonds and one 2-electron bond, which accounts for its paramagnetism and its formal bond order of 2. Chlorine dioxide and its heavier analogues bromine dioxide and iodine dioxide also contain three-electron bonds.

Molecules with odd-electron bonds are usually highly reactive. These types of bond are only stable between atoms with similar electronegativities.

Resonance

There are situations whereby a single Lewis structure is insufficient to explain the electron configuration in a molecule, hence a superposition of structures are needed. The same two atoms in such molecules can be bonded differently in different structures (a single bond in one, a double bond in another, or even none at all), resulting in a non-integer bond order. The nitrate ion is one such example with three equivalent structures. The bond between the nitrogen and each oxygen is a double bond in one structure and a single bond in the other two, so that the average bond order for each N–O interaction is 2 + 1 + 1/3 = 4/3.

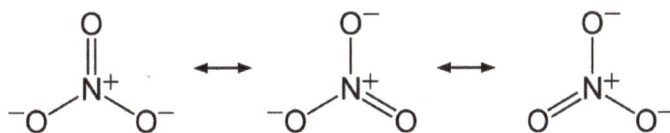

Aromaticity

In organic chemistry, when a molecule with a planar ring obeys Hückel's rule, where the number of π electrons fit the formula $4n + 2$ (where n is an integer), it attains extra stability and symmetry. In benzene, the prototypical aromatic compound, there are 6 π bonding electrons ($n = 1$, $4n + 2 = 6$). These occupy three delocalized π molecular orbitals (molecular orbital theory) or form conjugate π bonds in two resonance structures that linearly combine (valence bond theory), creating a regular hexagon exhibiting a greater stabilization than the hypothetical 1,3,5-cyclohexatriene.

In the case of heterocyclic aromatics and substituted benzenes, the electronegativity differences between different parts of the ring may dominate the chemical behaviour of aromatic ring bonds, which otherwise are equivalent.

Hypervalence

Certain molecules such as xenon difluoride and sulfur hexafluoride have higher co-ordination numbers than would be possible due to strictly covalent bonding according to the octet rule. This is explained by the three-center four-electron bond ("3c–4e") model which interprets the molecular wavefunction in terms of non-bonding highest occupied molecular orbitals in molecular orbital theory and ionic-covalent resonance in valence bond theory.

Electron-deficiency

In three-center two-electron bonds ("3c–2e") three atoms share two electrons in bonding. This type of bonding occurs in electron deficient compounds like diborane. Each such bond (2 per molecule in diborane) contains a pair of electrons which connect the boron atoms to each other in a banana shape, with a proton (nucleus of a hydrogen atom) in the middle of the bond, sharing electrons with both boron atoms. In certain cluster compounds, so-called four-center two-electron bonds also have been postulated.

Quantum Mechanical Description

After the development of quantum mechanics, two basic theories were proposed to provide a quantum description of chemical bonding: valence bond (VB) theory and molecular orbital (MO) theory. A more recent quantum description is given in terms of atomic contributions to the electronic density of states.

Covalency from Atomic Contribution to the Electronic Density of States

In COOP, COHP and BCOOP, evaluation of bond covalency is dependent on the basis set. To overcome this issue, an alternative formulation of the bond covalency can be provided in this way.

The center mass $cm(n,l,m_l,m_s)$ of an atomic orbital $|n,l,m_l,m_s\rangle$, with quantum numbers n, l, m_l, m_s, for atom A is defined as

$$cm^A(n,l,m_l,m_s) = \frac{\int_{E_0}^{E_1} E g^A_{|n,l,m_l,m_s\rangle}(E)\,dE}{\int_{E_0}^{E_1} g^A_{|n,l,m_l,m_s\rangle}(E)\,dE}$$

where $g^A_{|n,l,ml,ms\rangle}(E)$ is the contribution of the atomic orbital $|n,l,m_l,m_s\rangle$ of the atom A to the total electronic density of states $g(E)$ of the solid

$$g(E) = \sum_A \sum_{n,l} \sum_{m_l,m_s} g^A_{|n,l,m_l,m_s\rangle}(E)$$

where the outer sum runs over all atoms A of the unit cell. The energy window $[E_0, E_1]$ is chosen in such a way that it encompasses all relevant bands participating in the bond. If the range to select is unclear, it can be identified in practice by examining the molecular orbitals that describe the electron density along the considered bond.

The relative position $C_{n_A l_A, n_B l_B}$ of the center mass of $|n_A, l_A\rangle$ levels of atom A with respect to the center mass of $|n_B, l_B\rangle$ levels of atom B is given as

$$C_{n_A l_A, n_B l_B} = -\left| cm^A(n_A, l_A) - cm^B(n_B, l_B) \right|$$

where the contributions of the magnetic and spin quantum numbers are summed. According to this definition, the relative position of the A levels with respect to the B levels is

$$C_{A,B} = -\left| cm^A - cm^B \right|$$

where, for simplicity, we may omit the dependence from the principal quantum number n in the notation referring to $C_{n_A l_A, n_B l_B}$.

In this formalism, the greater the value of $C_{A,B}$, the higher the overlap of the selected atomic bands, and thus the electron density described by those orbitals gives a more covalent A–B bond. The quantity $C_{A,B}$ is denoted as the *covalency* of the A–B bond, which is specified in the same units of the energy E.

Ionic Bonding

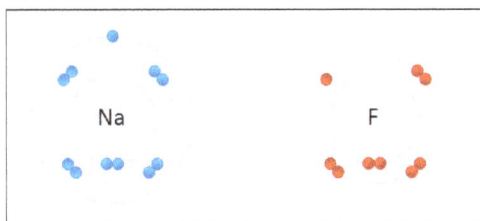

Sodium and fluorine undergoing a redox reaction to form sodium fluoride. Sodium loses its outer electron to give it a stable electron configuration, and this electron enters the fluorine atom exothermically. The oppositely charged ions – typically a great many of them – are then attracted to each other to form a solid.

Ionic bonding is a type of chemical bond that involves the electrostatic attraction between oppositely charged ions, and is the primary interaction occurring in ionic compounds. The ions are atoms that have gained one or more electrons (known as anions, which are negatively charged) and atoms that have lost one or more electrons (known as cations, which are positively charged). This transfer of electrons is known as electrovalence in contrast to covalence. In the simplest case, the cation is a metal atom and the anion is a nonmetal atom, but these ions can be of a more complex nature, e.g. molecular ions like NH_4^+ or SO_4^{2-}. In simpler words, an ionic bond is the transfer of electrons from a metal to a non-metal in order for both atoms to obtain a full valence shell.

It is important to recognize that *clean* ionic bonding – in which one atom or molecule completely share an electron from another – cannot exist: all ionic compounds have some degree of covalent bonding, or electron sharing. Thus, the term "ionic bonding" is given when the ionic character is greater than the covalent character—that is, a bond in which a large electronegativity difference exists between the two atoms, causing the bonding to be more polar (ionic) than in covalent bonding where electrons are shared more equally. Bonds with partially ionic and partially covalent character are called polar covalent bonds.

Ionic compounds conduct electricity when molten or in solution, typically as a solid. Ionic compounds generally have a high melting point, depending on the charge of the ions they consist of. The higher the charges the stronger the cohesive forces and the higher the melting point. They also tend to be soluble in water. Here, the opposite trend roughly holds: the weaker the cohesive forces, the greater the solubility.

Overview

Atoms that have an almost full or almost empty valence shells tend to be very reactive. Atoms that are strongly electronegative (as is the case with halogens) often only have one or two empty orbitals in their valence shell, and frequently bond with other molecules or gain electrons to form anions. Atoms that are weakly electronegative (such as alkali metals) have relatively few valence electrons that can easily be shared with atoms that are strongly electronegative. As a result, weakly electronegative atoms tend to distort their electrons cloud and form cations.

Formation

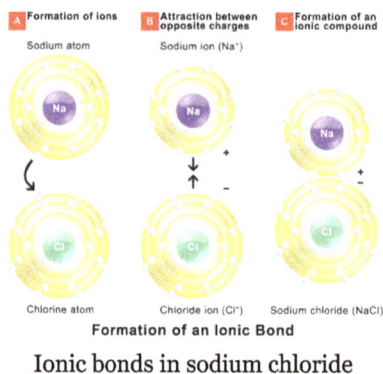

Formation of an Ionic Bond

Ionic bonds in sodium chloride

Ionic bonding can result from a redox reaction when atoms of an element (usually metal), whose ionization energy is low, give some of their electrons to achieve a stable electron configuration. In doing so, cations are formed. The atom of another element (usually nonmetal), whose electron affinity is positive, then accepts the electron(s), again to attain a stable electron configuration, and after accepting electron(s) the atom becomes an anion. Typically, the stable electron configuration is one of the noble gases for elements in the s-block and the p-block, and particular stable electron configurations for d-block and f-block elements. The electrostatic attraction between the anions and cations leads to the formation of a solid with a crystallographic lattice in which the ions are stacked in an alternating fashion. In such a lattice, it is usually not possible to distinguish discrete molecular units, so that the compounds formed are not molecular in nature. However, the ions themselves can be complex and form molecular ions like the acetate anion or the ammonium cation.

For example, common table salt is sodium chloride. When sodium (Na) and chlorine (Cl) are combined, the sodium atoms each lose an electron, forming cations (Na^+), and the chlorine atoms each gain an electron to form anions (Cl$^-$). These ions are then attracted to each other in a 1:1 ratio to form sodium chloride (NaCl).

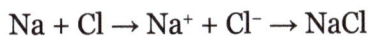

$$Na + Cl \rightarrow Na^+ + Cl^- \rightarrow NaCl$$

However, to maintain charge neutrality, strict ratios between anions and cations are observed so that ionic compounds, in general, obey the rules of stoichiometry despite not being molecular compounds. For compounds that are transitional to the alloys and possess mixed ionic and metallic bonding, this may not be the case anymore. Many sulfides, e.g., do form non-stoichiometric compounds.

Many ionic compounds are referred to as salts as they can also be formed by the neutralization reaction of an Arrhenius base like NaOH with an Arrhenius acid like HCl

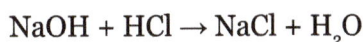

$$NaOH + HCl \rightarrow NaCl + H_2O$$

The salt NaCl is then said to consist of the acid rest Cl$^-$ and the base rest Na$^+$.

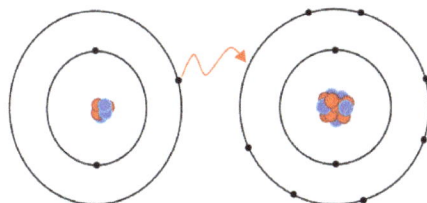

Representation of ionic bonding between lithium and fluorine to form lithium fluoride. Lithium has a low ionization energy and readily gives up its lone valence electron to a fluorine atom, which has a positive electron affinity and accepts the electron that was donated by the lithium atom. The end-result is that lithium is isoelectronic with helium and fluorine is isoelectronic with neon. Electrostatic interaction occurs between the two resulting ions, but typically aggregation is not limited to two of them. Instead, aggregation into a whole lattice held together by ionic bonding is the result.

The removal of electrons from the cation is endothermic, raising the system's overall energy. There may also be energy changes associated with breaking of existing bonds or the addition of more than one electron to form anions. However, the action of the anion's accepting the cation's valence electrons and the subsequent attraction of the ions to each other releases (lattice) energy and, thus, lowers the overall energy of the system.

Ionic bonding will occur only if the overall energy change for the reaction is favorable. In general, the reaction is exothermic, but, e.g., the formation of mercuric oxide (HgO) is endothermic. The charge of the resulting ions is a major factor in the strength of ionic bonding, e.g. a salt C$^+$A$^-$ is held together by electrostatic forces roughly four times weaker than C^{2+}A^{2-} according to Coulombs law, where C and A represent a generic cation and anion respectively. Of course the sizes of the ions and the particular packing of the lattice are ignored in this simple argument.

Structures

Ionic compounds in the solid state form lattice structures. The two principal factors in determining the form of the lattice are the relative charges of the ions and their relative sizes. Some structures are adopted by a number of compounds; for example, the structure of the rock salt sodium chloride is also adopted by many alkali halides, and binary oxides such as MgO. Pauling's rules provide guidelines for predicting and rationalizing the crystal structures of ionic crystals

Bond Strength

For a solid crystalline ionic compound the enthalpy change in forming the solid from gaseous ions is termed the lattice energy. The experimental value for the lattice energy can be determined using the Born-Haber cycle. It can also be calculated (predicted) using the Born-Landé equation as the sum of the electrostatic potential energy, calculated by summing interactions between cations and anions, and a short-range repulsive potential energy term. The electrostatic potential can be expressed in terms of the inter-ionic separation and a constant (Madelung constant) that takes account of the geometry of the crystal. The further away from the nucleus the weaker the shield. The Born-Landé equation gives a reasonable fit to the lattice energy of, e.g., sodium chloride, where the calculated (predicted) value is −756 kJ/mol, which compares to −787 kJ/mol using the Born-Haber cycle.

Polarization Effects

Ions in crystal lattices of purely ionic compounds are spherical; however, if the positive ion is small and/or highly charged, it will distort the electron cloud of the negative ion, an effect summarised in Fajans' rules. This polarization of the negative ion leads to a build-up of extra charge density between the two nuclei, i.e., to partial covalency. Larger negative ions are more easily polarized, but the effect is usually important only when positive ions with charges of 3+ (e.g., Al^{3+}) are involved. However, 2+ ions (Be^{2+}) or even 1+ (Li^+) show some polarizing power because their sizes are so small (e.g., LiI is ionic but has some covalent bonding present). Note that this is not the ionic polarization effect that refers to displacement of ions in the lattice due to the application of an electric field.

Comparison with Covalent Bonding

In ionic bonding, the atoms are bound by attraction of opposite ions, whereas, in covalent bonding, atoms are bound by sharing electrons to attain stable electron configurations. In covalent bonding, the molecular geometry around each atom is determined by valence shell electron pair repulsion VSEPR rules, whereas, in ionic materials, the geometry follows maximum packing rules. One could say that covalent bonding is more *directional* in the sense that the energy penalty for not adhering to the optimum bond angles is large, whereas ionic bonding has no such penalty. There are no shared electron pairs to repel each other, the ions should simply be packed as efficiently as possible. This often leads to much higher coordination numbers. In NaCl, each ion has 6 bonds and all bond angles are 90 degrees. In CsCl the coordination number is 8. By comparison carbon typically has a maximum of four bonds.

Purely ionic bonding cannot exist, as the proximity of the entities involved in the bonding allows some degree of sharing electron density between them. Therefore, all ionic bonding has some covalent character. Thus, bonding is considered ionic where the ionic character is greater than the covalent character. The larger the difference in electronegativity between the two types of atoms involved in the bonding, the more ionic (polar) it is. Bonds with partially ionic and partially covalent character are called polar covalent bonds. For example, Na−Cl and Mg−O interactions have a few percent covalency, while Si−O bonds are usually ~50% ionic and ~50% covalent. Pauling estimated that an electronegativity difference of 1.7 (on the Pauling scale) corresponds to 50% ionic character, so that a difference greater than 50% corresponds to a bond which is predominantly ionic. Ionic character in covalent bonds can be directly measured for atoms having quadrupolar nuclei (^{2}H, ^{14}N, ^{81,79}Br, ^{35,37}Cl or ^{127}I). These nuclei are generally objects of NQR nuclear quadrupole resonance and NMR nuclear magnetic resonance studies. Interactions between the nuclear quadrupole moments Q and the electric field gradients (EFG) are characterized via the nuclear quadrupole coupling constants $QCC = e^2q_{zz}Q/h$ where the eq_{zz} term corresponds to the principal component of the EFG tensor and e is the elementary charge. In turn, the electric field gradient opens the way to description of bonding modes in molecules when the QCC values are accurately determined by NMR or NQR methods.

In general, when ionic bonding occurs in the solid (or liquid) state, it is not possible to talk about a single "ionic bond" between two individual atoms, because the cohesive forces that keep the lattice together are of a more collective nature. This is quite different in the case of covalent bonding, where we can often speak of a distinct bond localized between two particular atoms. However, even if ionic bonding is combined with some covalency, the result is *not* necessarily discrete bonds of a localized character. In such cases, the resulting bonding often requires description in terms of a band structure consisting of gigantic molecular orbitals spanning the entire crystal. Thus, the bonding in the solid often retains its collective rather than localized nature. When the difference in electronegativity is decreased, the bonding may then lead to a semiconductor, a semimetal or eventually a metallic conductor with metallic bonding.

Metallic Bonding

Metallic bonding arises from the electrostatic attractive force between conduction electrons (in the form of an electron cloud of delocalized electrons) and positively charged metal ions. It may be described as the sharing of *free* electrons among a lattice of positively charged ions (cations). Metallic bonding accounts for many physical properties of metals, such as strength, ductility, thermal and electrical resistivity and conductivity, opacity, and luster.

Metallic bonding is not the only type of chemical bonding a metal can exhibit, even as a pure substance. For example, elemental gallium consists of covalently-bound pairs of atoms in both liquid and solid state—these pairs form a crystal lattice with metallic bonding between them. Another example of a metal–metal covalent bond is mercurous ion (Hg2+ 2).

History

As chemistry developed into a science it became clear that metals formed the large majority of the periodic table of the elements and great progress was made in the description of the salts that can be formed in reactions with acids. With the advent of electrochemistry it became clear that metals generally go into solution as positively charged ions and the oxidation reactions of the metals became well understood in the electrochemical series. A picture emerged of metals as positive ions held together by an ocean of negative electrons.

With the advent of quantum mechanics this picture was given more formal interpretation in the form of the free electron model and its further extension, the nearly free electron model. In both of these models the electrons are seen as a gas traveling through the lattice of the solid with an energy that is essentially isotropic in that it depends on the square of the magnitude, *not* the direction of the momentum vector k. In three-dimensional k-space, the set of points of the highest filled levels (the Fermi surface) should therefore be a sphere. In the nearly free correction of the model, box-like Brillouin zones are added to k-space by the periodic potential experienced from the (ionic) lattice, thus mildly breaking the isotropy.

The advent of X-ray diffraction and thermal analysis made it possible to study the structure of crystalline solids, including metals and their alloys, and the construction of phase diagrams became accessible. Despite all this progress the nature of intermetallic compounds and alloys largely remained a mystery and their study was often empirical. Chemists generally steered away from anything that did not seem to follow Dalton's laws of multiple proportions and the problem was considered the domain of a different science, metallurgy.

The almost-free electron model was eagerly taken up by some researchers in this field, notably Hume-Rothery, in an attempt to explain why certain intermetallic alloys with certain compositions would form and others would not. Initially his attempts were quite successful. His idea was to add electrons to inflate the spherical Fermi-balloon inside the series of Brillouin-boxes and determine when a certain box would be full. This indeed predicted a fairly large number of observed alloy compositions. Unfortunately, as soon as cyclotron resonance became available and the shape of the balloon could be determined, it was found that the assumption that the balloon was spherical did not hold at all, except perhaps in the case of caesium. This reduced many of the conclusions to examples of how a model can sometimes give a whole series of correct predictions, yet still be wrong.

The free-electron debacle showed researchers that the model assuming that the ions were in a sea of free electrons needed modification, and so a number of quantum mechanical models such as band structure calculations based on molecular orbitals or the density functional theory were developed. In these models, one either departs from the atomic orbitals of neutral atoms that share their electrons or (in the case of density functional theory) departs from the total electron density. The free-electron picture has, nevertheless, remained a dominant one in education.

The electronic band structure model became a major focus not only for the study of metals but even more so for the study of semiconductors. Together with the electronic states, the vibrational states were also shown to form bands. Rudolf Peierls showed that, in the case of a one-dimensional row of metallic atoms, say hydrogen, an instability had to arise that would lead to the breakup of such a chain into individual molecules. This sparked an interest in the general question: When is collective metallic bonding stable and when will a more localized form of bonding take its place? Much research went into the study of clustering of metal atoms.

As powerful as the concept of the band structure proved to be in the description of metallic bonding, it does have a drawback. It remains a one-electron approximation to a multitudinous many-body problem. In other words, the energy states of each electron are described as if all the other electrons simply form a homogeneous background. Researchers like Mott and Hubbard realized that this was perhaps appropriate for strongly delocalized s- and p-electrons but for d-electrons, and even more for f-electrons the interaction with electrons (and atomic displacements) in the local environment may become stronger than the delocalization that leads to broad bands. Thus, the transition from localized unpaired electrons to itinerant ones partaking in metallic bonding became more comprehensible.

The nature of metallic bonding

The combination of two phenomena gives rise to metallic bonding: delocalization of electrons and the availability of a far larger number of delocalized energy states than of delocalized electrons. The latter could be called electron deficiency.

In 2D

Graphene is an example of two-dimensional metallic bonding. Its metallic bonds are similar to aromatic bonding in benzene, naphthalene, anthracene, ovalene, and so on.

In 3D

Metal aromaticity in metal clusters is another example of delocalization, this time often in three-dimensional entities. Metals take the delocalization principle to its extreme and one could say that a crystal of a metal represents a single molecule over which all conduction electrons are delocalized in all three dimensions. This means that inside the metal one can generally not distinguish molecules, so that the metallic bonding is neither intra- nor intermolecular. 'Nonmolecular' would perhaps be a better term. Metallic bonding is mostly non-polar, because even in alloys there is little difference among the electronegativities of the atoms participating in the bonding interaction (and, in pure elemental metals, none at all). Thus, metallic bonding is an extremely delocalized communal form of covalent bonding. In a sense, metallic bonding is not a 'new' type of bonding at all, therefore, and it describes the bonding only as present in a *chunk* of condensed matter, be it crystalline solid, liquid, or even glass. Metallic vapors by contrast are often atomic (Hg) or at times contain molecules like Na_2 held together by a more conventional covalent bond. This is why it is not correct to speak of a single 'metallic bond'.

The delocalization is most pronounced for s- and p-electrons. For caesium it is so strong that the electrons are virtually free from the caesium atoms to form a gas constrained only by the surface of the metal. For caesium, therefore, the picture of Cs^+ ions held together by a negatively charged electron gas is not too inaccurate. For other elements the electrons are less free, in that they still experience the potential of the metal atoms, sometimes quite strongly. They require a more intricate quantum mechanical treatment (e.g., tight binding) in which the atoms are viewed as neutral, much like the carbon atoms in benzene. For d- and especially f-electrons the delocalization is not strong at all and this explains why these electrons are able to continue behaving as unpaired electrons that retain their spin, adding interesting magnetic properties to these metals.

Electron Deficiency and Mobility

Metal atoms contain few electrons in their valence shells relative to their periods or energy levels. They are electron deficient elements and the communal sharing does not change that. There remain far more available energy states than there are shared electrons. Both requirements for conductivity are therefore fulfilled: strong delocalization and partly filled energy bands. Such electrons can therefore easily change from one energy state into a slightly different one. Thus, not only do they become delocalized, forming a sea of electrons permeating the lattice, but they are also able to migrate through the lattice when an external electrical field is imposed, leading to electrical conductivity. Without the field, there are electrons moving equally in all directions. Under the field, some will adjust their state slightly, adopting a different wave vector. As a consequence, there will be more moving one way than the other and a net current will result.

The freedom of conduction electrons to migrate also give metal atoms, or layers of them, the capacity to slide past each other. Locally, bonds can easily be broken and replaced by new ones after the deformation. This process does not affect the communal metallic bonding very much. This gives rise to metals' typical characteristic phenomena of malleability and ductility. This is particularly true for pure elements. In the presence of dissolved impurities, the defects in the lattice that function as cleavage points may get blocked and the material becomes harder. Gold, for example, is very soft in pure form (24-karat), which is why alloys of 18-karat or lower are preferred in jewelry.

Metals are typically also good conductors of heat, but the conduction electrons only contribute partly to this phenomenon. Collective (i.e., delocalized) vibrations of the atoms known as phonons that travel through the solid as a wave, contribute strongly.

However, the latter also holds for a substance like diamond. It conducts heat quite well but *not* electricity. The latter is *not* a consequence of the fact that delocalization is absent in diamond, but simply that carbon is not electron deficient. The electron deficiency is an important point in distinguishing metallic from more conventional covalent bonding. Thus, we should amend the expression given above into: *Metallic bonding is an extremely delocalized communal form of electron deficient covalent bonding.*

Metallic Radius

Metallic radius is defined as one-half of the distance between the two adjacent metal ions in the metallic lattice. This radius depends on the nature of the atom as well as its environment—specifically, on the coordination number (CN), which in turn depends on the temperature and applied pressure.

When comparing periodic trends in the size of atoms it is often desirable to apply so-called Goldschmidt correction, which converts the radii to the values the atoms would have if they were 12-coordinated. Since metallic radii are always biggest for the highest coordination number, correction for less dense coordinations involves multiplying by x, where $0 < x < 1$. Specifically, for CN = 4, x = 0.88; for CN = 6, x = 0.96, and for CN = 8, x = 0.97. The correction is named after Victor Goldschmidt who obtained the numerical values quoted above.

The radii follow general periodic trends: they decrease across the period due to increase in the effective nuclear charge, which is not offset by the increased number of valence electrons. The radii also increase down the group due to increase in principal quantum number. Between rows 3 and 4, the lanthanide contraction is observed – there is very little increase of the radius down the group due to the presence of poorly shielding f orbitals.

Strength of the Bond

The atoms in metals have a strong attractive force between them. Much energy is required to overcome it. Therefore, metals often have high boiling points, with tungsten (5828 K) being extremely high. A remarkable exception are the elements of the zinc group: Zn, Cd, and Hg. Their electron configuration ends in $...ns^2$ and this comes to resemble a noble gas configuration like that of helium more and more when going down in the periodic table because the energy distance to the empty np orbitals becomes larger. These metals are therefore relatively volatile, and are avoided in ultra-high vacuum systems.

Otherwise, metallic bonding can be very strong, even in molten metals, such as Gallium. Even though gallium will melt from the heat of one's hand just above room temperature, its boiling point is not far from that of copper. Molten gallium is therefore a very nonvolatile liquid thanks to its strong metallic bonding.

The strong bonding of metals in the liquid form demonstrates that the energy of a metallic bond is not a strong function of the direction of the metallic bond; this lack of bond directionality is a direct consequence of electron delocalization, and is best understood in contrast to the directional bonding of covalent bonds. The energy of a metallic bond is thus mostly a function of the amount of electrons which surround the metallic atom, as exemplified by the Embedded atom model. This typically results in metals assuming relatively simple, close-packed crystal structures, such as FCC, BCC, and HCP.

Given high enough cooling rates and appropriate alloy composition, metallic bonding can occur even in glasses with an amorphous structure.

Much biochemistry is mediated by the weak interaction of metal ions and biomolecules. Such interactions and their associated conformational change has been measured using dual polarisation interferometry.

Solubility and Compound Formation

Metals are insoluble in water or organic solvents unless they undergo a reaction with them. Typically this is an oxidation reaction that robs the metal atoms of their itinerant electrons, destroying the metallic bonding. However metals are often readily soluble in each other while retaining the metallic character of their bonding. Gold, for example, dissolves easily in mercury, even at room temperature. Even in solid metals, the solubility can be extensive. If the structures of the two metals are the same, there can even be complete solid solubility, as in the case of electrum, the alloys of silver and gold. At times, however, two metals will form alloys with different structures than either of the two parents. One could call these materials metal compounds, but, because materials with metallic bonding are typically not molecular, Dalton's law of integral proportions is not valid and often a range of stoichiometric ratios can be achieved. It is better to abandon such concepts as 'pure substance' or 'solute' is such cases and speak of phases instead. The study of such phases has traditionally been more the domain of metallurgy than of chemistry, although the two fields overlap considerably.

Localization and Clustering: from Bonding to Bonds

The metallic bonding in complicated compounds does not necessarily involve all constituent elements equally. It is quite possible to have an element or more that do not partake at all. One could picture the conduction electrons flowing around them like a river around an island or a big rock. It is possible to observe which elements do partake, e.g., by looking at the core levels in an X-ray photoelectron spectroscopy (XPS) spectrum. If an element partakes, its peaks tend to be skewed.

Some intermetallic materials e.g. do exhibit metal clusters, reminiscent of molecules and these compounds are more a topic of chemistry than of metallurgy. The formation of the clusters could be seen as a way to 'condense out' (localize) the electron deficient bonding into bonds of a more localized nature. Hydrogen is an extreme example of this form of condensation. At high pressures it is a metal. The core of the planet Jupiter could be said to be held together by a combination of metallic bonding and high pressure induced by gravity. At lower pressures however the bonding becomes entirely localized into a regular covalent bond. The localization is so complete that the (more familiar) H_2 gas results. A similar argument holds for an element like boron. Though it is electron deficient compared to carbon, it does not form a metal. Instead it has a number of complicated structures in which icosahedral B_{12} clusters dominate. Charge density waves are a related phenomenon.

As these phenomena involve the movement of the atoms towards or away from each other, they can be interpreted as the coupling between the electronic and the vibrational states (i.e. the phonons) of the material. A different such electron-phonon interaction is thought to cause a very different result at low temperatures, that of superconductivity. Rather than blocking the mobility of the charge carriers by forming electron pairs in localized bonds, Cooper-pairs are formed that no longer experience any resistance to their mobility.

Optical Properties

The presence of an ocean of mobile charge carriers has profound effects on the optical properties of metals. They can only be understood by considering the electrons as a *collective* rather than considering the states of individual electrons involved in more conventional covalent bonds.

Light consists of a combination of an electrical and a magnetic field. The electrical field is usually able to excite an elastic response from the electrons involved in the metallic bonding. The result is that photons are not able to penetrate very far into the metal and are typically reflected. They bounce off, although some may also be absorbed. This holds equally for all photons of the visible spectrum, which is why metals are often silvery white or grayish with the characteristic specular reflection of metallic luster. The balance between reflection and absorption determines how white or how gray they are, although surface tarnish can obscure such observations. Silver, a very good metal with high conductivity is one of the whitest.

Notable exceptions are reddish copper and yellowish gold. The reason for their color is that there is an upper limit to the frequency of the light that metallic electrons can readily respond to, the plasmon frequency. At the plasmon frequency, the frequency-dependent dielectric function of the free electron gas goes from negative (reflecting) to positive (transmitting); higher frequency photons are not reflected at the surface, and do not contribute to the color of the metal. There are some materials like indium tin oxide (ITO) that are metallic conductors (actually degenerate semiconductors) for which this threshold is in the infrared, which is why they are transparent in the visible, but good mirrors in the IR.

For silver the limiting frequency is in the far UV, but for copper and gold it is closer to the visible. This explains the colors of these two metals. At the surface of a metal resonance effects known as surface plasmons can result. They are collective oscillations of the conduction electrons like a ripple in the electronic ocean. However, even if photons have enough energy they usually do not have enough momentum to set the ripple in motion. Therefore, plasmons are hard to excite on a bulk metal. This is why gold and copper still look like lustrous metals albeit with a dash of color. However, in colloidal gold the metallic bonding is confined to a tiny metallic particle, preventing the oscillation wave of the plasmon from 'running away'. The momentum selection rule is therefore broken, and the plasmon resonance causes an extremely intense absorption in the green with a resulting beautiful purple-red color. Such colors are orders of magnitude more intense than ordinary absorptions seen in dyes and the like that involve individual electrons and their energy states

Intermolecular Force

Intermolecular forces (IMFs) are forces of attraction or repulsion which act between neighboring particles (atoms, molecules, or ions). They are weak compared to the intramolecular forces, the forces which keep a molecule together. For example the covalent bond, involving the sharing of electron pairs between atoms is much stronger than the forces present between the neighboring molecules. They are an essential part of force fields frequently used in molecular mechanics.

The investigation of intermolecular forces starts from macroscopic observations which point out the existence and action of forces at a molecular level. These observations include non-ideal-gas thermodynamic behavior reflected by virial coefficients, vapor pressure, viscosity, superficial tension and absorption data. ' The first reference to the nature of microscopic forces is found in Alexis Clairaut's work *Theorie de la Figure de la Terre*. Other scientists who have contributed to the investigation of microscopic forces include: Laplace, Gauss, Maxwell and Boltzmann.

Attractive intermolecular forces are considered by the following types:

- Ion-induced dipole forces

- Ion-dipole forces

- van der Waals forces (Keesom force, Debye force, and London dispersion force)

Information on intermolecular force is obtained by macroscopic measurements of properties like viscosity, PVT data. The link to microscopic aspects is given by virial coefficients and Lennard-Jones potentials.

Dipole-dipole Interactions

Dipole-dipole interactions are electrostatic interactions between permanent dipoles in molecules. These interactions tend to align the molecules to increase attraction (reducing potential energy). An example of a dipole-dipole interaction can be seen in hydrogen chloride (HCl): the positive end of a polar molecule will attract the negative end of the other molecule and influence its position. Polar molecules have a net attraction between them. Examples of polar molecules include hydrogen chloride (HCl) and chloroform ($CHCl_3$).

$$\overset{\delta+}{H}-\overset{\delta-}{Cl}\cdots\overset{\delta+}{H}-\overset{\delta-}{Cl}$$

Often molecules contain dipolar groups, but have no overall dipole moment. This occurs if there is symmetry within the molecule that causes the dipoles to cancel each other out. This occurs in molecules such as tetrachloromethane and carbon dioxide. Note that the dipole-dipole interaction between two individual atoms is usually zero, since atoms rarely carry a permanent dipole.

Ion-dipole and Ion-induced Dipole Forces

Ion-dipole and ion-induced dipole forces are similar to dipole-dipole and induced-dipole interactions but involve ions, instead of only polar and non-polar molecules. Ion-dipole and ion-induced dipole forces are stronger than dipole-dipole interactions because the charge of any ion is much greater than the charge of a dipole moment. Ion-dipole bonding is stronger than hydrogen bonding.

An ion-dipole force consists of an ion and a polar molecule interacting. They align so that the positive and negative groups are next to one another, allowing for maximum attraction.

An ion-induced dipole force consists of an ion and a non-polar molecule interacting. Like a dipole-induced dipole force, the charge of the ion causes distortion of the electron cloud on the non-polar molecule.

Hydrogen Bonding

A hydrogen bond is the attraction between the lone pair of an electronegative atom and a hydrogen atom that is bonded to either nitrogen, oxygen, or fluorine. The hydrogen bond is often described as a strong electrostatic dipole-dipole interaction. However, it also has some features of covalent bonding: it is directional, stronger than a van der Waals interaction, produces interatomic distances shorter than the sum of van der Waals radius, and usually involves a limited number of interaction partners, which can be interpreted as a kind of valence.

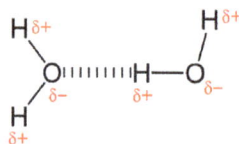

Intermolecular hydrogen bonding is responsible for the high boiling point of water (100 °C) compared to the other group 16 hydrides, which have no hydrogen bonds. Intramolecular hydrogen oxygen bonding is partly responsible for the secondary, tertiary, and quaternary structures of proteins and nucleic acids. It also plays an important role in the structure of polymers, both synthetic and natural.

Van der Waals Forces

The van der Waals forces arise from interaction between uncharged atoms or molecules, leading not only to such phenomena as the cohesion of condensed phases and physical adsorption of gases, but also to a universal force of attraction between macroscopic bodies.

Keesom (Permanent-permanent Dipoles) Interaction

The first contribution to van der Waals forces is due to electrostatic interactions between charges (in molecular ions), dipoles (for polar molecules), quadrupoles (all molecules with symmetry lower than cubic), and permanent multipoles. It is referred to as Keesom interactions(named after Willem Hendrik Keesom). These forces originate from the attraction between permanent dipoles (dipolar molecules) and are temperature dependent.

They consist of attractive interactions between dipoles that are ensemble averaged over different rotational orientations of the dipoles. It is assumed that the molecules are constantly rotating and never get locked into place. This is a good assumption, but at some point molecules do get locked into place. The energy of a Keesom interaction depends on the inverse sixth power of the distance, unlike the interaction energy of two spatially fixed dipoles, which depends on the inverse third power of the distance. The Keesom interaction can only occur among molecules that possess permanent dipole moments a.k.a. two polar molecules. Also Keesom interactions are very weak van der Waals interactions and do not occur in aqueous solutions that contain electrolytes. The angle averaged interaction is given by the following equation:

$$\frac{-m_1^2 m_2^2}{24\pi^2 \varepsilon_o^2 \varepsilon_r^2 k_b T r^6} = V$$

Where m = charge per length, ε_o = permitivity of free space, ε_r = dielectric constant of surrounding material, T = temperature, k_b = Boltzmann constant, and r = distance between molecules.

Debye (Permanent-induced Dipoles) Force

The second contribution is the induction (also known as polarization) or Debye force, arising from interactions between rotating permanent dipoles and from the polarizability of atoms and molecules (induced dipoles). These induced dipoles occur when one molecule with a permanent dipole repels another molecule's electrons. A molecule with permanent dipole can induce a dipole in a similar neighboring molecule and cause mutual attraction. Debye forces cannot occur between atoms. The forces between induced and permanent dipoles are not as temperature dependent as Keesom interactions because the induced dipole is free to shift and rotate around the non-polar molecule. The Debye induction effects and Keesom orientation effects are referred to as polar interactions.

The induced dipole forces appear from the induction (also known as polarization), which is the attractive interaction between a permanent multipole on one molecule with an induced (by the former di/multi-pole) multipole on another. This interaction is called the Debye force, named after Peter J.W. Debye.

One example of an induction-interaction between permanent dipole and induced dipole is the interaction between HCl and Ar. In this system, Ar experiences a dipole as its electrons are attracted (to the H side of HCl) or repelled (from the Cl side) by HCl. The angle averaged interaction is given by the following equation.

$$\frac{-m_1^2 \alpha_2}{16\pi^2 \varepsilon_o^2 \varepsilon_r^2 r^6} = V$$

Where α = polarizability

This kind of interaction can be expected between any polar molecule and non-polar/symmetrical molecule. The induction-interaction force is far weaker than dipole-dipole interaction, but stronger than the London dispersion force.

London Dispersion Force (Dipole-induced Dipoles Interaction)

The third and dominant contribution is the dispersion or London force (fluctuating dipole-induced dipole), which arises due to the non-zero instantaneous dipole moments of all atoms and molecules. Such polarization can be induced either by a polar molecule or by the repulsion of negatively charged electron clouds in non-polar molecules. Thus, London interactions are caused by random fluctuations of electron density in an electron cloud. An atom with a large number of electrons will have a greater associated London force than an atom with fewer electrons. The dispersion (London) force is the most important component because all materials are polarizable, whereas Keesom and Debye forces require permanent dipoles. The London interaction is universal and is present in atom-atom interactions as well. For various reasons, London interactions (dispersion) have been considered relevant for interactions between macroscopic bodies in condensed systems. Hamaker developed the theory of van der Waals between macroscopic bodies in 1937 and showed that the additivity of these interactions renders them considerably more long-range.

Relative Strength of Forces

Bond type	Dissociation energy (kcal/mol)
Ionic Lattice Energy	250–4000
Covalent Bond Energy	30–260
Hydrogen Bonds	1–12 (about 5 in water)
Dipole–Dipole	0.5–2
London Dispersion Forces	<1 to 15 (estimated from the enthalpies of vaporization of hydrocarbons)

Note: this comparison is only approximate – the actual relative strengths will vary depending on the molecules involved. Ionic and covalent bonding will always be stronger than intermolecular forces in any given substance.

Effect on the Behavior of Gases

Intermolecular forces are repulsive at short distances and attractive at long distances. In a gas, the repulsive force chiefly has the effect of keeping two molecules from occupying the same volume. This gives a real gas a tendency to occupy a larger volume than an ideal gas at the same temperature and pressure. The attractive force draws molecules closer together and gives a real gas a tendency to occupy a smaller volume than an ideal gas. Which interaction is more important depends on temperature and pressure.

In a gas, the distances between molecules are generally large, so intermolecular forces have only a small effect. The attractive force is not overcome by the repulsive force, but by the thermal energy of the molecules. Temperature is the measure of thermal energy, so increasing temperature reduces the influence of the attractive force. In contrast, the influence of the repulsive force is essentially unaffected by temperature.

When a gas is compressed to increase its density, the influence of the attractive force increases. If the gas is made sufficiently dense, the attractions can become large enough to overcome the tendency of thermal motion to cause the molecules to spread out. Then the gas can condense to form a solid or liquid (i.e., a condensed phase). Lower temperature favors the formation of a condensed phase. In a condensed phase, there is very nearly a balance between the attractive and repulsive forces.

Quantum Mechanical Theories

Intermolecular forces observed between atoms and molecules can be described phenomenologically as occurring between permanent and instantaneous dipoles, as outlined above. Alternatively, one may seek a fundamental, unifying theory that is able to explain the various types of interactions such as hydrogen bonding, van der Waals forces and dipole-dipole interactions. Typically, this is done by applying the ideas of quantum mechanics to molecules, and Rayleigh–Schrödinger perturbation theory has been especially effective in this regard. When applied to existing quantum chemistry methods, such a quantum mechanical explanation of intermolecular interactions, this provides an array of approximate methods that can be used to analyze intermolecular interactions.

Molecular Solid

A molecular solid is a solid composed of molecules held together by the van der Waals forces. Because these dipole forces are weaker than covalent or ionic bonds, molecular solids are soft and have relatively low melting temperature. Molecular solids can be amorphous solids or crystals. Examples of molecular solids include hydrocarbons, ice, sugar, fullerenes, sulfur, and solid carbon dioxide. Pure molecular solids are electrical insulators but they can be made conductive by doping.

Other classes of solid include ionic solids, held together by ionic bonding; glasses, ceramics, and polymeric solids, held together by covalent bonds; and metals, held together by metallic bonding.

Structure and Composition

Melting points of some molecular solids

Formula	T_m °C
H_2	−259.1
F_2	−219.6
O_2	−218.8
N_2	−210.0
CH_4	−182.4
C_2H_6	−181.8
C_3H_8	−165.0
C_4H_{10}	−138.3

C_5H_{12}	−129.8
Cl_2	−101.6
C_6H_{14}	−95.3
HBr	−86.8
HF	−80.0
NH_3	−80.0
HI	−50.8
$C_{10}H_{22}$	−29.7
HCl	−27.3
Br_2	−7.2
H_2O	0.0
C_6H_6	5.5
I_2	113.7
S_8	119.0
C_6Cl_6	220.0

The term "molecular solid" may refer not to a certain chemical composition, but to a specific form of a material. For example, solid phosphorus can crystallize in different allotropes called "white", "red", and "black" phosphorus. White phosphorus forms molecular crystals composed of tetrahedral P_4 molecules. Heating at ambient pressure to 250 °C or exposing to sunlight converts white phosphorus to red phosphorus where the P_4 tetrahedra are no longer isolated, but are connected by covalent bonds into polymer-like chains. Heating white phosphorus under high (GPa) pressures converts it to black phosphorus which has a layered, graphite-like structure.

The structural transitions in phosphorus are reversible: upon releasing high pressure, black phosphorus gradually converts into the red allotrope, and by vaporizing red phosphorus at 490 °C in inert atmosphere and condensing the vapor, covalent red phosphorus can be transformed back into the white molecular solid.

White, red, violet, and black phosphorus samples

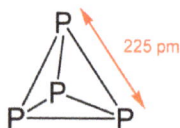

Structure unit of white phosphorus

Structures of red

violet

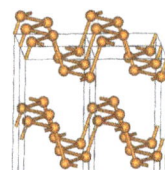

and black phosphorus

Similarly, yellow arsenic is a molecular solid composed of As_4 units; it is metastable and gradually transforms into gray arsenic upon heating or illumination. Some forms of sulfur and selenium are composed of S_8 (or Se_8) units and are molecular solids at ambient conditions, but they can convert into covalent allotropes having atomic chains extending all through the crystal.

Changes in the chemical composition can have even stronger effects on the bonding in solids. For example, whereas both hydrogen and lithium belong to the first group of the periodic table, LiCl is ionic and HCl is a molecular solid.

Examples of Molecular Solids

Several classes of molecular solids can be distinguished. The vast majority of molecular solids can be attributed to organic compounds containing carbon and hydrogen, such as hydrocarbons (C_nH_m) and diamondoids. Spherical molecules consisting of different number of carbon atoms, that is fullerenes, are another important class. Less numerous, yet distinctive molecular solids are halogens (e.g. Cl_2) and their compounds with hydrogen (HCl), as well as light chalcogens (O_2) and pnictogens (N_2).

Properties

Weakness of intermolecular forces results in low melting temperatures of molecular solids. Whereas the characteristic melting point of metals and ionic solids is ~1000 °C, most molecular solids melt well below ~300 °C, thus many corresponding substances are either liquid (ice) or gaseous (oxygen) at room temperature. Molecular solids also have relatively low density and hardness. This is because of the light elements involved and relatively long and thus weak inter-molecular bonds. Because of the charge neutrality of the constituent molecules and long distance between them, molecular solids are electrical insulators.

The above tendencies can be illustrated on example of different allotropes of phosphorus. White phosphorus, a molecular solid, has a relatively low density of 1.82 g/cm³ and melting point of 44.1 °C; it is a soft material which can be cut with a knife. When it is converted to the covalent red phosphorus, the density goes to 2.2–2.4 g/cm³ and melting point to 590 °C, and when white phosphorus is transformed into the (also covalent) black phosphorus, the density becomes 2.69–3.8 g/cm³ and melting temperature ~200 °C. Both red and black phosphorus forms are significantly harder than white phosphorus, and whereas white phosphorus is an insulator, the black allotrope, which consists of layers extending over the whole crystal, does conduct electricity.

Conductivity of molecular solids can be illustrated on example of fullerene. Its solid is an insulator because all valence electrons of carbon atoms are involved into the covalent bonds within the individual carbon molecules. However, inserting (intercalating) alkali metal atoms between the fullerene molecules provides extra electrons, which can be easily ionized from the metal atoms and make material conductive and even superconductive.

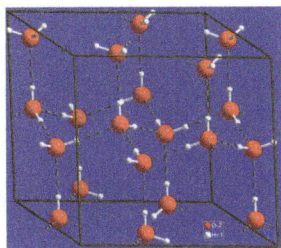

Crystal structure of hexagonal ice. Gray dashed lines indicate hydrogen bonds.

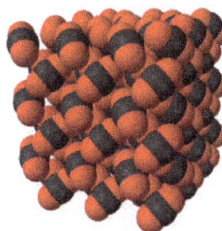

Unit cell of solid carbon dioxide, a molecular solid containing discrete CO_2 molecules

References

- Majer, V. and Svoboda, V. (1985) Enthalpies of Vaporization of Organic Compounds, Blackwell Scientific Publications, Oxford. ISBN 0632015292

- Margenau, H. and Kestner, N. (1969) Theory of intermolecular forces, International Series of Monographs in Natural Philosophy, Pergamon Press, ISBN 1483119289

- Stranks, D. R.; Heffernan, M. L.; Lee Dow, K. C.; McTigue, P. T.; Withers, G. R. A. (1970). Chemistry: A structural view. Carlton, Vic.: Melbourne University Press. p. 184. ISBN 0-522-83988-6.

- March, Jerry (1992). Advanced Organic Chemistry: Reactions, Mechanisms, and Structure. John Wiley & Sons. ISBN 0-471-60180-2.

- Maksic, Zvonimir (1990). "The Concept of the Chemical Bond in Solids". Theoretical Models of Chemical Bonding. New York: Springer-Verlag. pp. 417–452. ISBN 0-387-51553-4.

- Holleman, Arnold F; Wiberg, Egon; Wiberg, Nils (1985). "Arsen". Lehrbuch der Anorganischen Chemie (in German) (91–100 ed.). Walter de Gruyter. pp. 675–681. ISBN 3-11-007511-3.

Crystals and its Structure

A crystal is a solid material. The components of a crystal are always arranged in a systematic structure. Crystal, crystal structure, crystal twinning, atomic packing factor and crystal structures in the periodic table are some of the topics elucidated in this section. The chapter serves as a source to understand the basic concept of crystals.

Crystal

A crystal or crystalline solid is a solid material whose constituents (such as atoms, molecules or ions) are arranged in a highly ordered microscopic structure, forming a crystal lattice that extends in all directions. In addition, macroscopic single crystals are usually identifiable by their geometrical shape, consisting of flat faces with specific, characteristic orientations. The scientific study of crystals and crystal formation is known as crystallography. The process of crystal formation via mechanisms of crystal growth is called crystallization or solidification.

A crystal of amethyst quartz

Examples of large crystals include snowflakes, diamonds, and table salt. Most inorganic solids are not crystals but polycrystals, i.e. many microscopic crystals fused together into a single solid. Examples of polycrystals include most metals, rocks, ceramics, and ice. A third category of solids is amorphous solids, where the atoms have no periodic structure whatsoever. Examples of amorphous solids include glass, wax, and many plastics.

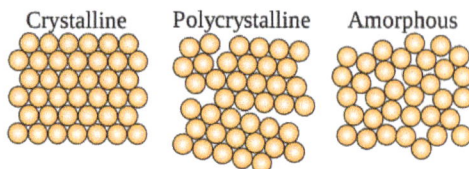

Microscopically, a single crystal has atoms in a near-perfect periodic arrangement; a polycrystal is composed of many microscopic crystals (called "crystallites" or "grains"); and an amorphous solid (such as glass) has no periodic arrangement even microscopically.

Crystal Structure (Microscopic)

Halite (Table salt, NaCl): Microscopic and Macroscopic

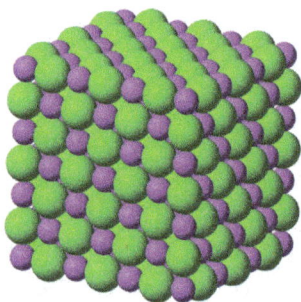

Microscopic structure of a halite crystal. (Purple is sodium ion, green is chlorine ion.) There is cubic symmetry in the atoms' arrangement.

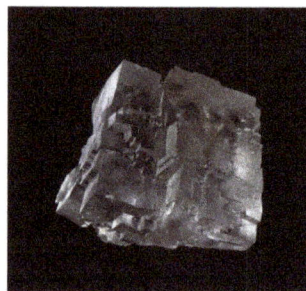

Macroscopic (~16cm) halite crystal. The right-angles between crystal faces are due to the cubic symmetry of the atoms' arrangement.

The scientific definition of a "crystal" is based on the microscopic arrangement of atoms inside it, called the crystal structure. A crystal is a solid where the atoms form a periodic arrangement.

Not all solids are crystals. For example, when liquid water starts freezing, the phase change begins with small ice crystals that grow until they fuse, forming a *polycrystalline* structure. In the final block of ice, each of the small crystals (called "crystallites" or "grains") is a true crystal with a periodic arrangement of atoms, but the whole polycrystal does *not* have a periodic arrangement of atoms, because the periodic pattern is broken at the grain boundaries. Most macroscopic inorganic solids are polycrystalline, including almost all metals, ceramics, ice, rocks, etc. Solids that are neither crystalline nor polycrystalline, such as glass, are called *amorphous solids*, also called glassy, vitreous, or noncrystalline. These have no periodic order, even microscopically. There are distinct differences between crystalline solids and amorphous solids: most notably, the process of forming a glass does not release the latent heat of fusion, but forming a crystal does.

A crystal structure (an arrangement of atoms in a crystal) is characterized by its *unit cell*, a small imaginary box containing one or more atoms in a specific spatial arrangement. The unit cells are stacked in three-dimensional space to form the crystal.

The symmetry of a crystal is constrained by the requirement that the unit cells stack perfectly with no gaps. There are 219 possible crystal symmetries, called crystallographic space groups. These are grouped into 7 crystal systems, such as cubic crystal system (where the crystals may form cubes or rectangular boxes, such as halite shown at right) or hexagonal crystal system (where the crystals may form hexagons, such as ordinary water ice).

Crystal Faces and Shapes

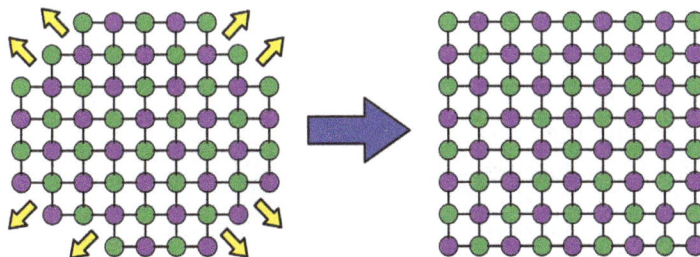

As a halite crystal is growing, new atoms can very easily attach to the parts of the surface with rough atomic-scale structure and many dangling bonds. Therefore, these parts of the crystal grow out very quickly (yellow arrows). Eventually, the whole surface consists of smooth, stable faces, where new atoms cannot as easily attach themselves.

Crystals are commonly recognized by their shape, consisting of flat faces with sharp angles. These shape characteristics are not *necessary* for a crystal—a crystal is scientifically defined by its microscopic atomic arrangement, not its macroscopic shape—but the characteristic macroscopic shape is often present and easy to see.

Euhedral crystals are those with obvious, well-formed flat faces. Anhedral crystals do not, usually because the crystal is one grain in a polycrystalline solid.

The flat faces (also called facets) of a euhedral crystal are oriented in a specific way relative to the underlying atomic arrangement of the crystal: they are planes of relatively low Miller index. This occurs because some surface orientations are more stable than others (lower surface energy). As a crystal grows, new atoms attach easily to the rougher and less stable parts of the surface, but less easily to the flat, stable surfaces. Therefore, the flat surfaces tend to grow larger and smoother, until the whole crystal surface consists of these plane surfaces.

One of the oldest techniques in the science of crystallography consists of measuring the three-dimensional orientations of the faces of a crystal, and using them to infer the underlying crystal symmetry.

A crystal's habit is its visible external shape. This is determined by the crystal structure (which restricts the possible facet orientations), the specific crystal chemistry and bonding (which may favor some facet types over others), and the conditions under which the crystal formed.

Occurrence in Nature

Ice crystals.

Fossil shell with calcite crystals.

Rocks

By volume and weight, the largest concentrations of crystals in the Earth are part of its solid bedrock. Crystals found in rocks typically range in size from a fraction of a millimetre to several centimetres across, although exceptionally large crystals are occasionally found. As of 1999, the world's largest known naturally occurring crystal is a crystal of beryl from Malakialina, Madagascar, 18 m (59 ft) long and 3.5 m (11 ft) in diameter, and weighing 380,000 kg (840,000 lb).

Some crystals have formed by magmatic and metamorphic processes, giving origin to large masses of crystalline rock. The vast majority of igneous rocks are formed from molten magma and the degree of crystallization depends primarily on the conditions under which they solidified. Such rocks as granite, which have cooled very slowly and under great pressures, have completely crystallized; but many kinds of lava were poured out at the surface and cooled very rapidly, and in this latter group a small amount of amorphous or glassy matter is common. Other crystalline rocks, the metamorphic rocks such as marbles, mica-schists and quartzites, are recrystallized. This means that they were at first fragmental rocks like limestone, shale and sandstone and have never been in a molten condition nor entirely in solution, but the high temperature and pressure conditions of metamorphism have acted on them by erasing their original structures and inducing recrystallization in the solid state.

Other rock crystals have formed out of precipitation from fluids, commonly water, to form druses or quartz veins. The evaporites such as halite, gypsum and some limestones have been deposited from aqueous solution, mostly owing to evaporation in arid climates.

Ice

Water-based ice in the form of snow, sea ice and glaciers is a very common manifestation of crystalline or polycrystalline matter on Earth. A single snowflake is typically a single crystal, while an ice cube is a polycrystal.

Organigenic Crystals

Many living organisms are able to produce crystals, for example calcite and aragonite in the case of most molluscs or hydroxylapatite in the case of vertebrates.

Polymorphism and Allotropy

The same group of atoms can often solidify in many different ways. Polymorphism is the ability of a solid to exist in more than one crystal form. For example, water ice is ordinarily found in the hexagonal form Ice I_h, but can also exist as the cubic Ice I_c, the rhombohedral ice II, and many other forms. The different polymorphs are usually called different *phases*.

In addition, the same atoms may be able to form noncrystalline phases. For example, water can also form amorphous ice, while SiO_2 can form both fused silica (an amorphous glass) and quartz (a crystal). Likewise, if a substance can form crystals, it can also form polycrystals.

For pure chemical elements, polymorphism is known as allotropy. For example, diamond and graphite are two crystalline forms of carbon, while amorphous carbon is a noncrystalline form. Polymorphs, despite having the same atoms, may have wildly different properties. For example, diamond is among the hardest substances known, while graphite is so soft that it is used as a lubricant.

Polyamorphism is a similar phenomenon where the same atoms can exist in more than one amorphous solid form.

Crystallization

Vertical cooling crystallizer in a beet sugar factory.

Crystallization is the process of forming a crystalline structure from a fluid or from materials dissolved in a fluid. (More rarely, crystals may be deposited directly from gas; see thin-film deposition and epitaxy.)

Crystallization is a complex and extensively-studied field, because depending on the conditions, a single fluid can solidify into many different possible forms. It can form a single crystal, perhaps with various possible phases, stoichiometries, impurities, defects, and habits. Or, it can form a polycrystal, with various possibilities for the size, arrangement, orientation, and phase of its grains. The final form of the solid is determined by the conditions under which the fluid is being solidified, such as the chemistry of the fluid, the ambient pressure, the temperature, and the speed with which all these parameters are changing.

Specific industrial techniques to produce large single crystals (called *boules*) include the Czochralski process and the Bridgman technique. Other less exotic methods of crystallization may be used, depending on the physical properties of the substance, including hydrothermal synthesis, sublimation, or simply solvent-based crystallization.

Large single crystals can be created by geological processes. For example, selenite crystals in ex-cess of 10 meters are found in the Cave of the Crystals in Naica, Mexico.

Crystals can also be formed by biological processes. Conversely, some organisms have special techniques to *prevent* crystallization from occurring, such as antifreeze proteins.

Defects, Impurities, and Twinning

An *ideal* crystal has every atom in a perfect, exactly repeating pattern. However, in reality, most crystalline materials have a variety of crystallographic defects, places where the crystal's pattern is interrupted. The types and structures of these defects may have a profound effect on the properties of the materials.

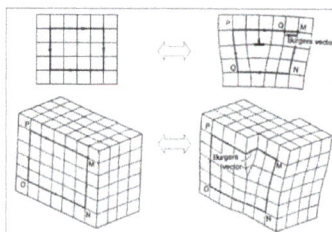

Two types of crystallographic defects. Top right: edge dislocation. Bottom right: screw dislocation.

A few examples of crystallographic defects include vacancy defects (an empty space where an atom should fit), interstitial defects (an extra atom squeezed in where it does not fit), and dislocations. Dislocations are especially important in materials science, because they help determine the mechanical strength of materials.

Another common type of crystallographic defect is an impurity, meaning that the "wrong" type of atom is present in a crystal. For example, a perfect crystal of diamond would only contain carbon atoms, but a real crystal might perhaps contain a few boron atoms as well. These boron impurities change the diamond's color to slightly blue. Likewise, the only difference between ruby and sapphire is the type of impurities present in a corundum crystal.

Twinned pyrite crystal group.

In semiconductors, a special type of impurity, called a dopant, drastically changes the crystal's electrical properties. Semiconductor devices, such as transistors, are made possible largely by putting different semiconductor dopants into different places, in specific patterns.

Twinning is a phenomenon somewhere between a crystallographic defect and a grain boundary. Like a grain boundary, a twin boundary has different crystal orientations on its two sides. But unlike a grain boundary, the orientations are not random, but related in a specific, mirror-image way.

Mosaicity is a spread of crystal plane orientations. A mosaic crystal is supposed to consist of smaller crystalline units that are somewhat misaligned with respect to each other.

Chemical Bonds

In general, solids can be held together by various types of chemical bonds, such as metallic bonds, ionic bonds, covalent bonds, van der Waals bonds, and others. None of these are necessarily crystalline or non-crystalline. However, there are some general trends as follows.

Metals are almost always polycrystalline, though there are exceptions like amorphous metal and single-crystal metals. The latter are grown synthetically. (A microscopically-small piece of metal may naturally form into a single crystal, but larger pieces generally do not.) Ionically bonded solids are usually crystalline or polycrystalline. In practice, large salt crystals can be created by solidification of a molten fluid, or by crystallization out of a solution. Covalently bonded crystals are also very common, notable examples being diamond, quartz, and graphite. Polymer materials generally will form crystalline regions, but the lengths of the molecules usually prevent complete crystallization—and sometimes polymers are completely amorphous. Weak van der Waals forces also help hold together certain crystals, including graphite.

Quasicrystals

A quasicrystal consists of arrays of atoms that are ordered but not strictly periodic. They have many attributes in common with ordinary crystals, such as displaying a discrete pattern in x-ray diffraction, and the ability to form shapes with smooth, flat faces.

Quasicrystals are most famous for their ability to show five-fold symmetry, which is impossible for an ordinary periodic crystal.

The International Union of Crystallography has redefined the term "crystal" to include both ordinary periodic crystals and quasicrystals ("any solid having an essentially discrete diffraction diagram").

Quasicrystals, first discovered in 1982, are quite rare in practice. Only about 100 solids are known to form quasicrystals, compared to about 400,000 periodic crystals known in 2004. The 2011 Nobel Prize in Chemistry was awarded to Dan Shechtman for the discovery of quasicrystals.

Special Properties from Anisotropy

Crystals can have certain special electrical, optical, and mechanical properties that glass and polycrystals normally cannot. These properties are related to the anisotropy of the crystal, i.e. the lack of rotational symmetry in its atomic arrangement. One such property is the piezoelectric effect, where a voltage across the crystal can shrink or stretch it. Another is birefringence, where a double image appears when looking through a crystal. Moreover, various properties of a crystal, including electrical conductivity, electrical permittivity, and Young's modulus, may be different in different directions in a crystal. For example, graphite crystals consist of a stack of sheets, and although each individual sheet is mechanically very strong, the sheets are rather loosely bound to each other. Therefore, the mechanical strength of the material is quite different depending on the direction of stress.

Not all crystals have all of these properties. Conversely, these properties are not quite exclusive to crystals. They can appear in glasses or polycrystals that have been made anisotropic by working or stress—for example, stress-induced birefringence.

Crystallography

Crystallography is the science of measuring the crystal structure (in other words, the atomic arrangement) of a crystal. One widely used crystallography technique is X-ray diffraction. Large numbers of known crystal structures are stored in crystallographic databases.

Crystal Structure

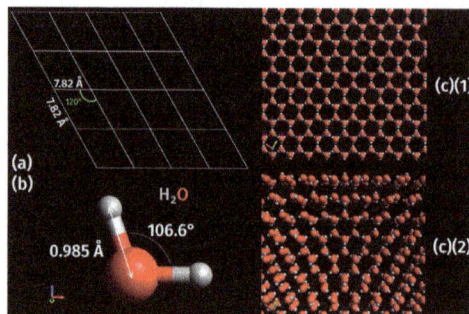

The (3-D) crystal structure of H_2O ice Ih (c) consists of bases of H_2O ice molecules (b) located on lattice points within the (2-D) hexagonal space lattice (a). The values for the H–O–H angle and O–H distance have come from *Physics of Ice* with uncertainties of ±1.5° and ±0.005 Å, respectively. The white box in (c) is the unit cell defined by Bernal and Fowler

In crystallography, crystal structure is a description of the ordered arrangement of atoms, ions or molecules in a crystalline material. Ordered structures occur from the intrinsic nature of the constituent particles to form symmetric patterns that repeat along the principal directions of three-dimensional space in matter.

The smallest group of particles in the material that constitutes the repeating pattern is the unit cell of the structure. The unit cell completely defines the symmetry and structure of the entire crystal lattice, which is built up by repetitive translation of the unit cell along its principal axes. The repeating patterns are said to be located at the points of the Bravais lattice.

The lengths of the principal axes, or edges, of the unit cell and the angles between them are the lattice constants, also called *lattice parameters*. The symmetry properties of the crystal are described by the concept of space groups. All possible symmetric arrangements of particles in space may be described by the set of seven space groups.

The crystal structure and symmetry play a role in determining many physical properties, such as cleavage, electronic band structure, and optical transparency.

Unit Cell

The crystal structure of a material (the arrangement of atoms within a given type of crystal) can be described in terms of its unit cell. The unit cell is a small box containing one or more atoms arranged in 3 dimensions. The unit cells stacked in three-dimensional space describe the bulk arrangement of atoms of the crystal. The unit cell is represented in terms of its lattice parameters, which are the lengths of the cell edges (a, b and c) and the angles between them (alpha, beta and gamma), while the positions of the atoms inside the unit cell are described by the set of atomic positions (x_i, y_i, z_i) measured from a lattice point. Commonly, atomic positions are represented in terms of fractional coordinates, relative to the unit cell lengths.

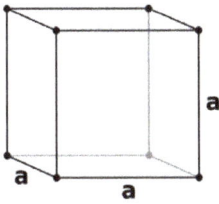

| Simple cubic (P) | Body-centered cubic (I) | Face-centered cubic (F) |

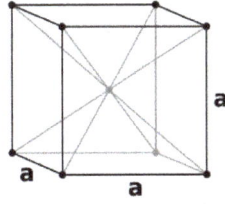

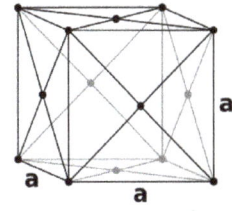

The atom positions within the unit cell can be calculated through application of symmetry operations to the asymmetric unit. The asymmetric unit refers to the smallest possible occupation of space within the unit cell. This does not, however imply that the entirety of the asymmetric unit must lie within the boundaries of the unit cell. Symmetric transformations of atom positions are calculated from the space group of the crystal structure, and this is usually a black box operation performed by computer programs. However, manual calculation of the atomic positions within the unit cell can be performed from the asymmetric unit, through the application of the symmetry operators described within the *International Tables for Crystallography: Volume A.*

Miller Indices

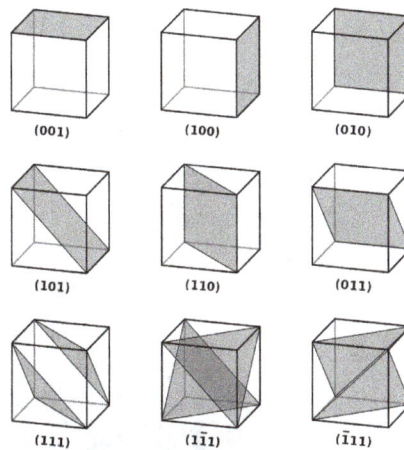

Planes with different Miller indices in cubic crystals

Vectors and planes in a crystal lattice are described by the three-value Miller index notation. It uses the indices ℓ, m, and n as directional parameters, which are separated by 90°, and are thus orthogonal.

By definition, the syntax (ℓmn) denotes a plane that intercepts the three points a_1/ℓ, a_2/m, and a_3/n, or some multiple thereof. That is, the Miller indices are proportional to the inverses of the intercepts of the plane with the unit cell (in the basis of the lattice vectors). If one or more of the indices is zero, it means that the planes do not intersect that axis (i.e., the intercept is "at infinity"). A plane containing a coordinate axis is translated so that it no longer contains that axis before its Miller indices are determined. The Miller indices for a plane are integers with no common factors. Negative indices are indicated with horizontal bars, as in (123). In an orthogonal coordinate system for a cubic cell, the Miller indices of a plane are the Cartesian components of a vector normal to the plane.

Considering only (ℓmn) planes intersecting one or more lattice points (the *lattice planes*), the distance d between adjacent lattice planes is related to the (shortest) reciprocal lattice vector orthogonal to the planes by the formula

Planes and Directions

The crystallographic directions are geometric lines linking nodes (atoms, ions or molecules) of a crystal. Likewise, the crystallographic planes are geometric *planes* linking nodes. Some directions and planes have a higher density of nodes. These high density planes have an influence on the behavior of the crystal as follows:

- Optical properties: Refractive index is directly related to density (or periodic density fluctuations).

- Adsorption and reactivity: Physical adsorption and chemical reactions occur at or near surface atoms or molecules. These phenomena are thus sensitive to the density of nodes.

- Surface tension: The condensation of a material means that the atoms, ions or molecules are more stable if they are surrounded by other similar species. The surface tension of an interface thus varies according to the density on the surface.

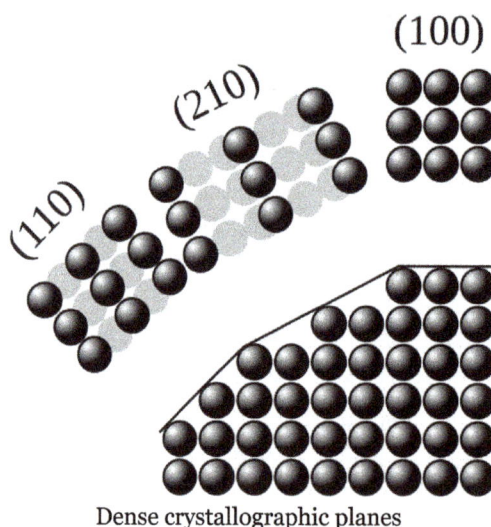

Dense crystallographic planes

- Microstructural defects: Pores and crystallites tend to have straight grain boundaries following higher density planes.

- Cleavage: This typically occurs preferentially parallel to higher density planes.

- Plastic deformation: Dislocation glide occurs preferentially parallel to higher density planes. The perturbation carried by the dislocation (Burgers vector) is along a dense direction. The shift of one node in a more dense direction requires a lesser distortion of the crystal lattice.

Some directions and planes are defined by symmetry of the crystal system. In monoclinic, rhombohedral, tetragonal, and trigonal/hexagonal systems there is one unique axis (sometimes called the principal axis) which has higher rotational symmetry than the other two axes. The basal plane is the plane perpendicular to the principal axis in these crystal systems. For triclinic, orthorhombic, and cubic crystal systems the axis designation is arbitrary and there is no principal axis.

Cubic Structures

Because of the symmetry of cubic crystals, it is possible to change the place and sign of the integers and have equivalent directions and planes:

- Coordinates in *angle brackets* such as ⟨100⟩ denote a *family* of directions that are equivalent due to symmetry operations, such as , , or the negative of any of those directions.

- Coordinates in *curly brackets* or *braces* such as {100} denote a family of plane normals that are equivalent due to symmetry operations, much the way angle brackets denote a family of directions.

For face-centered cubic (fcc) and body-centered cubic (bcc) lattices, the primitive lattice vectors are not orthogonal. However, in these cases the Miller indices are conventionally defined relative to the lattice vectors of the cubic supercell and hence are again simply the Cartesian directions.

Classification

The defining property of a crystal is its inherent symmetry, by which we mean that under certain 'operations' the crystal remains unchanged. All crystals have translational symmetry in three directions, but some have other symmetry elements as well. For example, rotating the crystal 180° about a certain axis may result in an atomic configuration that is identical to the original configuration. The crystal is then said to have a twofold rotational symmetry about this axis. In addition to rotational symmetries like this, a crystal may have symmetries in the form of mirror planes and translational symmetries, and also the so-called "compound symmetries," which are a combination of translation and rotation/mirror symmetries. A full classification of a crystal is achieved when all of these inherent symmetries of the crystal are identified.

Lattice Systems

These lattice systems are a grouping of crystal structures according to the axial system used to describe their lattice. Each lattice system consists of a set of three axes in a particular geometric arrangement. There are seven lattice systems. They are similar to but not quite the same as the seven crystal systems and the six crystal families.

Crystal family	Lattice system	Symmetry	14 Bravais Lattices			
			Primitive	Base-centered	Body-centered	Face-centered
	triclinic	C_i				

Crystal family	Lattice system	Symmetry	14 Bravais Lattices			
			Primitive	Base-centered	Body-centered	Face-centered
monoclinic		C_{2h}	$\beta \neq 90°$ $a \neq c$	$\beta \neq 90°$ $a \neq c$		
orthorhombic		D_{2h}	$a \neq b \neq c$	$a \neq b \neq c$	$a \neq b \neq c$	$a \neq b \neq c$
tetragonal		D_{4h}	$a \neq c$		$a \neq c$	
hexagonal	rhombohedral	D_{3d}	$\alpha \neq 90°$			
	hexagonal	D_{6h}	$\gamma = 120°$			
cubic		O_h				

The simplest and most symmetric, the cubic (or isometric) system, has the symmetry of a cube, that is, it exhibits four threefold rotational axes oriented at 109.5° (the tetrahedral angle) with respect to each other. These threefold axes lie along the body diagonals of the cube. The other six lattice systems, are hexagonal, tetragonal, rhombohedral (often confused with the trigonal crystal system), orthorhombic, monoclinic and triclinic.

Interplanar Spacing

The spacing d between adjacent (hkl) lattice planes is given by:

- Cubic:

$$\frac{1}{d^2} = \frac{h^2 + k^2 + l^2}{a^2}$$

- Tetragonal:

$$\frac{1}{d^2} = \frac{h^2 + k^2}{a^2} + \frac{l^2}{c^2}$$

- Hexagonal:

$$\frac{1}{d^2} = \frac{4}{3}\left(\frac{h^2 + hk + k^2}{a^2}\right) + \frac{l^2}{c^2}$$

- Rhombohedral:

$$\frac{1}{d^2} = \frac{(h^2 + k^2 + l^2)\sin^2\alpha + 2(hk + kl + hl)(\cos^2\alpha - \cos\alpha)}{a^2(1 - 3\cos^2\alpha + 2\cos^3\alpha)}$$

- Orthorhombic:

$$\frac{1}{d^2} = \frac{h^2}{a^2} + \frac{k^2}{b^2} + \frac{l^2}{c^2}$$

- Monoclinic:

$$\frac{1}{d^2} = \left(\frac{h^2}{a^2} + \frac{k^2\sin^2\beta}{b^2} + \frac{l^2}{c^2} - \frac{2hl\cos\beta}{ac}\right)\csc^2\beta$$

- Triclinic:

$$\frac{1}{d^2} = \frac{\dfrac{h^2}{a^2}\sin^2\alpha + \dfrac{k^2}{b^2}\sin^2\beta + \dfrac{l^2}{c^2}\sin^2\gamma}{1 - \cos^2\alpha - \cos^2\beta - \cos^2\gamma + 2\cos\alpha\cos\beta\cos\gamma}$$

Atomic Coordination

By considering the arrangement of atoms relative to each other, their coordination numbers (or number of nearest neighbors), interatomic distances, types of bonding, etc., it is possible to form a general view of the structures and alternative ways of visualizing them.

Close Packing

The hcp lattice (left) and the fcc lattice (right)

The principles involved can be understood by considering the most efficient way of packing together equal-sized spheres and stacking close-packed atomic planes in three dimensions. For example, if plane A lies beneath plane B, there are two possible ways of placing an additional atom on top of layer B. If an additional layer was placed directly over plane A, this would give rise to the following series:

...ABABABAB...

This arrangement of atoms in a crystal structure is known as hexagonal close packing (hcp).

If, however, all three planes are staggered relative to each other and it is not until the fourth layer is positioned directly over plane A that the sequence is repeated, then the following sequence arises:

...ABCABCABC...

This type of structural arrangement is known as cubic close packing (ccp).

The unit cell of a ccp arrangement of atoms is the face-centered cubic (fcc) unit cell. This is not immediately obvious as the closely packed layers are parallel to the {111} planes of the fcc unit cell. There are four different orientations of the close-packed layers.

The 74% packing efficiency is the maximum density possible in unit cells constructed of spheres of only one size. Most crystalline forms of metallic elements are hcp, fcc, or bcc (body-centered cubic). The coordination number of atoms in hcp and fcc structures is 12 and its atomic packing factor (APF) is the number mentioned above, 0.74. This can be compared to the APF of a bcc structure, which is 0.68.

Bravais Lattices

Bravais lattices, also referred to as *space lattices*, describe the geometric arrangement of the lattice points, and therefore the translational symmetry of the crystal. The three dimensions of space afford 14 distinct Bravais lattices describing the translational symmetry. All crystalline materials recognized today, not including quasicrystals, fit in one of these arrangements. The fourteen three-dimensional lattices, classified by crystal system, are shown above.

The crystal structure consists of the same group of atoms, the *basis*, positioned around each and every lattice point. This group of atoms therefore repeats indefinitely in three dimensions according to the arrangement of one of the Bravais lattices. The characteristic rotation and mirror symmetries of the unit cell is described by its crystallographic point group.

Point Groups

The crystallographic point group or *crystal class* is the mathematical group comprising the symmetry operations that leave at least one point unmoved and that leave the appearance of the crystal structure unchanged. These symmetry operations include

- *Reflection*, which reflects the structure across a *reflection plane*

- *Rotation*, which rotates the structure a specified portion of a circle about a *rotation axis*

- *Inversion*, which changes the sign of the coordinate of each point with respect to a *center of symmetry* or *inversion point*

- *Improper rotation*, which consists of a rotation about an axis followed by an inversion.

Rotation axes (proper and improper), reflection planes, and centers of symmetry are collectively called *symmetry elements*. There are 32 possible crystal classes. Each one can be classified into one of the seven crystal systems.

Space Groups

In addition to the operations of the point group, the space group of the crystal structure contains translational symmetry operations. These include:

- Pure *translations*, which move a point along a vector

- *Screw axes*, which rotate a point around an axis while translating parallel to the axis.

- *Glide planes*, which reflect a point through a plane while translating it parallel to the plane.

There are 230 distinct space groups.

Grain Boundaries

Grain boundaries are interfaces where crystals of different orientations meet. A grain boundary is a single-phase interface, with crystals on each side of the boundary being identical except in orientation. The term "crystallite boundary" is sometimes, though rarely, used. Grain boundary areas contain those atoms that have been perturbed from their original lattice sites, dislocations, and impurities that have migrated to the lower energy grain boundary.

Treating a grain boundary geometrically as an interface of a single crystal cut into two parts, one of which is rotated, we see that there are five variables required to define a grain boundary. The first two numbers come from the unit vector that specifies a rotation axis. The third number designates the angle of rotation of the grain. The final two numbers specify the plane of the grain boundary (or a unit vector that is normal to this plane).

Grain boundaries disrupt the motion of dislocations through a material, so reducing crystallite size is a common way to improve strength, as described by the Hall–Petch relationship. Since grain boundaries are defects in the crystal structure they tend to decrease the electrical and thermal conductivity of the material. The high interfacial energy and relatively weak bonding in most grain boundaries often makes them preferred sites for the onset of corrosion and for the precipitation of new phases from the solid. They are also important to many of the mechanisms of creep.

Grain boundaries are in general only a few nanometers wide. In common materials, crystallites are large enough that grain boundaries account for a small fraction of the material. However, very small grain sizes are achievable. In nanocrystalline solids, grain boundaries become a significant volume fraction of the material, with profound effects on such properties as diffusion and plasticity. In the limit of small crystallites, as the volume fraction of grain boundaries approaches 100%, the material ceases to have any crystalline character, and thus becomes an amorphous solid.

Defects and Impurities

Real crystals feature defects or irregularities in the ideal arrangements described above and it is these defects that critically determine many of the electrical and mechanical properties of real materials. When one atom substitutes for one of the principal atomic components within the crystal structure, alteration in the electrical and thermal properties of the material may ensue. Impurities may also manifest as spin impurities in certain materials. Research on magnetic impurities demonstrates that substantial alteration of certain properties such as specific heat may be affected by small concentrations of an impurity, as for example impurities in semiconducting ferromagnetic alloys may lead to different properties as first predicted in the late 1960s. Dislocations in the crystal lattice allow shear at lower stress than that needed for a perfect crystal structure.

Prediction of Structure

The difficulty of predicting stable crystal structures based on the knowledge of only the chemical composition has long been a stumbling block on the way to fully computational materials design. Now, with more powerful algorithms and high-performance computing, structures of medium complexity can be predicted using such approaches as evolutionary algorithms, random sampling, or metadynamics.

The crystal structures of simple ionic solids (e.g., NaCl or table salt) have long been rationalized in terms of Pauling's rules, first set out in 1929 by Linus Pauling, referred to by many since as the "father of the chemical bond". Pauling also considered the nature of the interatomic forces in metals, and concluded that about half of the five d-orbitals in the transition metals are involved in bonding, with the remaining nonbonding d-orbitals being responsible for the magnetic properties. He, therefore, was able to correlate the number of d-orbitals in bond formation with the bond length as well as many of the physical properties of the substance. He subsequently introduced the metallic orbital, an extra orbital necessary to permit uninhibited resonance of valence bonds among various electronic structures.

In the resonating valence bond theory, the factors that determine the choice of one from among alternative crystal structures of a metal or intermetallic compound revolve around the energy of resonance of bonds among interatomic positions. It is clear that some modes of resonance would make larger contributions (be more mechanically stable than others), and that in particular a simple ratio of number of bonds to number of positions would be exceptional. The resulting principle is that a special stability is associated with the simplest ratios or "bond numbers": $\frac{1}{2}$, $\frac{1}{3}$, $\frac{2}{3}$, $\frac{1}{4}$, $\frac{3}{4}$, etc. The choice of structure and the value of the axial ratio (which determines the relative bond lengths) are thus a result of the effort of an atom to use its valency in the formation of stable bonds with simple fractional bond numbers.

After postulating a direct correlation between electron concentration and crystal structure in beta-phase alloys, Hume-Rothery analyzed the trends in melting points, compressibilities and bond lengths as a function of group number in the periodic table in order to establish a system of valencies of the transition elements in the metallic state. This treatment thus emphasized the increasing bond strength as a function of group number. The operation of directional forces were emphasized in one article on the relation between bond hybrids and the metallic structures. The resulting correlation between electronic and crystalline structures is summarized by a single parameter, the weight of the d-electrons per hybridized metallic orbital. The "d-weight" calculates out to 0.5, 0.7 and 0.9 for the fcc, hcp and bcc structures respectively. The relationship between d-electrons and crystal structure thus becomes apparent.

In crystal structure predictions/simulations, the periodicity is usually applied, since the system is imagined as unlimited big in all directions. Starting from a triclinic structure with no further symmetry property assumed, the system may be driven to show some additional symmetry properties by applying Newton's Second Law on particles in the unit cell and a recently developed dynamical equation for the system period vectors (lattice parameters including angles), even if the system is subject to external stress.

Polymorphism

Quartz is one of the several thermodynamically stable crystalline forms of silica, SiO_2. The most important forms of silica include: α-quartz, β-quartz, tridymite, cristobalite, coesite, and stishovite.

Polymorphism is the occurrence of multiple crystalline forms of a material. It is found in many crystalline materials including polymers, minerals, and metals. According to Gibbs' rules of phase equilibria, these unique crystalline phases are dependent on intensive variables such as pressure and temperature. Polymorphism is related to allotropy, which refers to elemental solids. The complete morphology of a material is described by polymorphism and other variables such as crystal habit, amorphous fraction or crystallographic defects. Polymorphs have different stabilities and may spontaneously convert from a metastable form (or thermodynamically unstable form) to the stable form at a particular temperature. They also exhibit different melting points, solubilities, and X-ray diffraction patterns.

One good example of this is the quartz form of silicon dioxide, or SiO_2. In the vast majority of silicates, the Si atom shows tetrahedral coordination by 4 oxygens. All but one of the crystalline forms involve tetrahedral $\{SiO_4\}$ units linked together by shared vertices in different arrangements. In different minerals the tetrahedra show different degrees of networking and polymerization. For example, they occur singly, joined together in pairs, in larger finite clusters including rings, in chains, double chains, sheets, and three-dimensional frameworks. The minerals are classified into groups based on these structures. In each of its 7 thermodynamically stable crystalline forms or polymorphs of crystalline quartz, only 2 out of 4 of each the edges of the $\{SiO_4\}$ tetrahedra are shared with others, yielding the net chemical formula for silica: SiO_2.

Another example is elemental tin (Sn), which is malleable near ambient temperatures but is brittle when cooled. This change in mechanical properties due to existence of its two major allotropes, α- and β-tin. The two allotropes that are encountered at normal pressure and temperature, α-tin and β-tin, are more commonly known as *gray tin* and *white tin* respectively. Two more allotropes, γ and σ, exist at temperatures above 161 °C and pressures above several GPa. White tin is metallic, and is the stable crystalline form at or above room temperature. Below 13.2 °C, tin exists in the gray form, which has a diamond cubic crystal structure, similar to diamond, silicon or germanium. Gray tin has no metallic properties at all, is a dull gray powdery material, and has few uses, other than a few specialized semiconductor applications. Although the α–β transformation temperature of tin is nominally 13.2 °C, impurities (e.g. Al, Zn, etc.) lower the transition temperature well below 0 °C, and upon addition of Sb or Bi the transformation may not occur at all.

Physical Properties

Twenty of the 32 crystal classes are piezoelectric, and crystals belonging to one of these classes (point groups) display piezoelectricity. All piezoelectric classes lack a center of symmetry. Any material develops a dielectric polarization when an electric field is applied, but a substance that has such a natural charge separation even in the absence of a field is called a polar material. Whether or not a material is polar is determined solely by its crystal structure. Only 10 of the 32 point groups are polar. All polar crystals are pyroelectric, so the 10 polar crystal classes are sometimes referred to as the pyroelectric classes.

There are a few crystal structures, notably the perovskite structure, which exhibit ferroelectric behavior. This is analogous to ferromagnetism, in that, in the absence of an electric field during production, the ferroelectric crystal does not exhibit a polarization. Upon the application of an electric field of sufficient magnitude, the crystal becomes permanently polarized. This polarization can be reversed by a sufficiently large counter-charge, in the same way that a ferromagnet can be reversed. However, although they are called ferroelectrics, the effect is due to the crystal structure (not the presence of a ferrous metal).

Crystallographic Defect

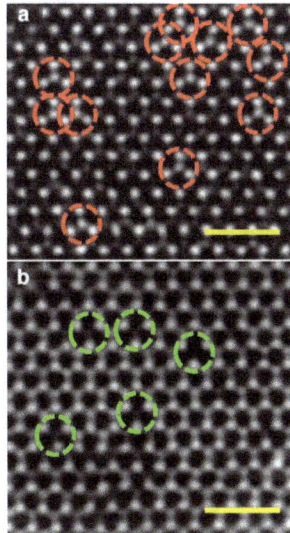

Electron microscopy of antisites (a, Mo substitutes for S) and vacancies (b, missing S atoms) in a monolayer of molybdenum disulfide. Scale bar: 1 nm.

Crystalline solids exhibit a periodic crystal structure. The positions of atoms or molecules occur on repeating fixed distances, determined by the unit cell parameters. However, the arrangement of atoms or molecules in most crystalline materials is not perfect. The regular patterns are interrupted by crystallographic defects.

Point Defects

Point defects are defects that occur only at or around a single lattice point. They are not extended in space in any dimension. Strict limits for how small a point defect is are generally not defined explicitly, typically, however, these defects involve at most a few extra or missing atoms. Larger defects in an ordered structure are usually considered dislocation loops. For historical reasons, many point defects, especially in ionic crystals, are called *centers*: for example a vacancy in many ionic solids is called a luminescence center, a color center, or F-center. These dislocations permit ionic transport through crystals leading to electrochemical reactions. These are frequently specified using Kröger–Vink Notation.

- Vacancy defects are lattice sites which would be occupied in a perfect crystal, but are vacant. If a neighboring atom moves to occupy the vacant site, the vacancy moves in the opposite direction to the site which used to be occupied by the moving atom. The stability of the surrounding crystal structure guarantees that the neighboring atoms will not simply collapse around the vacancy. In some materials, neighboring atoms actually move away from a vacancy, because they experience attraction from atoms in the surroundings. A vacancy (or pair of vacancies in an ionic solid) is sometimes called a Schottky defect.

- Interstitial defects are atoms that occupy a site in the crystal structure at which there is usually not an atom. They are generally high energy configurations. Small atoms in some crystals can occupy interstices without high energy, such as hydrogen in palladium.

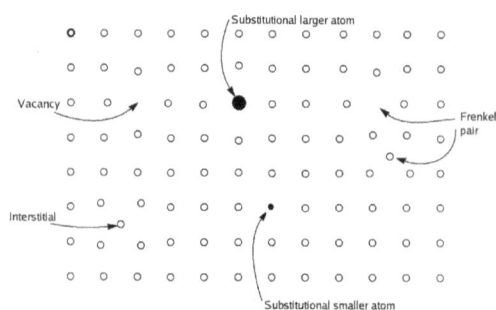

Schematic illustration of some simple point defect types in a monatomic solid

- A nearby pair of a vacancy and an interstitial is often called a Frenkel defect or Frenkel pair. This is caused when an ion moves into an interstitial site and creates a vacancy.

- Due to fundamental limitations of material purification methods, materials are never 100% pure, which by definition induces defects in crystal structure. In the case of an impurity, the atom is often incorporated at a regular atomic site in the crystal structure. This is neither a vacant site nor is the atom on an interstitial site and it is called a *substitutional* defect. The atom is not supposed to be anywhere in the crystal, and is thus an impurity. In some cases where the radius of the substitutional atom (ion) is substantially smaller than that of the atom (ion) it is replacing, its equilibrium position can be shifted away from the lattice site. These types of substitutional defects are often referred to as off-center ions. There are two different types of substitutional defects: Isovalent substitution and aliovalent substitution. Isovalent substitution is where the ion that is substituting the original ion is of the same oxidation state as the ion it is replacing. Aliovalent substitution is where the ion that is substituting the original ion is of a different oxidation state than the ion it is replacing. Aliovalent substitutions change the overall charge within the ionic compound, but the ionic compound must be neutral. Therefore, a charge compensation mechanism is required. Hence either one of the metals is partially or fully oxidised or reduced, or ion vacancies are created.

- Antisite defects occur in an ordered alloy or compound when atoms of different type exchange positions. For example, some alloys have a regular structure in which every other atom is a different species; for illustration assume that type A atoms sit on the corners of a cubic lattice, and type B atoms sit in the center of the cubes. If one cube has an A atom at its center, the atom is on a site usually occupied by a B atom, and is thus an antisite defect. This is neither a vacancy nor an interstitial, nor an impurity.

- Topological defects are regions in a crystal where the normal chemical bonding environment is topologically different from the surroundings. For instance, in a perfect sheet of graphite (graphene) all atoms are in rings containing six atoms. If the sheet contains regions where the number of atoms in a ring is different from six, while the total number of atoms remains the same, a topological defect has formed. An example is the Stone Wales defect in nanotubes, which consists of two adjacent 5-membered and two 7-membered atom rings.

- Also amorphous solids may contain defects. These are naturally somewhat hard to define, but sometimes their nature can be quite easily understood. For instance, in ideally bonded amorphous silica all Si atoms have 4 bonds to O atoms and all O atoms have 2 bonds to Si atom. Thus e.g. an O atom with only one Si bond (a dangling bond) can be considered a defect in silica. Moreover, defects can also be defined in amorphous solids based on empty or densely packed local atomic neighbourhoods, and the properties of such 'defects' can be shown to be similar to normal vacancies and interstitials in crystals,.

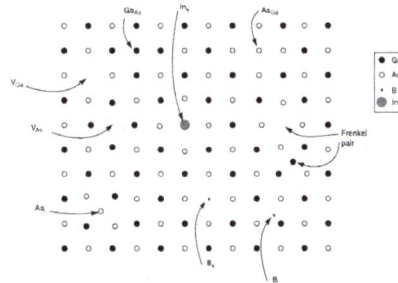

Schematic illustration of defects in a compound solid, using GaAs as an example.

- Complexes can form between different kinds of point defects. For example, if a vacancy encounters an impurity, the two may bind together if the impurity is too large for the lattice. Interstitials can form 'split interstitial' or 'dumbbell' structures where two atoms effectively share an atomic site, resulting in neither atom actually occupying the site.

Line Defects

Line defects can be described by gauge theories.

Dislocations are linear defects around which some of the atoms of the crystal lattice are misaligned. There are two basic types of dislocations, the *edge* dislocation and the *screw* dislocation. "Mixed" dislocations, combining aspects of both types, are also common.

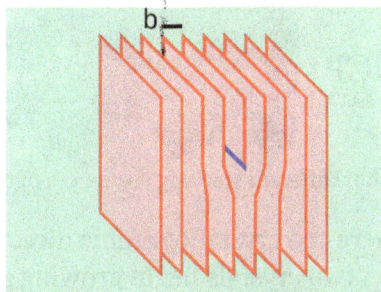

An *edge dislocation* is shown. The dislocation line is presented in blue, the Burgers vector b in black.

Edge dislocations are caused by the termination of a plane of atoms in the middle of a crystal. In such a case, the adjacent planes are not straight, but instead bend around the edge of the terminating plane so that the crystal structure is perfectly ordered on either side. The analogy with a stack of paper is apt: if a half a piece of paper is inserted in a stack of paper, the defect in the stack is only noticeable at the edge of the half sheet.

The screw dislocation is more difficult to visualise, but basically comprises a structure in which a helical path is traced around the linear defect (dislocation line) by the atomic planes of atoms in the crystal lattice.

The presence of dislocation results in lattice strain (distortion). The direction and magnitude of such distortion is expressed in terms of a Burgers vector (b). For an edge type, b is perpendicular to the dislocation line, whereas in the cases of the screw type it is parallel. In metallic materials, b is aligned with close-packed crystallographic directions and its magnitude is equivalent to one interatomic spacing.

Dislocations can move if the atoms from one of the surrounding planes break their bonds and re-bond with the atoms at the terminating edge.

It is the presence of dislocations and their ability to readily move (and interact) under the influence of stresses induced by external loads that leads to the characteristic malleability of metallic materials.

Dislocations can be observed using transmission electron microscopy, field ion microscopy and atom probe techniques. Deep level transient spectroscopy has been used for studying the electrical activity of dislocations in semiconductors, mainly silicon.

Disclinations are line defects corresponding to "adding" or "subtracting" an angle around a line. Basically, this means that if you track the crystal orientation around the line defect, you get a rotation. Usually, they were thought to play a role only in liquid crystals, but recent developments suggest that they might have a role also in solid materials, e.g. leading to the self-healing of cracks.

Planar Defects

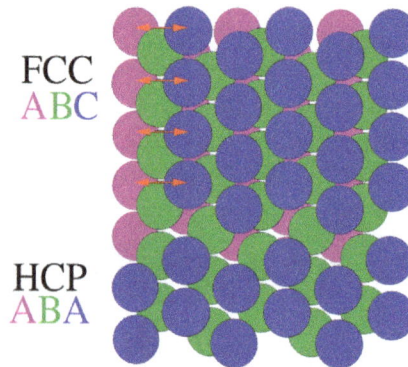

FCC
ABC

HCP
ABA

Origin of stacking faults: Different stacking sequences of close-packed crystals

- Grain boundaries occur where the crystallographic direction of the lattice abruptly changes. This usually occurs when two crystals begin growing separately and then meet.

- Antiphase boundaries occur in ordered alloys: in this case, the crystallographic direction remains the same, but each side of the boundary has an opposite phase: For example, if the ordering is usually ABABABAB (hexagonal close-packed crystal), an antiphase boundary takes the form of ABABBABA.

- Stacking faults occur in a number of crystal structures, but the common example is in close-packed structures. They are formed by a local deviation of the stacking sequence of layers in a crystal. An example would be the ABABCABAB stacking sequence.

- A twin boundary is a defect that introduces a plane of mirror symmetry in the ordering of a crystal. For example, in cubic close-packed crystals, the stacking sequence of a twin boundary would be ABCABCBACBA.

- On surfaces of single crystals, steps between atomically flat terraces can also be regarded as planar defects. It has been shown that such defects and their geometry have significant influence on the adsorption of organic molecules

Bulk Defects

- three-dimensional macroscopic or bulk defects, such as pores, cracks, or inclusions

- Voids — small regions where there are no atoms, and which can be thought of as clusters of vacancies

- Impurities can cluster together to form small regions of a different phase. These are often called precipitates.

Mathematical Classification Methods

A successful mathematical classification method for physical lattice defects, which works not only with the theory of dislocations and other defects in crystals but also, e.g., for disclinations in liquid crystals and for excitations in superfluid ^{3}He, is the topological homotopy theory.

Computer Simulation Methods

Density functional theory, classical molecular dynamics and kinetic Monte Carlo simulations are widely used to study the properties of defects in solids with computer simulations. Simulating jamming of hard spheres of different sizes and/or in containers with non-commeasurable sizes using the Lubachevsky–Stillinger algorithm can be an effective technique for demonstrating some types of crystallographic defects.

Crystal Twinning

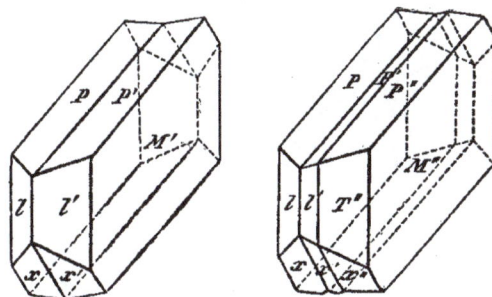

Diagram of twinned crystals of Albite. On the more perfect cleavage, which is parallel to the basal plane (P), is a system of fine striations, parallel to the second cleavage (M).

Crystal twinning occurs when two separate crystals share some of the same crystal lattice points in a symmetrical manner. The result is an intergrowth of two separate crystals in a variety of specific configurations. A twin boundary or composition surface separates the two crystals. Crystallographers classify twinned crystals by a number of twin laws. These twin laws are specific to the crystal system. The type of twinning can be a diagnostic tool in mineral identification.

Twinning can often be a problem in X-ray crystallography, as a twinned crystal does not produce a simple diffraction pattern.

Types of Twinning

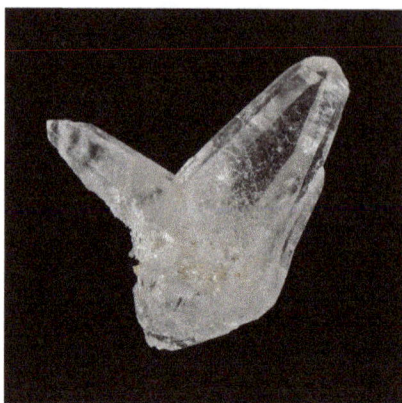

Quartz - Japan twin

Simple twinned crystals may be contact twins or penetration twins. Contact twins share a single composition surface often appearing as mirror images across the boundary. Plagioclase, quartz, gypsum, and spinel often exhibit contact twinning. Merohedral twinning occurs when the lattices of the contact twins superimpose in three dimensions, such as by relative rotation of one twin from the other. An example is metazeunerite. In penetration twins the individual crystals have the appearance of *passing through* each other in a symmetrical manner. Orthoclase, staurolite, pyrite, and fluorite often show penetration twinning.

Galvanized surface with macroscopic crystalline features. Twin boundaries are visible as striations within each crystallite, most prominently in the bottom-left and top-right.

If several twin crystal parts are aligned by the same twin law they are referred to as multiple or repeated twins. If these multiple twins are aligned in parallel they are called polysynthetic twins. When the multiple twins are not parallel they are cyclic twins. Albite, calcite, and pyrite often show polysynthetic twinning. Closely spaced polysynthetic twinning is often observed as striations or fine parallel lines on the crystal face. Rutile, aragonite, cerussite, and chrysoberyl often exhibit cyclic twinning, typically in a radiating pattern. But in general based on relationship between twin axis and twin plane there are 3 types of twinning exists, such as; 1-parallel twinning:-when twin axis and compositional plane lies parallel to each other. 2-normal twin:-when the twin plane and compositional plane lies noramally. 3-complex twin:-combination of parallel twinning and normal twinning on one compositional plane.

Modes of Formation

There are three modes of formation of twinned crystals. Growth twins are the result of an interruption or change in the lattice during formation or growth due to a possible deformation from a larger substituting ion. Annealing or deformation twins are the result of a change in crystal system during cooling as one *form* becomes unstable and the crystal structure must re-organize or *transform* into another more stable form. Deformation or gliding twins are the result of stress on the crystal after the crystal has formed. If a FCC metal like aluminum experiences extreme stresses, it will experience twinning as seen in the case of explosions. Deformation twinning is a common result of regional metamorphism.

Crystals that grow adjacent to each other may be aligned to resemble twinning. This parallel growth simply reduces system energy and is not twinning.

Twin Boundaries

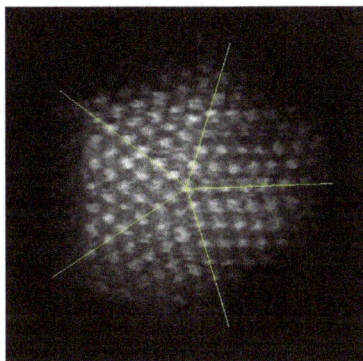

Fivefold twinning in a gold nanoparticle (electron microscope image).

Twin boundaries occur when two crystals of the same type intergrow, so that only a slight misorientation exists between them. It is a highly symmetrical interface, often with one crystal the mirror image of the other; also, atoms are shared by the two crystals at regular intervals. This is also a much lower-energy interface than the grain boundaries that form when crystals of arbitrary orientation grow together.

Twin boundaries are partly responsible for shock hardening and for many of the changes that occur in cold work of metals with limited slip systems or at very low temperatures. They also occur due to martensitic transformations: the motion of twin boundaries is responsible for the pseudoelastic and shape-memory behavior of nitinol, and their presence is partly responsible for the hardness due to quenching of steel. In certain types of high strength steels, very fine deformation twins act as primary obstacles against dislocation motion. These steels are referred to as 'TWIP' steels, where TWIP stands for TWinning Induced Plasticity.

Deformation Twins

Of the three common crystalline structures BCC, FCC, and HCP, the HCP structure is the most likely to form deformation twins when strained, because they rarely have a sufficient number of slip systems for an arbitrary shape change. High strain rates, low Stacking-fault energy and low temperatures facilitate deformation twinning.

Atomic Packing Factor

In crystallography, atomic packing factor (APF), packing efficiency or packing fraction is the fraction of volume in a crystal structure that is occupied by constituent particles. It is dimensionless and always less than unity. In atomic systems, by convention, the APF is determined by assuming that atoms are rigid spheres. The radius of the spheres is taken to be the maximal value such that the atoms do not overlap. For one-component crystals (those that contain only one type of particle), the packing fraction is represented mathematically by

$$APF = \frac{N_{particle}V_{particle}}{V_{unit\ cell}}$$

where $N_{particle}$ is the number of particles in the unit cell, $V_{particle}$ is the volume of each particle, and $V_{unit\ cell}$ is the volume occupied by the unit cell. It can be proven mathematically that for one-com-ponent structures, the most dense arrangement of atoms has an APF of about 0.74, obtained by the close-packed structures. For multiple-component structures, the APF can exceed 0.74.

Single Component Crystal Structures

Common sphere packings taken on by atomic systems are listed below with their corresponding packing fraction.

- Hexagonal close-packed (hcp): 0.74
- Face-centered cubic (fcc): 0.74 (also called cubic close-packed, ccp)
- Body-centered cubic (bcc): 0.68
- Simple cubic: 0.52
- Diamond cubic: 0.34

The majority of metals take on either the hcp, ccp or bcc structure.

Body-centered Cubic

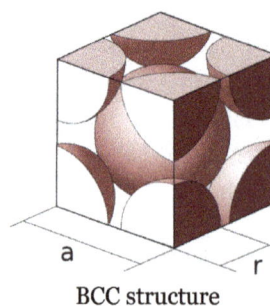

BCC structure

The primitive unit cell for the body-centered cubic crystal structure contains several fractions taken from nine atoms: one on each corner of the cube and one atom in the center. Because the volume of each of the eight corner atoms is shared between eight adjacent cells, each BCC cell contains the equivalent volume of two atoms (one central and one on the corner).

Each corner atom touches the center atom. A line that is drawn from one corner of the cube through the center and to the other corner passes through $4r$, where r is the radius of an atom. By geometry, the length of the diagonal is $a\sqrt{3}$. Therefore, the length of each side of the BCC structure can be related to the radius of the atom by

$$a = \frac{4r}{\sqrt{3}}.$$

Knowing this and the formula for the volume of a sphere, it becomes possible to calculate the APF as follows:

$$\text{APF} = \frac{N_{\text{atoms}} V_{\text{atom}}}{V_{\text{crystal}}} = \frac{2 \cdot \frac{4}{3}\pi r^3}{\left(\frac{4r}{\sqrt{3}}\right)^3}$$

$$= \frac{\pi \sqrt{3}}{8} \approx 0.68017476.$$

Hexagonal Close-packed

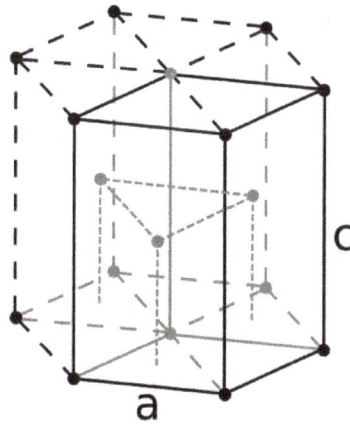

HCP structure

For the hexagonal close-packed structure the derivation is similar. Here the unit cell (equivalent to 3 primitive unit cells) is a hexagonal prism containing six atoms. Let a be the side length of its base and c be its height. The volume of the unit cell of hcp can be taken as $24\sqrt{2}r^3$. Then:

$$a = 2r, \qquad c = 4\sqrt{\tfrac{2}{3}}r.$$

It is then possible to calculate the APF as follows:

$$\text{APF} = \frac{N_{\text{atoms}} V_{\text{atom}}}{V_{\text{crystal}}} = \frac{6 \cdot \frac{4}{3}\pi r^3}{\frac{3\sqrt{3}}{2}a^2 c}$$

$$= \frac{6 \cdot \frac{4}{3}\pi r^3}{\frac{3\sqrt{3}}{2}(2r)^2 \sqrt{\frac{2}{3}} \cdot 4r} = \frac{6 \cdot \frac{4}{3}\pi r^3}{\frac{3\sqrt{3}}{2}\sqrt{\frac{2}{3}} \cdot 16r^3}$$

$$= \frac{\pi}{\sqrt{18}} = \frac{\pi}{3\sqrt{2}} \approx 0.74048048.$$

Periodic Table (Crystal Structure)

For elements that are solid at standard temperature and pressure the table gives the crystalline structure of the most thermodynamically stable form(s) in those conditions. In all other cases the structure given is for the element at its melting point. Data is presented only for the first 112 elements (hydrogen through copernicium; it is not available for any further ones), and predictions are given for elements that have never been produced in bulk (astatine, francium, and elements 100–112).

Table

Crystal structure of elements in the periodic table

1 H HEX																	2 He HCP
3 Li BCC	4 Be HCP											5 B RHO	6 C HEX	7 N HEX	8 O SC	9 F SC	10 Ne FCC
11 Na BCC	12 Mg HCP											13 Al FCC	14 Si DC	15 P ORTH	16 S ORTH	17 Cl ORTH	18 Ar FCC
19 K BCC	20 Ca FCC	21 Sc HCP	22 Ti HCP	23 V BCC	24 Cr BCC	25 Mn BCC	26 Fe BCC	27 Co HCP	28 Ni FCC	29 Cu FCC	30 Zn HCP	31 Ga ORTH	32 Ge DC	33 As RHO	34 Se HEX	35 Br ORTH	36 Kr FCC
37 Rb BCC	38 Sr FCC	39 Y HCP	40 Zr HCP	41 Nb BCC	42 Mo BCC	43 Tc HCP	44 Ru HCP	45 Rh FCC	46 Pd FCC	47 Ag FCC	48 Cd HCP	49 In TETR	50 Sn TETR	51 Sb RHO	52 Te HEX	53 I ORTH	54 Xe FCC
55 Cs BCC	56 Ba BCC	*	72 Hf HCP	73 Ta BCC/tetr	74 W BCC	75 Re HCP	76 Os HCP	77 Ir FCC	78 Pt FCC	79 Au FCC	80 Hg RHO	81 Tl HCP	82 Pb FCC	83 Bi RHO	84 Po SC/RHO	85 At [FCC]	86 Rn FCC
87 Fr [BCC]	88 Ra BCC	**	104 Rf [HCP]	105 Db [BCC]	106 Sg [BCC]	107 Bh [HCP]	108 Hs [HCP]	109 Mt [FCC]	110 Ds [BCC]	111 Rg [BCC]	112 Cn [HCP]	113 Nh	114 Fl	115 Mc	116 Lv	117 Ts	118 Og

	57 La DHCP	58 Ce DHCP/FCC	59 Pr DHCP	60 Nd DHCP	61 Pm DHCP	62 Sm RHO	63 Eu BCC	64 Gd HCP	65 Tb HCP	66 Dy HCP	67 Ho HCP	68 Er HCP	69 Tm HCP	70 Yb FCC	71 Lu HCP
*															
**	89 Ac FCC	90 Th FCC	91 Pa TETR	92 U ORTH	93 Np ORTH	94 Pu MON	95 Am DHCP	96 Cm DHCP	97 Bk DHCP	98 Cf DHCP	99 Es FCC	100 Fm [FCC]	101 Md [FCC]	102 No [FCC]	103 Lr [HCP]

Legend:

.../... mixed structure

[...] predicted structure

BCC: body centered cubic

FCC: face centered cubic (cubic close packed)

HCP: hexagonal close packed

DHCP: double hexagonal close packed

ORTH: orthorhombic

TETR: tetragonal

RHO: rhombohedral

HEX: hexagonal

SC: simple cubic

DC: diamond cubic

MON: monoclinic

Unusual Structures

Element	crystal system	coordination number	notes
Mn	cubic	distorted bcc – unit cell contains Mn atoms in 4 different environments.	

Zn	hexagonal	distorted from ideal hcp. 6 nearest neighbors in same plane- 6 in adjacent planes 14% farther away	
Ga	orthorhombic	each Ga atom has one nearest neighbour at 244 pm, 2 at 270 pm, 2 at 273 pm, 2 at 279 pm.	The structure is related to Iodine.
Cd	hexagonal	distorted from ideal hcp. 6 nearest neighbours in the same plane- 6 in adjacent planes 15% farther away	
In	tetragonal	slightly distorted fcc structure	
Sn	tetragonal	4 neighbours at 302 pm; 2 at 318 pm; 4 at 377 pm; 8 at 441 pm	white tin form (thermodynamical stable above 286.4 K)
Sb	rhombohedral	puckered sheet; each Sb atom has 3 neighbours in the same sheet at 290.8pm; 3 in adjacent sheet at 335.5 pm.	grey metallic form.
Hg	rhombohedral	6 nearest neighbours at 234 K and 1 atm (it is liquid at room temperature and thus has no crystal structure at ambient conditions!)	this structure can be considered to be a distorted hcp lattice with the nearest neighbours in the same plane being approx 16% farther away
Bi	rhombohedral	puckered sheet; each Bi atom has 3 neighbours in the same sheet at 307.2 pm; 3 in adjacent sheet at 352.9 pm.	Bi, Sb and grey As have the same space group in their crystal
Po	cubic	6 nearest neighbours	simple cubic lattice. The atoms in the unit cell are at the corner of a cube.
Sm	trigonal	12 nearest neighbours	complex hcp with 9 layer repeat, ABCBCACAB....
Pa	tetragonal	body centred tetragonal unit cell, which can be considered to be a distorted bcc	
U	orthorhombic	strongly distorted hcp structure. Each atom has four near neighbours, 2 at 275.4 pm, 2 at 285.4 pm. The next four at distances 326.3 pm and four more at 334.2 pm.	
Np	orthorhombic	highly distorted bcc structure. Lattice parameters: a=666.3 pm, b=472.3 pm, c=488.7 pm	

Pu	monoclinic	slightly distorted hexagonal structure. 16 atoms per unit cell. Lattice parameters: a= 618.3 pm, b=482.2 pm, c=1096.3 pm, β= 101.79 °	

Usual Crystal Structures

Close Packed Metal Structures

Many metals adopt close packed structures i.e. hexagonal close packed and face centred cubic structures (cubic close packed). A simple model for both of these is to assume that the metal atoms are spherical and are packed together in the most efficient way (close packing or closest packing). In closest packing every atom has 12 equidistant nearest neighbours, and therefore a coordination number of 12. If the close packed structures are considered as being built of layers of spheres then the difference between hexagonal close packing and face centred cubic is how each layer is positioned relative to others. Whilst there are many ways that can be envisaged for a regular buildup of layers:

- hexagonal close packing has alternate layers positioned directly above/below each other, A,B,A,B, ... (also termed $P6_3/mmc$, Pearson symbol hP2, strukturbericht A3) .

- face centered cubic has every third layer directly above/below each other,A,B,C,A,B,C,....... (also termed cubic close packing, Fm3m, Pearson symbol cF4, strukturbericht A1) .

- double hexagonal close packing has layers directly above/below each other, A,B,A,C,A,B,A,C,.... of period length 4 like an alternative mixture of fcc and hcp packing (also termed $P6_3/mmc$, Pearson Symbol hP4, strukturbericht A3').

- α-Sm packing has a period of 9 layers A,B,A,B,C,B,C,A,C,.... (R3m, Pearson Symbol hR3, strukturbericht C19).

Hexagonal Close Packed

In the ideal hcp structure the unit cell axial ratio is $2\sqrt{\dfrac{2}{3}}$ ~ 1.633, However, there are deviations

from this in some metals where the unit cell is distorted in one direction but the structure still retains the hcp space group—remarkable all the elements have a ratio of lattice parameters c/a < 1.633 (best are Mg and Co and worst Be with c/a ~ 1.568). In others like Zn and Cd the deviations from the ideal change the symmetry of the structure and these have a lattice parameter ratio c/a > 1.85.

Face Centered Cubic (Cubic Close Packed)

More content relating to number of planes within structure and implications for glide/slide e.g. ductility.

Double Hexagonal Close Packed

Similar to the ideal hcp structure, the perfect dhcp structure should have a lattice parameter ratio

of $\dfrac{c}{a} = 4\sqrt{\dfrac{2}{3}}$ ~ 3.267. In the real dhcp structures of 5 lanthanides (including β-Ce) $\dfrac{c}{2a}$ variates between

1.596 (Pm) and 1.6128 (Nd). For the 4 known actinides dhcp lattices the corresponding number vary between 1.620 (Bk) and 1.625 (Cf).

Body Centred Cubic

This is not a close packed structure. In this each metal atom is at the centre of a cube with 8 nearest neighbors, however the 6 atoms at the centres of the adjacent cubes are only approximately 15% further away so the coordination number can therefore be considered to be 14 when these are included. Note that if the body centered cubic unit cell is compressed along one 4 fold axis the structure becomes face centred cubic (cubic close packed).

Lattice Energy

The lattice energy of a crystalline solid is usually defined as the energy of formation of the crystal from infinitely-separated ions and as such is invariably negative. The precise value of the lattice energy may not be determined experimentally, because of the impossibility of preparing an adequate amount of gaseous ions or atoms and measuring the energy released during their condensation to form the solid. However, the value of the lattice energy may either be derived theoretically from electrostatics or from a thermodynamic cycling reaction, the Born–Haber cycle.

Historical Development

The concept of lattice energy was originally developed for rocksalt-structured and sphalerite-structured compounds like NaCl and ZnS, where the ions occupy high-symmetry crystal lattice sites. In the case of NaCl, the lattice energy is the energy released by the reaction

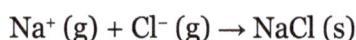

$$Na^+ (g) + Cl^- (g) \rightarrow NaCl (s)$$

which would amount to -786 kJ/mol.

Some older textbooks define lattice energy with the opposite sign, i.e. the energy required to convert the crystal into infinitely separated gaseous ions in vacuum, an endothermic process. Following this convention, the lattice energy of NaCl would be +786 kJ/mol. The lattice energy for ionic crystals such as sodium chloride, metals such as iron, or covalently linked materials such as diamond is considerably greater in magnitude than for solids such as sugar or iodine, whose neutral molecules interact only by weaker dipole-dipole or van der Waals forces.

Theoretical Treatments

The relationship between the molar lattice energy and the molar lattice enthalpy is given by the following equation:

$$\Delta_G U = \Delta_G H - p \Delta V_m,$$

where $\ddot{A}_G U$ is the molar lattice energy, $\ddot{A}_G H$ the molar lattice enthalpy and $\ddot{A}V_m$ the change of the volume per mol. Therefore the lattice enthalpy further takes into account that work has to be performed against an outer pressure p. Lattice Energy of an ionic compound depends upon charge of the ion and size of the ions.Moreover factors such as packing of ions doesn't matter efficiently

Born–landé Equation

In 1918 Born and Landé proposed that the lattice energy could be derived from the electric potential of the ionic lattice and a repulsive potential energy term.

$$E = -\frac{N_A M z^+ z^- q_e^2}{4\pi\varepsilon_0 r_0}\left(1-\frac{1}{n}\right),$$

where

N_A is the Avogadro constant;

M is the Madelung constant, relating to the geometry of the crystal;

z^+ is the charge number of cation;

z^- is the charge number of anion;

q_e is the elementary charge, equal to 1.6022×10^{-19} C;

ε_0 is the permittivity of free space, equal to 8.854×10^{-12} C² J⁻¹ m⁻¹;

r_0 is the distance to closest ion; and

n is the Born exponent, a number between 5 and 12, determined experimentally by measuring the compressibility of the solid, or derived theoretically.

The Born–Landé equation gives a reasonable fit to the lattice energy.

Compound	Calculated Lattice Energy	Experimental Lattice Energy
NaCl	−756 kJ/mol	−787 kJ/mol
LiF	−1007 kJ/mol	−1046 kJ/mol
CaCl$_2$	−2170 kJ/mol	−2255 kJ/mol

From the Born–Landé equation it can be seen that the lattice energy of a compound is dependent on a number of factors

- as the charges on the ions increase the lattice energy increases (becomes more negative),

- when ions are closer together the lattice energy increases (becomes more negative)

Barium oxide (BaO), for instance, which has the NaCl structure and therefore the same Madelung constant, has a bond radius of 275 picometers and a lattice energy of -3054 kJ/mol, while sodium chloride (NaCl) has a bond radius of 283 picometers and a lattice energy of -786 kJ/mol.

Kapustinskii Equation

The Kapustinskii equation can be used as a simpler way of deriving lattice energies where high precision is not required.

Effect of Polarisation

For ionic compounds with ions occupying lattice sites with crystallographic point groups C_1, C_{1h}, C_n or C_{nv} (n = 2, 3, 4 or 6) the concept of the lattice energy and the Born–Haber cycle has to be extended. In these cases the polarization energy E_{pol} associated with ions on polar lattice sites has to be included in the Born–Haber cycle and the solid formation reaction has to start from the already polarized species. As an example, one may consider the case of iron-pyrite FeS_2, where sulfur ions occupy lattice site of point symmetry group C_3. The lattice energy defining reaction then reads

$$Fe^{2+} (g) + 2 \text{ pol } S^- (g) \rightarrow FeS_2 (s)$$

where pol S^- stands for the polarized, gaseous sulfur ion. It has been shown that the neglection of the effect led to 15% difference between theoretical and experimental thermodynamic cycle energy of FeS_2 that reduced to only 2%, when the sulfur polarization effects were included.

Born–landé Equation

The Born–Landé equation is a means of calculating the lattice energy of a crystalline ionic compound. In 1918 Max Born and Alfred Landé proposed that the lattice energy could be derived from the electrostatic potential of the ionic lattice and a repulsive potential energy term.

$$E = -\frac{N_A M z^+ z^- e^2}{4\pi\epsilon_0 r_0}\left(1 - \frac{1}{n}\right)$$

where:

- N_A = Avogadro constant;

- M = Madelung constant, relating to the geometry of the crystal;

- z^+ = charge number of cation

- z^- = charge number of anion

- e = elementary charge, 1.6022×10^{-19} C

- ε_0 = permittivity of free space

    $4\pi\varepsilon_0 = 1.112\times10^{-10}$ C^2/(J·m)

- r_0 = distance to closest ion

- n = Born exponent, typically a number between 5 and 12, determined experimentally by measuring the compressibility of the solid, or derived theoretically.

Derivation

The ionic lattice is modeled as an assembly of hard elastic spheres which are compressed together by the mutual attraction of the electrostatic charges on the ions. They achieve the observed equilibrium distance apart due to a balancing short range repulsion.

Electrostatic Potential

The electrostatic potential energy, E_{pair}, between a pair of ions of equal and opposite charge is:

$$E_{pair} = -\frac{z^2 e^2}{4\pi\epsilon_0 r}$$

where

z = magnitude of charge on one ion

e = elementary charge, 1.6022×10^{-19} C

ϵ_0 = permittivity of free space

$4\pi\epsilon_0$ = 1.112×10^{-10} C²/(J m)

r = distance separating the ion centers

For a simple lattice consisting ions with equal and opposite charge in a 1:1 ratio, interactions between one ion and all other lattice ions need to be summed to calculate E_M, sometimes called the Madelung or lattice energy:

$$E_M = -\frac{z^2 e^2 M}{4\pi\epsilon_0 r}$$

where

M = Madelung constant, which is related to the geometry of the crystal

r = closest distance between two ions of opposite charge

Repulsive Term

Born and Lande suggested that a repulsive interaction between the lattice ions would be proportional to $1/r^n$ so that the repulsive energy term, E_R, would be expressed:

$$E_R = \frac{B}{r^n}$$

where

B = constant scaling the strength of the repulsive interaction

r = closest distance between two ions of opposite charge

n = Born exponent, a number between 5 and 12 expressing the steepness of the repulsive barrier

Total Energy

The total intensive potential energy of an ion in the lattice can therefore be expressed as the sum of the Madelung and repulsive potentials:

$$E(r) = -\frac{z^2 e^2 M}{4\pi\epsilon_0 r} + \frac{B}{r^n}$$

Minimizing this energy with respect to r yields the equilibrium separation r_0 in terms of the unknown constant B:

$$\frac{dE}{dr} = \frac{z^2 e^2 M}{4\pi\epsilon_0 r^2} - \frac{nB}{r^{n+1}}$$

$$0 = \frac{z^2 e^2 M}{4\pi\epsilon_0 r_0^2} - \frac{nB}{r_0^{n+1}}$$

$$r_0 = \left(\frac{4\pi\epsilon_0 nB}{z^2 e^2 M}\right)^{\frac{1}{n-1}}$$

$$B = \frac{z^2 e^2 M}{4\pi\epsilon_0 n} r_0^{n-1}$$

Evaluating the minimum intensive potential energy and substituting the expression for B in terms of r_0 yields the Born–Landé equation:

$$E(r_0) = -\frac{Mz^2 e^2}{4\pi\epsilon_0 r_0}\left(1 - \frac{1}{n}\right)$$

Calculated Lattice Energies

The Born–Landé equation gives a reasonable fit to the lattice energy

Compound	Calculated Lattice Energy	Experimental Lattice Energy

NaCl	−756 kJ/mol	−787 kJ/mol
LiF	−1007 kJ/mol	−1046 kJ/mol
CaCl$_2$	−2170 kJ/mol	−2255 kJ/mol

Born Exponent

The Born exponent is typically between 5 and 12. Approximate experimental values are listed below:

Ion configuration	He	Ne	Ar, Cu+	Kr, Ag+	Xe, Au+
n	5	7	9	10	12

References

- Encyclopaedia of Physics (2nd Edition), R.G. Lerner, G.L. Trigg, VHC publishers, 1991, ISBN (Verlagsgesellschaft) 3-527-26954-1, ISBN (VHC Inc.) 0-89573-752-3

- Donald E. Sands (1994). "§4-2 Screw axes and §4-3 Glide planes". Introduction to Crystallography (Reprint of WA Benjamin corrected 1975 ed.). Courier-Dover. pp. 70–71. ISBN 0486678393.

- Holleman, Arnold F.; Wiberg, Egon; Wiberg, Nils (1985). "Tin". Lehrbuch der Anorganischen Chemie (in German) (91–100 ed.). Walter de Gruyter. pp. 793–800. ISBN 3-11-007511-3.

- Schwartz, Mel (2002). "Tin and Alloys, Properties". Encyclopedia of Materials, Parts and Finishes (2nd ed.). CRC Press. ISBN 1-56676-661-3.

- Ellis, Arthur B.; et al. (1995). Teaching General Chemistry: A Materials Science Companion (3rd ed.). Washington, DC: American Chemical Society. ISBN 084122725X.

- Moore, Lesley E.; Smart, Elaine A. (2005). Solid State Chemistry: An Introduction (3rd ed.). Boca Raton, FL: Taylor & Francis, CRC. p. 8. ISBN 0748775161.

- I.D. Brown, The chemical Bond in Inorganic Chemistry, IUCr monographs in crystallography, Oxford University Press, 2002, ISBN 0-19-850870-0

- Roger Norris, Lawrie Ryan, David Acaster (2011). Cambridge International AS and A Level Chemistry Coursebook (1st ed.). Cambridge: Cambridge University Press. p. 254. ISBN 978-0-521-12661-8.

Methods for Characterizing Solid

X-ray crystallography is a method that is used to determine the atomic and molecular structure of a crystal. The methods that are used for characterizing solid apart from x-ray crystallography are neutron diffraction, electron microscope, X-ray absorption spectroscopy, powder diffraction etc. This section discusses the methods of characterizing solids in a critical manner providing key analysis to the subject matter.

X-ray Crystallography

X-ray crystallography is a tool used for identifying the atomic and molecular structure of a crystal, in which the crystalline atoms cause a beam of incident X-rays to diffract into many specific directions. By measuring the angles and intensities of these diffracted beams, a crystallographer can produce a three-dimensional picture of the density of electrons within the crystal. From this electron density, the mean positions of the atoms in the crystal can be determined, as well as their chemical bonds, their disorder and various other information.

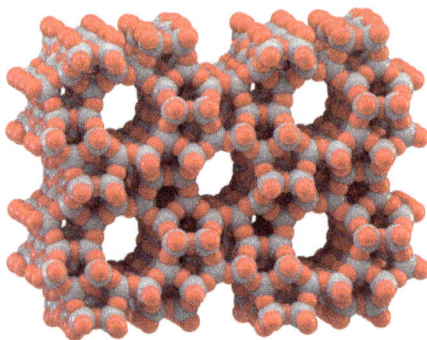

X-ray crystallography provides an atomistic model of zeolite, an aluminosilicate.

Since many materials can form crystals—such as salts, metals, minerals, semiconductors, as well as various inorganic, organic and biological molecules—X-ray crystallography has been fundamental in the development of many scientific fields. In its first decades of use, this method determined the size of atoms, the lengths and types of chemical bonds, and the atomic-scale differences among various materials, especially minerals and alloys. The method also revealed the structure and function of many biological molecules, including vitamins, drugs, proteins and nucleic acids such as DNA. X-ray crystallography is still the chief method for characterizing the atomic structure of new materials and in discerning materials that appear similar by other experiments. X-ray crystal structures can also account for unusual electronic or elastic properties of a material, shed light on chemical interactions and processes, or serve as the basis for designing pharmaceuticals against diseases.

In a single-crystal X-ray diffraction measurement, a crystal is mounted on a goniometer. The goniometer is used to position the crystal at selected orientations. The crystal is illuminated with a finely focused monochromatic beam of X-rays, producing a diffraction pattern of regularly spaced spots known as *reflections*. The two-dimensional images taken at different orientations are converted into a three-dimensional model of the density of electrons within the crystal using the mathematical method of Fourier transforms, combined with chemical data known for the sample. Poor resolution (fuzziness) or even errors may result if the crystals are too small, or not uniform enough in their internal makeup.

X-ray crystallography is related to several other methods for determining atomic structures. Similar diffraction patterns can be produced by scattering electrons or neutrons, which are likewise interpreted by Fourier transformation. If single crystals of sufficient size cannot be obtained, various other X-ray methods can be applied to obtain less detailed information; such methods include fiber diffraction, powder diffraction and (if the sample is not crystallized) small-angle X-ray scattering (SAXS). If the material under investigation is only available in the form of nanocrystalline powders or suffers from poor crystallinity, the methods of electron crystallography can be applied for determining the atomic structure.

For all above mentioned X-ray diffraction methods, the scattering is elastic; the scattered X-rays have the same wavelength as the incoming X-ray. By contrast, *inelastic* X-ray scattering methods are useful in studying excitations of the sample, rather than the distribution of its atoms.

History

Early Scientific History of Crystals and X-Rays

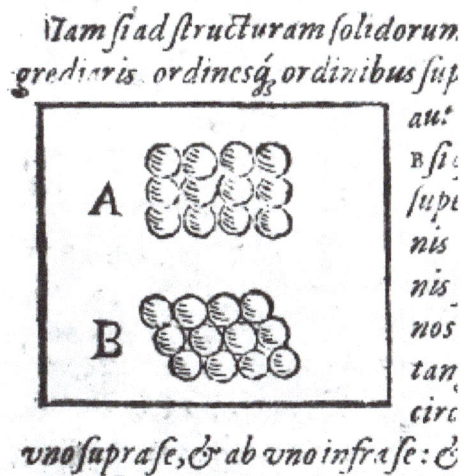

Drawing of square (Figure A, above) and hexagonal (Figure B, below) packing from Kepler's work, *Strena seu de Nive Sexangula*.

Crystals, though long admired for their regularity and symmetry, were not investigated scientifically until the 17th century. Johannes Kepler hypothesized in his work *Strena seu de Nive Sexangula* (A New Year's Gift of Hexagonal Snow) (1611) that the hexagonal symmetry of snowflake crystals was due to a regular packing of spherical water particles.

The Danish scientist Nicolas Steno (1669) pioneered experimental investigations of crystal symmetry. Steno showed that the angles between the faces are the same in every exemplar of a particular type of crystal, and René Just Haüy (1784) discovered that every face of a crystal can be described by simple stacking patterns of blocks of the same shape and size. Hence, William Hallowes Miller in 1839 was able to give each face a unique label of three small integers, the Miller indices which remain in use today for identifying crystal faces. Haüy's study led to the correct idea that crystals are a regular three-dimensional array (a Bravais lattice) of atoms and molecules; a single unit cell is repeated indefinitely along three principal directions that are not necessarily perpendicular. In the 19th century, a complete catalog of the possible symmetries of a crystal was worked out by Johan Hessel, Auguste Bravais, Evgraf Fedorov, Arthur Schönflies and (belatedly) William Barlow (1894). From the available data and physical reasoning, Barlow proposed several crystal structures in the 1880s that were validated later by X-ray crystallography; however, the available data were too scarce in the 1880s to accept his models as conclusive.

As shown by X-ray crystallography, the hexagonal symmetry of snowflakes results from the tetrahedral arrangement of hydrogen bonds about each water molecule. The water molecules are arranged similarly to the silicon atoms in the tridymite polymorph of SiO_2. The resulting crystal structure has hexagonal symmetry when viewed along a principal axis.

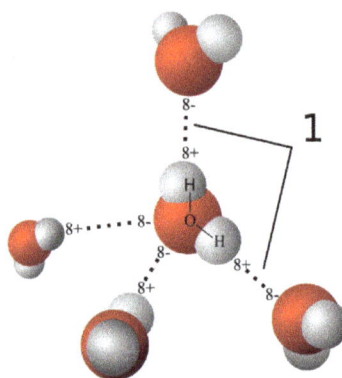

X-ray crystallography shows the arrangement of water molecules in ice, revealing the hydrogen bonds (1) that hold the solid together. Few other methods can determine the structure of matter with such precision (*resolution*).

Wilhelm Röntgen discovered X-rays in 1895, just as the studies of crystal symmetry were being concluded. Physicists were initially uncertain of the nature of X-rays, but soon suspected (correctly) that they were waves of electromagnetic radiation, in other words, another form of light. At that time, the wave model of light—specifically, the Maxwell theory of electromagnetic radiation—was well accepted among scientists, and experiments by Charles Glover Barkla showed that X-rays exhibited phenomena associated with electromagnetic waves, including transverse polarization and spectral lines akin to those observed in the visible wavelengths. Single-slit experiments in the laboratory of Arnold Sommerfeld suggested that X-rays had a wavelength of about 1 angstrom. However, X-rays are composed of photons, and thus are not only waves of electromagnetic radiation but also exhibit particle-like properties. Albert Einstein introduced the photon concept in 1905, but it was not broadly accepted until 1922, when Arthur Compton confirmed it by the scattering of X-rays from electrons. Therefore, these particle-like properties of X-rays, such as their ionization of gases, caused William Henry Bragg to argue in 1907 that X-rays were *not* electromagnetic radiation. Nevertheless, Bragg's view was not broadly accepted and the observation of X-ray diffraction by Max von Laue in 1912 confirmed for most scientists that X-rays were a form of electromagnetic radiation.

X-ray Diffraction

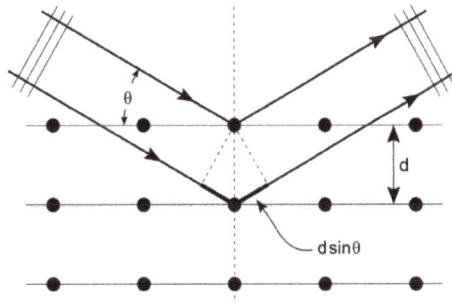

The incoming beam (coming from upper left) causes each scatterer to re-radiate a small portion of its intensity as a spherical wave. If scatterers are arranged symmetrically with a separation d, these spherical waves will be in sync (add constructively) only in directions where their path-length difference $2d \sin \theta$ equals an integer multiple of the wavelength λ. In that case, part of the incoming beam is deflected by an angle 2θ, producing a *reflection* spot in the diffraction pattern.

Crystals are regular arrays of atoms, and X-rays can be considered waves of electromagnetic radiation. Atoms scatter X-ray waves, primarily through the atoms' electrons. Just as an ocean wave striking a lighthouse produces secondary circular waves emanating from the lighthouse, so an X-ray striking an electron produces secondary spherical waves emanating from the electron. This phenomenon is known as elastic scattering, and the electron (or lighthouse) is known as the *scatterer*. A regular array of scatterers produces a regular array of spherical waves. Although these waves cancel one another out in most directions through destructive interference, they add constructively in a few specific directions, determined by Bragg's law:

$$2d \sin \theta = n\lambda$$

Here d is the spacing between diffracting planes, θ is the incident angle, n is any integer, and λ is the wavelength of the beam. These specific directions appear as spots on the diffraction pattern called *reflections*. Thus, X-ray diffraction results from an electromagnetic wave (the X-ray) impinging on a regular array of scatterers (the repeating arrangement of atoms within the crystal).

X-rays are used to produce the diffraction pattern because their wavelength λ is typically the same order of magnitude (1–100 angstroms) as the spacing d between planes in the crystal. In principle, any wave impinging on a regular array of scatterers produces diffraction, as predicted first by Francesco Maria Grimaldi in 1665. To produce significant diffraction, the spacing between the scatterers and the wavelength of the impinging wave should be similar in size. For illustration, the diffraction of sunlight through a bird's feather was first reported by James Gregory in the later 17th century. The first artificial diffraction gratings for visible light were constructed by David Rittenhouse in 1787, and Joseph von Fraunhofer in 1821. However, visible light has too long a wavelength (typically, 5500 angstroms) to observe diffraction from crystals. Prior to the first X-ray diffraction experiments, the spacings between lattice planes in a crystal were not known with certainty.

The idea that crystals could be used as a diffraction grating for X-rays arose in 1912 in a conversation between Paul Peter Ewald and Max von Laue in the English Garden in Munich. Ewald had proposed a resonator model of crystals for his thesis, but this model could not be validated using visible light, since the wavelength was much larger than the spacing between the resonators. Von Laue realized that electromagnetic radiation of a shorter wavelength was needed to observe such small spacings, and suggested that X-rays might have a wavelength comparable to the unit-cell spacing in crystals. Von Laue worked with two technicians, Walter Friedrich and his assistant Paul Knipping, to shine a beam of X-rays through a copper sulfate crystal and record its diffraction on a photographic plate. After being developed, the plate showed a large number of well-defined spots arranged in a pattern of intersecting circles around the spot produced by the central beam. Von Laue developed a law that connects the scattering angles and the size and orientation of the unit-cell spacings in the crystal, for which he was awarded the Nobel Prize in Physics in 1914.

Scattering

As described in the mathematical derivation below, the X-ray scattering is determined by the density of electrons within the crystal. Since the energy of an X-ray is much greater than that of a valence electron, the scattering may be modeled as Thomson scattering, the interaction of an electromagnetic ray with a free electron. This model is generally adopted to describe the polarization of the scattered radiation.

The intensity of Thomson scattering for one particle with mass m and charge q is:

$$I_o = I_e \left(\frac{q^4}{m^2 c^4} \right) \frac{1 + \cos^2 2\theta}{2} = I_e 7.94.10^{-26} \frac{1 + \cos^2 2\theta}{2} = I_e f$$

Hence the atomic nuclei, which are much heavier than an electron, contribute negligibly to the scattered X-rays.

Development from 1912 to 1920

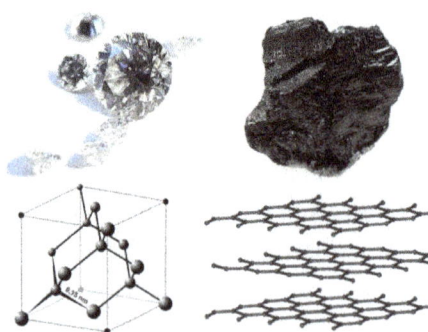

Although diamonds (top left) and graphite (top right) are identical in chemical composition—being both pure carbon—X-ray crystallography revealed the arrangement of their atoms (bottom) accounts for their different properties. In diamond, the carbon atoms are arranged tetrahedrally and held together by single covalent bonds, making it strong in all directions. By contrast, graphite is composed of stacked sheets. Within the sheet, the bonding is covalent and has hexagonal symmetry, but there are no covalent bonds between the sheets, making graphite easy to cleave into flakes.

After Von Laue's pioneering research, the field developed rapidly, most notably by physicists William Lawrence Bragg and his father William Henry Bragg. In 1912–1913, the younger Bragg developed Bragg's law, which connects the observed scattering with reflections from evenly spaced planes within the crystal. The Braggs, father and son, shared the 1915 Nobel Prize in Physics for their work in crystallography. The earliest structures were generally simple and marked by one-dimensional symmetry. However, as computational and experimental methods improved over the next decades, it became feasible to deduce reliable atomic positions for more complicated two- and three-dimensional arrangements of atoms in the unit-cell.

The potential of X-ray crystallography for determining the structure of molecules and minerals—then only known vaguely from chemical and hydrodynamic experiments—was realized immediately. The earliest structures were simple inorganic crystals and minerals, but even these revealed fundamental laws of physics and chemistry. The first atomic-resolution structure to be "solved" (i.e., determined) in 1914 was that of table salt. The distribution of electrons in the table-salt structure showed that crystals are not necessarily composed of covalently bonded molecules, and proved the existence of ionic compounds. The structure of diamond was solved in the same year, proving the tetrahedral arrangement of its chemical bonds and showing that the length of C–C single bond was 1.52 angstroms. Other early structures included copper, calcium fluoride (CaF_2, also known as *fluorite*), calcite ($CaCO_3$) and pyrite (FeS_2) in 1914; spinel ($MgAl_2O_4$) in 1915; the rutile and anatase forms of titanium dioxide (TiO_2) in 1916; pyrochroite $Mn(OH)_2$ and, by extension, brucite $Mg(OH)_2$ in 1919;. Also in 1919 sodium nitrate ($NaNO_3$) and caesium dichloroiodide ($CsICl_2$) were determined by Ralph Walter Graystone Wyckoff, and the wurtzite (hexagonal ZnS) structure became known in 1920.

The structure of graphite was solved in 1916 by the related method of powder diffraction, which was developed by Peter Debye and Paul Scherrer and, independently, by Albert Hull in 1917. The structure of graphite was determined from single-crystal diffraction in 1924 by two groups independently. Hull also used the powder method to determine the structures of various metals, such as iron and magnesium.

Cultural and Aesthetic Importance of X-ray Crystallography

In what has been called his scientific autobiography, *The Development of X-ray Analysis*, Sir William Lawrence Bragg mentioned that he believed the field of crystallography was particularly welcoming to women because the techno-aesthetics of the molecular structures resembled textiles and household objects. Bragg was known to compare crystal formation to "curtains, wallpapers, mosaics, and roses."

In 1951, the Festival Pattern Group at the Festival of Britain hosted a collaborative group of textile manufacturers and experienced crystallographers to design lace and prints based on the X-ray crystallography of insulin, china clay, and hemoglobin. One of the leading scientists of the project was Dr. Helen Megaw (1907–2002), the Assistant Director of Research at the Cavendish Laboratory in Cambridge at the time. Megaw is credited as one of the central figures who took inspiration from crystal diagrams and saw their potential in design. In 2008, the Wellcome Collection in London curated an exhibition on the Festival Pattern Group called "From Atom to Patterns."

Contributions to Chemistry and Material Science

X-ray crystallography has led to a better understanding of chemical bonds and non-covalent interactions. The initial studies revealed the typical radii of atoms, and confirmed many theoretical models of chemical bonding, such as the tetrahedral bonding of carbon in the diamond structure, the octahedral bonding of metals observed in ammonium hexachloroplatinate (IV), and the resonance observed in the planar carbonate group and in aromatic molecules. Kathleen Lonsdale's 1928 structure of hexamethylbenzene established the hexagonal symmetry of benzene and showed a clear difference in bond length between the aliphatic C–C bonds and aromatic C–C bonds; this finding led to the idea of resonance between chemical bonds, which had profound consequences for the development of chemistry. Her conclusions were anticipated by William Henry Bragg, who published models of naphthalene and anthracene in 1921 based on other molecules, an early form of molecular replacement.

Also in the 1920s, Victor Moritz Goldschmidt and later Linus Pauling developed rules for eliminating chemically unlikely structures and for determining the relative sizes of atoms. These rules led to the structure of brookite (1928) and an understanding of the relative stability of the rutile, brookite and anatase forms of titanium dioxide.

The distance between two bonded atoms is a sensitive measure of the bond strength and its bond order; thus, X-ray crystallographic studies have led to the discovery of even more exotic types of bonding in inorganic chemistry, such as metal-metal double bonds, metal-metal quadruple bonds, and three-center, two-electron bonds. X-ray crystallography—or, strictly speaking, an inelastic Compton scattering experiment—has also provided evidence for the partly covalent character of hydrogen bonds. In the field of organometallic chemistry, the X-ray structure of ferrocene initiated scientific studies of sandwich compounds, while that of Zeise's salt stimulated research into "back bonding" and metal-pi complexes. Finally, X-ray crystallography had a pioneering role in the development of supramolecular chemistry, particularly in clarifying the structures of the crown ethers and the principles of host-guest chemistry.

In material sciences, many complicated inorganic and organometallic systems have been analyzed using single-crystal methods, such as fullerenes, metalloporphyrins, and other complicated compounds. Single-crystal diffraction is also used in the pharmaceutical industry, due to recent problems with polymorphs. The major factors affecting the quality of single-crystal structures are the crystal's size and regularity; recrystallization is a commonly used technique to improve these factors in small-molecule crystals. The Cambridge Structural Database contains over 800,000 structures as of September 2016; over 99% of these structures were determined by X-ray diffraction.

Mineralogy and Metallurgy

Since the 1920s, X-ray diffraction has been the principal method for determining the arrangement of atoms in minerals and metals. The application of X-ray crystallography to mineralogy began with the structure of garnet, which was determined in 1924 by Menzer. A systematic X-ray crystallographic study of the silicates was undertaken in the 1920s. This study showed that, as the Si/O ratio is altered, the silicate crystals exhibit significant changes in their atomic arrangements. Machatschki extended these insights to minerals in which aluminium substitutes for the silicon atoms of the silicates. The first application of X-ray crystallography to metallurgy likewise occurred in the mid-1920s. Most notably, Linus Pauling's structure of the alloy Mg_2Sn led to his theory of the stability and structure of complex ionic crystals.

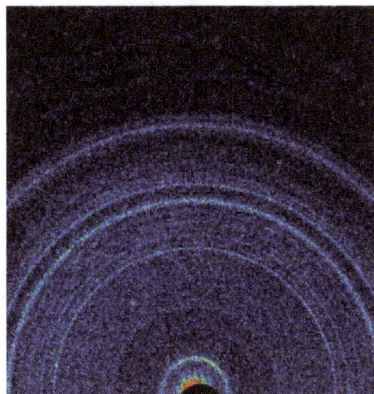

First X-ray diffraction view of Martian soil – CheMin analysis reveals feldspar, pyroxenes, olivine and more (Curiosity rover at "Rocknest", October 17, 2012).

On October 17, 2012, the Curiosity rover on the planet Mars at "Rocknest" performed the first X-ray diffraction analysis of Martian soil. The results from the rover's CheMin analyzer revealed the presence of several minerals, including feldspar, pyroxenes and olivine, and suggested that the Martian soil in the sample was similar to the "weathered basaltic soils" of Hawaiian volcanoes.

Early Organic and Small Biological Molecules

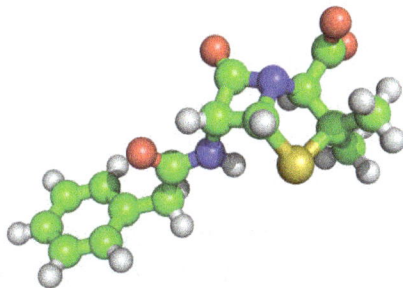

The three-dimensional structure of penicillin, solved by Dorothy Crowfoot Hodgkin in 1945. The green, white, red, yellow and blue spheres represent atoms of carbon, hydrogen, oxygen, sulfur and nitrogen, respectively.

The first structure of an organic compound, hexamethylenetetramine, was solved in 1923. This was followed by several studies of long-chain fatty acids, which are an important component of biological membranes. In the 1930s, the structures of much larger molecules with two-dimensional complexity began to be solved. A significant advance was the structure of phthalocyanine, a large planar molecule that is closely related to porphyrin molecules important in biology, such as heme, corrin and chlorophyll.

X-ray crystallography of biological molecules took off with Dorothy Crowfoot Hodgkin, who solved the structures of cholesterol (1937), penicillin (1946) and vitamin B12 (1956), for which she was awarded the Nobel Prize in Chemistry in 1964. In 1969, she succeeded in solving the structure of insulin, on which she worked for over thirty years.

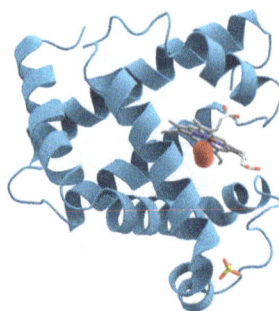

Ribbon diagram of the structure of myoglobin, showing colored alpha helices. Such proteins are long, linear molecules with thousands of atoms; yet the relative position of each atom has been determined with sub-atomic resolution by X-ray crystallography. Since it is difficult to visualize all the atoms at once, the ribbon shows the rough path of the protein polymer from its N-terminus (blue) to its C-terminus (red).

Biological Macromolecular Crystallography

Crystal structures of proteins (which are irregular and hundreds of times larger than cholesterol) began to be solved in the late 1950s, beginning with the structure of sperm whale myoglobin by Sir John Cowdery Kendrew, for which he shared the Nobel Prize in Chemistry with Max Perutz in 1962. Since that success, over 86817 X-ray crystal structures of proteins, nucleic acids and other biological molecules have been determined. For comparison, the nearest competing method in terms of structures analyzed is nuclear magnetic resonance (NMR) spectroscopy, which has resolved 9561 chemical structures. Moreover, crystallography can solve structures of arbitrarily large molecules, whereas solution-state NMR is restricted to relatively small ones (less than 70 kDa). X-ray crystallography is now used routinely by scientists to determine how a pharmaceutical drug interacts with its protein target and what changes might improve it. However, intrinsic membrane proteins remain challenging to crystallize because they require detergents or other means to solubilize them in isolation, and such detergents often interfere with crystallization. Such membrane proteins are a large component of the genome, and include many proteins of great physiological importance, such as ion channels and receptors. Helium cryogenics are used to prevent radiation damage in protein crystals.

On the other end of the size scale, relative small molecules are able to lure the resolving power of X-ray crystallography. The structure assigned in 1991 to the antibiotic isolated from a marina organism, diazonamide A – $C_{40}H_{34}Cl_2N_6O_6$ with M = 765.65 – proved to be incorrect by the classical proof of structure: a synthetic sample was not identical to the natural product. The mistake was possible because of the inability of X-ray crystallography to distinguish between the correct -OH / >NH and the interchanged -NH$_2$ / -O- groups in the incorrect structure.

Relationship to Other Scattering Techniques

Elastic vs. Inelastic Scattering

X-ray crystallography is a form of elastic scattering; the outgoing X-rays have the same energy, and thus same wavelength, as the incoming X-rays, only with altered direction. By contrast, *inelastic scattering* occurs when energy is transferred from the incoming X-ray to the crystal, e.g., by exciting an inner-shell electron to a higher energy level. Such inelastic scattering reduces the energy (or increases the wavelength) of the outgoing beam. Inelastic scattering is useful for probing such excitations of matter, but not in determining the distribution of scatterers within the matter, which is the goal of X-ray crystallography.

X-rays range in wavelength from 10 to 0.01 nanometers; a typical wavelength used for crystallography is 1 Å (0.1 nm), which is on the scale of covalent chemical bonds and the radius of a single atom. Longer-wavelength photons (such as ultraviolet radiation) would not have sufficient resolution to determine the atomic positions. At the other extreme, shorter-wavelength photons such as gamma rays are difficult to produce in large numbers, difficult to focus, and interact too strongly with matter, producing particle-antiparticle pairs. Therefore, X-rays are the "sweetspot" for wavelength when determining atomic-resolution structures from the scattering of electromagnetic radiation.

Other X-ray Techniques

Other forms of elastic X-ray scattering include powder diffraction, Small-Angle X-ray Scattering (SAXS) and several types of X-ray fiber diffraction, which was used by Rosalind Franklin in determining the double-helix structure of DNA. In general, single-crystal X-ray diffraction offers more structural information than these other techniques; however, it requires a sufficiently large and regular crystal, which is not always available.

These scattering methods generally use *monochromatic* X-rays, which are restricted to a single wavelength with minor deviations. A broad spectrum of X-rays (that is, a blend of X-rays with different wavelengths) can also be used to carry out X-ray diffraction, a technique known as the Laue method. This is the method used in the original discovery of X-ray diffraction. Laue scattering provides much structural information with only a short exposure to the X-ray beam, and is therefore used in structural studies of very rapid events (Time resolved crystallography). However, it is not as well-suited as monochromatic scattering for determining the full atomic structure of a crystal and therefore works better with crystals with relatively simple atomic arrangements.

The Laue back reflection mode records X-rays scattered backwards from a broad spectrum source. This is useful if the sample is too thick for X-rays to transmit through it. The diffracting planes in the crystal are determined by knowing that the normal to the diffracting plane bisects the angle between the incident beam and the diffracted beam. A Greninger chart can be used to interpret the back reflection Laue photograph.

Electron and Neutron Diffraction

Other particles, such as electrons and neutrons, may be used to produce a diffraction pattern. Although electron, neutron, and X-ray scattering are based on different physical processes, the resulting diffraction patterns are analyzed using the same coherent diffraction imaging techniques.

As derived below, the electron density within the crystal and the diffraction patterns are related by a simple mathematical method, the Fourier transform, which allows the density to be calculated relatively easily from the patterns. However, this works only if the scattering is *weak*, i.e., if the scattered beams are much less intense than the incoming beam. Weakly scattered beams pass through the remainder of the crystal without undergoing a second scattering event. Such re-scattered waves are called "secondary scattering" and hinder the analysis. Any sufficiently thick crystal will produce secondary scattering, but since X-rays interact relatively weakly with the electrons, this is generally not a significant concern. By contrast, electron beams may produce strong secondary scattering even for relatively thin crystals (>100 nm). Since this thickness corresponds to the diameter of many viruses, a promising direction is the electron diffraction of isolated macromolecular assemblies, such as viral capsids and molecular machines, which may be carried out with a cryo-electron microscope. Moreover, the strong interaction of electrons with matter (about 1000 times stronger than for X-rays) allows determination of the atomic structure of extremely small volumes. The field of applications for electron crystallography ranges from bio molecules like membrane proteins over organic thin films to the complex structures of (nanocrystalline) intermetallic compounds and zeolites.

Neutron diffraction is an excellent method for structure determination, although it has been difficult to obtain intense, monochromatic beams of neutrons in sufficient quantities. Traditionally, nuclear reactors have been used, although sources producing neutrons by spallation are becoming increasingly available. Being uncharged, neutrons scatter much more readily from the atomic nuclei rather than from the electrons. Therefore, neutron scattering is very useful for observing the positions of light atoms with few electrons, especially hydrogen, which is essentially invisible in the X-ray diffraction. Neutron scattering also has the remarkable property that the solvent can be made invisible by adjusting the ratio of normal water, H_2O, and heavy water, D_2O.

Methods

Overview of Single-crystal X-ray Diffraction

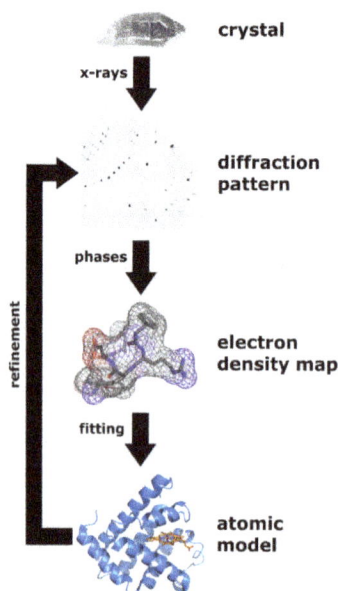

Workflow for solving the structure of a molecule by X-ray crystallography.

The oldest and most precise method of X-ray crystallography is *single-crystal X-ray diffraction*, in which a beam of X-rays strikes a single crystal, producing scattered beams. When they land on a piece of film or other detector, these beams make a *diffraction pattern* of spots; the strengths and angles of these beams are recorded as the crystal is gradually rotated. Each spot is called a *reflec-*

tion, since it corresponds to the reflection of the X-rays from one set of evenly spaced planes within the crystal. For single crystals of sufficient purity and regularity, X-ray diffraction data can determine the mean chemical bond lengths and angles to within a few thousandths of an angstrom and to within a few tenths of a degree, respectively. The atoms in a crystal are not static, but oscillate about their mean positions, usually by less than a few tenths of an angstrom. X-ray crystallography allows measuring the size of these oscillations.

Procedure

The technique of single-crystal X-ray crystallography has three basic steps. The first—and often most difficult—step is to obtain an adequate crystal of the material under study. The crystal should be sufficiently large (typically larger than 0.1 mm in all dimensions), pure in composition and regular in structure, with no significant internal imperfections such as cracks or twinning.

In the second step, the crystal is placed in an intense beam of X-rays, usually of a single wavelength (*monochromatic X-rays*), producing the regular pattern of reflections. As the crystal is gradually rotated, previous reflections disappear and new ones appear; the intensity of every spot is recorded at every orientation of the crystal. Multiple data sets may have to be collected, with each set covering slightly more than half a full rotation of the crystal and typically containing tens of thousands of reflections.

In the third step, these data are combined computationally with complementary chemical information to produce and refine a model of the arrangement of atoms within the crystal. The final, refined model of the atomic arrangement—now called a *crystal structure*—is usually stored in a public database.

Limitations

As the crystal's repeating unit, its unit cell, becomes larger and more complex, the atomic-level picture provided by X-ray crystallography becomes less well-resolved (more "fuzzy") for a given number of observed reflections. Two limiting cases of X-ray crystallography—"small-molecule" (which includes continuous inorganic solids) and "macromolecular" crystallography—are often discerned. *Small-molecule crystallography* typically involves crystals with fewer than 100 atoms in their asymmetric unit; such crystal structures are usually so well resolved that the atoms can be discerned as isolated "blobs" of electron density. By contrast, *macromolecular crystallography* often involves tens of thousands of atoms in the unit cell. Such crystal structures are generally less well-resolved (more "smeared out"); the atoms and chemical bonds appear as tubes of electron density, rather than as isolated atoms. In general, small molecules are also easier to crystallize than macromolecules; however, X-ray crystallography has proven possible even for viruses with hundreds of thousands of atoms. Though normally x-ray crystallography can only be performed if the sample is in crystal form, new research has been done into sampling non-crystalline forms of samples.

Crystallization

Although crystallography can be used to characterize the disorder in an impure or irregular crystal, crystallography generally requires a pure crystal of high regularity to solve the structure of a complicated arrangement of atoms. Pure, regular crystals can sometimes be obtained from natural or synthetic materials, such as samples of metals, minerals or other macroscopic materials. The regularity of such crystals can sometimes be improved with macromolecular crystal annealing and other methods. However, in many cases, obtaining a diffraction-quality crystal is the chief barrier to solving its atomic-resolution structure.

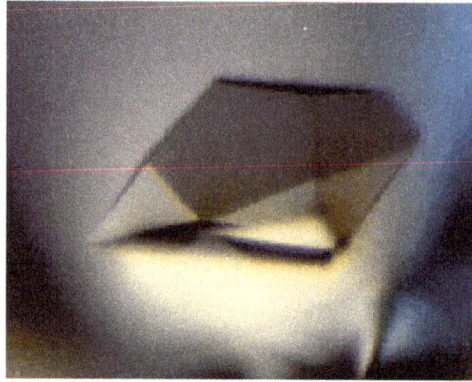

A protein crystal seen under a microscope. Crystals used in X-ray crystallography may be smaller than a millimeter across.

Small-molecule and macromolecular crystallography differ in the range of possible techniques used to produce diffraction-quality crystals. Small molecules generally have few degrees of conformational freedom, and may be crystallized by a wide range of methods, such as chemical vapor deposition and recrystallization. By contrast, macromolecules generally have many degrees of freedom and their crystallization must be carried out so as to maintain a stable structure. For example, proteins and larger RNA molecules cannot be crystallized if their tertiary structure has been unfolded; therefore, the range of crystallization conditions is restricted to solution conditions in which such molecules remain folded.

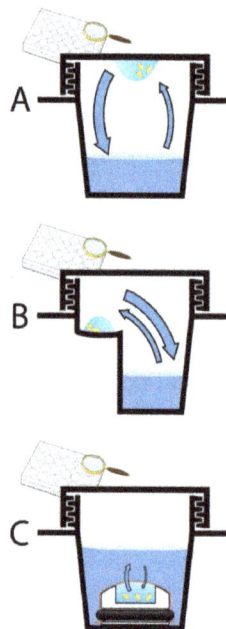

Three methods of preparing crystals, A: Hanging drop. B: Sitting drop. C: Microdialysis

Protein crystals are almost always grown in solution. The most common approach is to lower the solubility of its component molecules very gradually; if this is done too quickly, the molecules will precipitate from solution, forming a useless dust or amorphous gel on the bottom of the container. Crystal growth in solution is characterized by two steps: *nucleation* of a microscopic crystallite (possibly having only 100 molecules), followed by *growth* of that crystallite, ideally to a diffraction-quality crystal. The solution conditions that favor the first step (nucleation) are not always the same conditions that favor the second step (subsequent growth). The crystallographer's goal is to identify solution conditions that favor the development of a single, large crystal, since larger crystals offer improved resolution of the molecule. Consequently, the solution conditions should *disfavor* the first step (nucleation) but *favor* the second (growth), so that only one large crystal forms per droplet. If nucleation is favored too much, a shower of small crystallites will form in the droplet, rather than one large crystal; if favored too little, no crystal will form whatsoever. Other approaches involves, crystallizing proteins under oil, where aqueous protein solutions are dispensed under liquid oil, and water evaporates through the layer of oil. Different oils have different evaporation permeabilities, therefore yielding changes in concentration rates from different percipient/protein mixture. The technique relies on bringing the protein directly into the nucleation zone by mixing protein with the appropriate amount of percipient to prevent the diffusion of water out of the drop.

It is extremely difficult to predict good conditions for nucleation or growth of well-ordered crystals. In practice, favorable conditions are identified by *screening*; a very large batch of the molecules is prepared, and a wide variety of crystallization solutions are tested. Hundreds, even thousands, of solution conditions are generally tried before finding the successful one. The various conditions can use one or more physical mechanisms to lower the solubility of the molecule; for example, some may change the pH, some contain salts of the Hofmeister series or chemicals that lower the dielectric constant of the solution, and still others contain large polymers such as polyethylene glycol that drive the molecule out of solution by entropic effects. It is also common to try several temperatures for encouraging crystallization, or to gradually lower the temperature so that the solution becomes supersaturated. These methods require large amounts of the target molecule, as they use high concentration of the molecule(s) to be crystallized. Due to the difficulty in obtaining such large quantities (milligrams) of crystallization-grade protein, robots have been developed that are capable of accurately dispensing crystallization trial drops that are in the order of 100 nanoliters in volume. This means that 10-fold less protein is used per experiment when compared to crystallization trials set up by hand (in the order of 1 microliter).

Several factors are known to inhibit or mar crystallization. The growing crystals are generally held at a constant temperature and protected from shocks or vibrations that might disturb their crystallization. Impurities in the molecules or in the crystallization solutions are often inimical to crystallization. Conformational flexibility in the molecule also tends to make crystallization less likely, due to entropy. Ironically, molecules that tend to self-assemble into regular helices are often unwilling to assemble into crystals. Crystals can be marred by twinning, which can occur when a unit cell can pack equally favorably in multiple orientations; although recent advances in computational methods may allow solving the structure of some twinned crystals. Having failed to crystallize a target molecule, a crystallographer may try again with a slightly modified version of the molecule; even small changes in molecular properties can lead to large differences in crystallization behavior.

Data Collection

Mounting the Crystal

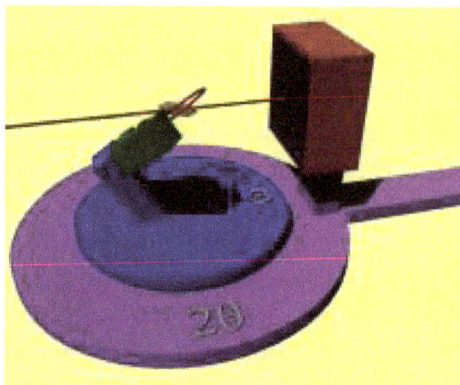

Animation showing the five motions possible with a four-circle kappa goniometer. The rotations about each of the four angles φ, κ, ω and 2θ leave the crystal within the X-ray beam, but change the crystal orientation. The detector (red box) can be slid closer or further away from the crystal, allowing higher resolution data to be taken (if closer) or better discernment of the Bragg peaks (if further away).

The crystal is mounted for measurements so that it may be held in the X-ray beam and rotated. There are several methods of mounting. In the past, crystals were loaded into glass capillaries with the crystallization solution (the mother liquor). Nowadays, crystals of small molecules are typically attached with oil or glue to a glass fiber or a loop, which is made of nylon or plastic and attached to a solid rod. Protein crystals are scooped up by a loop, then flash-frozen with liquid nitrogen. This freezing reduces the radiation damage of the X-rays, as well as the noise in the Bragg peaks due to thermal motion (the Debye-Waller effect). However, untreated protein crystals often crack if flash-frozen; therefore, they are generally pre-soaked in a cryoprotectant solution before freezing. Unfortunately, this pre-soak may itself cause the crystal to crack, ruining it for crystallography. Generally, successful cryo-conditions are identified by trial and error.

The capillary or loop is mounted on a goniometer, which allows it to be positioned accurately within the X-ray beam and rotated. Since both the crystal and the beam are often very small, the crystal must be centered within the beam to within ~25 micrometers accuracy, which is aided by a camera focused on the crystal. The most common type of goniometer is the "kappa goniometer", which offers three angles of rotation: the ω angle, which rotates about an axis perpendicular to the beam; the κ angle, about an axis at ~50° to the ω axis; and, finally, the φ angle about the loop/capillary axis. When the κ angle is zero, the ω and φ axes are aligned. The κ rotation allows for convenient mounting of the crystal, since the arm in which the crystal is mounted may be swung out towards the crystallographer. The oscillations carried out during data collection (mentioned below) involve the ω axis only. An older type of goniometer is the four-circle goniometer, and its relatives such as the six-circle goniometer.

X-ray Sources

Rotating Anode

Small scale can be done on a local X-ray tube source, typically coupled with an image plate detector. These have the advantage of being (relatively) inexpensive and easy to maintain, and allow for quick screening and collection of samples. However, the wavelength light produced is limited by anode material, typically copper. Further, intensity is limited by the power applied and cooling capacity available to avoid melting the anode. In such systems, electrons are boiled off of a cathode and accelerated through a strong electric potential of ~50 kV; having reached a high speed, the electrons collide with a metal plate, emitting *bremsstrahlung* and some strong spectral lines corresponding to the excitation of inner-shell electrons of the metal. The most common metal used is copper, which can be kept cool easily, due to its high thermal conductivity, and which produces strong K_α and K_β lines. The K_β line is sometimes suppressed with a thin (~10 μm) nickel foil. The simplest and cheapest variety of sealed X-ray tube has a stationary anode (the Crookes tube) and run with ~2 kW of electron beam power. The more expensive variety has a rotating-anode type source that run with ~14 kW of e-beam power.

X-rays are generally filtered (by use of X-Ray Filters) to a single wavelength (made monochromatic) and collimated to a single direction before they are allowed to strike the crystal. The filtering not only simplifies the data analysis, but also removes radiation that degrades the crystal without contributing useful information. Collimation is done either with a collimator (basically, a long tube) or with a clever arrangement of gently curved mirrors. Mirror systems are preferred for small crystals (under 0.3 mm) or with large unit cells (over 150 Å)

Synchrotron Radiation

Synchrotron radiation are some of the brightest lights on earth. It is the single most powerful tool available to X-ray crystallographers. It is made of X-ray beams generated in large machines called synchrotrons. These machines accelerate electrically charged particles, often electrons, to nearly the speed of light and confine them in a (roughly) circular loop using magnetic fields.

Synchrotrons are generally national facilities, each with several dedicated beamlines where data is collected without interruption. Synchrotrons were originally designed for use by high-energy physicists studying subatomic particles and cosmic phenomena. The largest component of each synchrotron is its electron storage ring. This ring is actually not a perfect circle, but a many-sided polygon. At each corner of the polygon, or sector, precisely aligned magnets bend the electron stream. As the electrons' path is bent, they emit bursts of energy in the form of X-rays.

Using synchrotron radiation frequently has specific requirements for X-ray crystallography. The intense ionizing radiation can cause radiation damage to samples, particularly macromolecular crystals. Cryo crystallography protects the sample from radiation damage, by freezing the crystal at liquid nitrogen temperatures (~100 K). However, synchrotron radiation frequently has the advantage of user selectable wavelengths, allowing for anomalous scattering experiments which maximizes anomalous signal. This is critical in experiments such as SAD and MAD.

Free Electron Laser

Recently, free electron lasers have been developed for use in X-ray crystallography. These are the brightest X-ray sources currently available; with the X-rays coming in femtosecond bursts. The intensity of the source is such that atomic resolution diffraction patterns can be resolved for crystals otherwise too small for collection. However, the intense light source also destroys the sample, requiring multiple crystals to be shot. As each crystal is randomly oriented in the beam, hundreds of thousands of individual diffraction images must be collected in order to get a complete data-set. This method, serial femtosecond crystallography, has been used in solving the structure of a number of protein crystal structures, sometimes noting differences with equivalent structures collected from synchrotron sources.

Recording the Reflections

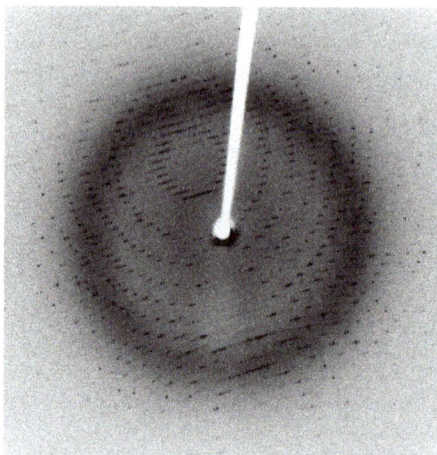

An X-ray diffraction pattern of a crystallized enzyme. The pattern of spots (*reflections*) and the relative strength of each spot (*intensities*) can be used to determine the structure of the enzyme.

When a crystal is mounted and exposed to an intense beam of X-rays, it scatters the X-rays into a pattern of spots or *reflections* that can be observed on a screen behind the crystal. A similar pattern may be seen by shining a laser pointer at a compact disc. The relative intensities of these spots provide the information to determine the arrangement of molecules within the crystal in atomic detail. The intensities of these reflections may be recorded with photographic film, an area detector or with a charge-coupled device (CCD) image sensor. The peaks at small angles correspond to low-resolution data, whereas those at high angles represent high-resolution data; thus, an upper limit on the eventual resolution of the structure can be determined from the first few images. Some measures of diffraction quality can be determined at this point, such as the mosaicity of the crystal and its overall disorder, as observed in the peak widths. Some pathologies of the crystal that would render it unfit for solving the structure can also be diagnosed quickly at this point.

One image of spots is insufficient to reconstruct the whole crystal; it represents only a small slice of the full Fourier transform. To collect all the necessary information, the crystal must be rotated step-by-step through 180°, with an image recorded at every step; actually, slightly more than 180° is required to cover reciprocal space, due to the curvature of the Ewald sphere. However, if the crystal has a higher symmetry, a smaller angular range such as 90° or 45° may be recorded. The rotation axis should be changed at least once, to avoid developing a "blind spot" in reciprocal space close to the rotation axis. It is customary to rock the crystal slightly (by 0.5–2°) to catch a broader region of reciprocal space.

Multiple data sets may be necessary for certain phasing methods. For example, MAD phasing requires that the scattering be recorded at least three (and usually four, for redundancy) wavelengths of the incoming X-ray radiation. A single crystal may degrade too much during the collection of one data set, owing to radiation damage; in such cases, data sets on multiple crystals must be taken.

Data Analysis

Crystal Symmetry, Unit Cell, and Image Scaling

The recorded series of two-dimensional diffraction patterns, each corresponding to a different crystal orientation, is converted into a three-dimensional model of the electron density; the conversion uses the mathematical technique of Fourier transforms, which is explained below. Each spot corresponds to a different type of variation in the electron density; the crystallographer must determine *which* variation corresponds to *which* spot (*indexing*), the relative strengths of the spots in different images (*merging and scaling*) and how the variations should be combined to yield the total electron density (*phasing*).

Data processing begins with *indexing* the reflections. This means identifying the dimensions of the unit cell and which image peak corresponds to which position in reciprocal space. A byproduct of indexing is to determine the symmetry of the crystal, i.e., its *space group*. Some space groups can be eliminated from the beginning. For example, reflection symmetries cannot be observed in chiral molecules; thus, only 65 space groups of 230 possible are allowed for protein molecules which are almost always chiral. Indexing is generally accomplished using an *autoindexing* routine. Having assigned symmetry, the data is then *integrated*. This converts the hundreds of images containing the thousands of reflections into a single file, consisting of (at the very least) records of the Miller index of each reflection, and an intensity for each reflection (at this state the file often also includes error estimates and measures of partiality (what part of a given reflection was recorded on that image)).

A full data set may consist of hundreds of separate images taken at different orientations of the crystal. The first step is to merge and scale these various images, that is, to identify which peaks appear in two or more images (*merging*) and to scale the relative images so that they have a consistent intensity scale. Optimizing the intensity scale is critical because the relative intensity of the peaks is the key information from which the structure is determined. The repetitive technique of crystallographic data collection and the often high symmetry of crystalline materials cause the diffractometer to record many symmetry-equivalent reflections multiple times. This allows calculating the symmetry-related R-factor, a reliability index based upon how similar are the measured intensities of symmetry-equivalent reflections, thus assessing the quality of the data.

Initial Phasing

The data collected from a diffraction experiment is a reciprocal space representation of the crystal lattice. The position of each diffraction 'spot' is governed by the size and shape of the unit cell, and the inherent symmetry within the crystal. The intensity of each diffraction 'spot' is recorded, and this intensity is proportional to the square of the *structure factor* amplitude. The structure factor is a complex number containing information relating to both the amplitude and phase of a wave. In order to obtain an interpretable *electron density map*, both amplitude and phase must be known (an electron density map allows a crystallographer to build a starting model of the molecule). The phase cannot be directly recorded during a diffraction experiment: this is known as the phase problem. Initial phase estimates can be obtained in a variety of ways:

- *Ab initio* phasing or direct methods – This is usually the method of choice for small molecules (<1000 non-hydrogen atoms), and has been used successfully to solve the phase problems for small proteins. If the resolution of the data is better than 1.4 Å (140 pm), direct methods can be used to obtain phase information, by exploiting known phase relationships between certain groups of reflections.

- Molecular replacement – if a related structure is known, it can be used as a search model in molecular replacement to determine the orientation and position of the molecules within the unit cell. The phases obtained this way can be used to generate *electron density maps.*

- Anomalous X-ray scattering (*MAD or SAD phasing*) – the X-ray wavelength may be scanned past an absorption edge of an atom, which changes the scattering in a known way. By recording full sets of reflections at three different wavelengths (far below, far above and in the middle of the absorption edge) one can solve for the substructure of the anomalously diffracting atoms and hence the structure of the whole molecule. The most popular method of incorporating anomalous scattering atoms into proteins is to express the protein in a methionine auxotroph (a host incapable of synthesizing methionine) in a media rich in seleno-methionine, which contains selenium atoms. A MAD experiment can then be conducted around the absorption edge, which should then yield the position of any methionine residues within the protein, providing initial phases.

- Heavy atom methods (multiple isomorphous replacement) – If electron-dense metal atoms can be introduced into the crystal, direct methods or Patterson-space methods can be used to determine their location and to obtain initial phases. Such heavy atoms can be introduced either by soaking the crystal in a heavy atom-containing solution, or by co-crystallization (growing the crystals in the presence of a heavy atom). As in MAD phasing, the changes in the scattering amplitudes can be interpreted to yield the phases. Although this is the original method by which protein crystal structures were solved, it has largely been superseded by MAD phasing with selenomethionine.

Model Building and Phase Refinement

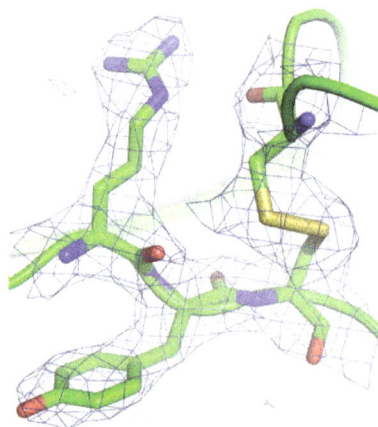

A protein crystal structure at 2.7 Å resolution. The mesh encloses the region in which the electron density exceeds a given threshold. The straight segments represent chemical bonds between the non-hydrogen atoms of an arginine (upper left), a tyrosine (lower left), a disulfide bond (upper right, in yellow), and some peptide groups (running left-right in the middle). The two curved green tubes represent spline fits to the polypeptide backbone.

Having obtained initial phases, an initial model can be built. This model can be used to refine the phases, leading to an improved model, and so on. Given a model of some atomic positions, these positions and their respective Debye-Waller factors (or B-factors, accounting for the thermal motion of the atom) can be refined to fit the observed diffraction data, ideally yielding a better set of phases. A new model can then be fit to the new electron density map and a further round of refinement is carried out. This continues until the correlation between the diffraction data and the model is maximized. The agreement is measured by an R-factor defined as

$$R = \frac{\sum_{\text{all reflections}} |F_o - F_c|}{\sum_{\text{all reflections}} |F_o|}$$

where F is the structure factor. A similar quality criterion is R_{free}, which is calculated from a subset (~10%) of reflections that were not included in the structure refinement. Both R factors depend on the resolution of the data. As a rule of thumb, R_{free} should be approximately the resolution in angstroms divided by 10; thus, a data-set with 2 Å resolution should yield a final R_{free} ~ 0.2. Chemical bonding features such as stereochemistry, hydrogen bonding and distribution of bond lengths and angles are complementary measures of the model quality. Phase bias is a serious problem in such iterative model building. *Omit maps* are a common technique used to check for this.

It may not be possible to observe every atom of the crystallized molecule – it must be remembered that the resulting electron density is an average of all the molecules within the crystal. In some cases, there is too much residual disorder in those atoms, and the resulting electron density for atoms existing in many conformations is smeared to such an extent that it is no longer detectable in the electron density map. Weakly scattering atoms such as hydrogen are routinely invisible. It is also possible for a single atom to appear multiple times in an electron density map, e.g., if a protein sidechain has multiple (<4) allowed conformations. In still other cases, the crystallographer may detect that the covalent structure deduced for the molecule was incorrect, or changed. For example, proteins may be cleaved or undergo post-translational modifications that were not detected prior to the crystallization.

Deposition of the Structure

Once the model of a molecule's structure has been finalized, it is often deposited in a crystallographic database such as the Cambridge Structural Database (for small molecules), the Inorganic Crystal Structure Database (ICSD) (for inorganic compounds) or the Protein Data Bank (for protein structures). Many structures obtained in private commercial ventures to crystallize medicinally relevant proteins are not deposited in public crystallographic databases.

Diffraction Theory

The main goal of X-ray crystallography is to determine the density of electrons $f(r)$ throughout the crystal, where r represents the three-dimensional position vector within the crystal. To do this, X-ray scattering is used to collect data about its Fourier transform $F(q)$, which is inverted mathematically to obtain the density defined in real space, using the formula

$$f(\mathbf{r}) = \frac{1}{(2\pi)^3} \int F(\mathbf{q}) e^{i\mathbf{q}\cdot\mathbf{r}} d\mathbf{q}$$

where the integral is taken over all values of q. The three-dimensional real vector q represents a point in reciprocal space, that is, to a particular oscillation in the electron density as one moves in the direction in which q points. The length of q corresponds to 2π divided by the wavelength of the oscillation. The corresponding formula for a Fourier transform will be used below

$$F(\mathbf{q}) = \int f(\mathbf{r}) e^{-i\mathbf{q}\cdot\mathbf{r}} d\mathbf{r}$$

where the integral is summed over all possible values of the position vector r within the crystal.

The Fourier transform $F(q)$ is generally a complex number, and therefore has a magnitude $|F(q)|$ and a phase $\varphi(q)$ related by the equation

$$F(\mathbf{q}) = |F(\mathbf{q})| e^{i\phi(\mathbf{q})}$$

The intensities of the reflections observed in X-ray diffraction give us the magnitudes $|F(q)|$ but not the phases $\varphi(q)$. To obtain the phases, full sets of reflections are collected with known alterations to the scattering, either by modulating the wavelength past a certain absorption edge or by adding strongly scattering (i.e., electron-dense) metal atoms such as mercury. Combining the magnitudes and phases yields the full Fourier transform $F(q)$, which may be inverted to obtain the electron density $f(r)$.

Crystals are often idealized as being *perfectly* periodic. In that ideal case, the atoms are positioned on a perfect lattice, the electron density is perfectly periodic, and the Fourier transform $F(q)$ is zero except when q belongs to the reciprocal lattice (the so-called *Bragg peaks*). In reality, however, crystals are not perfectly periodic; atoms vibrate about their mean position, and there may be disorder of various types, such as mosaicity, dislocations, various point defects, and heterogeneity in the conformation of crystallized molecules. Therefore, the Bragg peaks have a finite width and there may be significant *diffuse scattering*, a continuum of scattered X-rays that fall between the Bragg peaks.

Intuitive Understanding by Bragg's Law

An intuitive understanding of X-ray diffraction can be obtained from the Bragg model of diffraction. In this model, a given reflection is associated with a set of evenly spaced sheets running through the crystal, usually passing through the centers of the atoms of the crystal lattice. The orientation of a particular set of sheets is identified by its three Miller indices (h, k, l), and let their spacing be noted by d. William Lawrence Bragg proposed a model in which the incoming X-rays are scattered specularly (mirror-like) from each plane; from that assumption, X-rays scattered from adjacent planes will combine constructively (constructive interference) when the angle θ between the plane and the X-ray results in a path-length difference that is an integer multiple n of the X-ray wavelength λ.

$$2d \sin\theta = n\lambda$$

A reflection is said to be *indexed* when its Miller indices (or, more correctly, its reciprocal lattice vector components) have been identified from the known wavelength and the scattering angle 2θ. Such indexing gives the unit-cell parameters, the lengths and angles of the unit-cell, as well as its space group. Since Bragg's law does not interpret the relative intensities of the reflections, however, it is generally inadequate to solve for the arrangement of atoms within the unit-cell; for that, a Fourier transform method must be carried out.

Scattering as a Fourier Transform

The incoming X-ray beam has a polarization and should be represented as a vector wave; however, for simplicity, let it be represented here as a scalar wave. We also ignore the complication of the time dependence of the wave and just concentrate on the wave's spatial dependence. Plane waves can be represented by a wave vector k_{in}, and so the strength of the incoming wave at time $t=0$ is given by

$$A e^{i k_{in} \cdot r}$$

At position r within the sample, let there be a density of scatterers $f(r)$; these scatterers should produce a scattered spherical wave of amplitude proportional to the local amplitude of the incoming wave times the number of scatterers in a small volume dV about r

$$\text{amplitude of scattered wave} = A e^{i k \cdot r} S f(r) dV$$

where S is the proportionality constant.

Let's consider the fraction of scattered waves that leave with an outgoing wave-vector of k_{out} and strike the screen at r_{screen}. Since no energy is lost (elastic, not inelastic scattering), the wavelengths are the same as are the magnitudes of the wave-vectors $|k_{in}| = |k_{out}|$. From the time that the photon is scattered at r until it is absorbed at r_{screen}, the photon undergoes a change in phase

$$e^{i k_{out} \cdot (r_{screen} - r)}$$

The net radiation arriving at r_{screen} is the sum of all the scattered waves throughout the crystal

$$A S \int dr f(\mathbf{r}) e^{i k_{in} \cdot r} e^{i k_{out} \cdot (r_{screen} - r)} = A S e^{i k_{out} \cdot r_{screen}} \int dr f(\mathbf{r}) e^{i(k_{in} - k_{out}) \cdot r}$$

which may be written as a Fourier transform

$$A S e^{i k_{out} \cdot r_{screen}} \int dr f(\mathbf{r}) e^{-i q \cdot r} = A S e^{i k_{out} \cdot r_{screen}} F(\mathbf{q})$$

where $q = k_{out} - k_{in}$. The measured intensity of the reflection will be square of this amplitude

$$A^2 S^2 |F(\mathbf{q})|^2$$

Friedel and Bijvoet Mates

For every reflection corresponding to a point q in the reciprocal space, there is another reflection of the same intensity at the opposite point -q. This opposite reflection is known as the *Friedel mate* of the original reflection. This symmetry results from the mathematical fact that the density of electrons $f(r)$ at a position r is always a real number. As noted above, $f(r)$ is the inverse transform of its Fourier transform $F(q)$; however, such an inverse transform is a complex number in general. To ensure that $f(r)$ is real, the Fourier transform $F(q)$ must be such that the Friedel mates $F(-q)$ and $F(q)$ are complex conjugates of one another. Thus, $F(-q)$ has the same magnitude as $F(q)$ but they have the opposite phase, i.e., $\varphi(q) = -\varphi(q)$

$$F(-\mathbf{q}) = |F(-\mathbf{q})| e^{i\phi(-\mathbf{q})} = F^*(\mathbf{q}) = |F(\mathbf{q})| e^{-i\phi(\mathbf{q})}$$

The equality of their magnitudes ensures that the Friedel mates have the same intensity $|F|^2$. This symmetry allows one to measure the full Fourier transform from only half the reciprocal space, e.g., by rotating the crystal slightly more than 180° instead of a full 360° revolution. In crystals with significant symmetry, even more reflections may have the same intensity (Bijvoet mates); in such cases, even less of the reciprocal space may need to be measured. In favorable cases of high symmetry, sometimes only 90° or even only 45° of data are required to completely explore the reciprocal space.

The Friedel-mate constraint can be derived from the definition of the inverse Fourier transform

$$f(\mathbf{r}) = \int \frac{d\mathbf{q}}{(2\pi)^3} F(\mathbf{q}) e^{i\mathbf{q}\cdot\mathbf{r}} = \int \frac{d\mathbf{q}}{(2\pi)^3} |F(\mathbf{q})| e^{i\phi(\mathbf{q})} e^{i\mathbf{q}\cdot\mathbf{r}}$$

Since Euler's formula states that $e^{ix} = \cos(x) + i\sin(x)$, the inverse Fourier transform can be separated into a sum of a purely real part and a purely imaginary part

$$f(\mathbf{r}) = \int \frac{d\mathbf{q}}{(2\pi)^3} |F(\mathbf{q})| e^{i(\phi+\mathbf{q}\cdot\mathbf{r})} = \int \frac{d\mathbf{q}}{(2\pi)^3} |F(\mathbf{q})|$$

$$\cos(\phi+\mathbf{q}\cdot\mathbf{r}) + i\int \frac{d\mathbf{q}}{(2\pi)^3} |F(\mathbf{q})| \sin(\phi+\mathbf{q}\cdot\mathbf{r}) = I_{\cos} + iI_{\sin}$$

The function $f(r)$ is real if and only if the second integral $I_{\sin}$ is zero for all values of r. In turn, this is true if and only if the above constraint is satisfied

$$I_{\sin} = \int \frac{d\mathbf{q}}{(2\pi)^3} |F(\mathbf{q})| \sin(\phi+\mathbf{q}\cdot\mathbf{r}) = \int \frac{d\mathbf{q}}{(2\pi)^3} |F(-\mathbf{q})| \sin(-\phi-\mathbf{q}\cdot\mathbf{r}) = -I_{\sin}$$

since $I_{\sin} = -I_{\sin}$ implies that $I_{\sin} = 0$.

Ewald's Sphere

Each X-ray diffraction image represents only a slice, a spherical slice of reciprocal space, as may be seen by the Ewald sphere construction. Both k_{out} and k_{in} have the same length, due to the elastic scattering, since the wavelength has not changed. Therefore, they may be represented as two radial vectors in a sphere in reciprocal space, which shows the values of q that are sampled in a given diffraction image. Since there is a slight spread in the incoming wavelengths of the incoming X-ray beam, the values of $|F(q)|$ can be measured only for q vectors located between the two spheres corresponding to those radii. Therefore, to obtain a full set of Fourier transform data, it is necessary to rotate the crystal through slightly more than 180°, or sometimes less if sufficient symmetry is present. A full 360° rotation is not needed because of a symmetry intrinsic to the Fourier transforms of real functions (such as the electron density), but "slightly more" than 180° is needed to cover all of reciprocal space within a given resolution because of the curvature of the Ewald sphere. In practice, the crystal is rocked by a small amount (0.25-1°) to incorporate reflections near the boundaries of the spherical Ewald's shells.

Patterson Function

A well-known result of Fourier transforms is the autocorrelation theorem, which states that the autocorrelation c(r) of a function $f(r)$

$$c(\mathbf{r}) = \int d\mathbf{x} f(\mathbf{x}) f(\mathbf{x} + \mathbf{r}) = \int \frac{d\mathbf{q}}{(2\pi)^3} C(\mathbf{q}) e^{i\mathbf{q} \cdot \mathbf{r}}$$

has a Fourier transform $C(q)$ that is the squared magnitude of $F(q)$

$$C(\mathbf{q}) = |F(\mathbf{q})|^2$$

Therefore, the autocorrelation function c(r) of the electron density (also known as the *Patterson function*) can be computed directly from the reflection intensities, without computing the phases. In principle, this could be used to determine the crystal structure directly; however, it is difficult to realize in practice. The autocorrelation function corresponds to the distribution of vectors between atoms in the crystal; thus, a crystal of N atoms in its unit cell may have $N(N-1)$ peaks in its Patterson function. Given the inevitable errors in measuring the intensities, and the mathematical difficulties of reconstructing atomic positions from the interatomic vectors, this technique is rarely used to solve structures, except for the simplest crystals.

Advantages of a Crystal

In principle, an atomic structure could be determined from applying X-ray scattering to non-crystalline samples, even to a single molecule. However, crystals offer a much stronger signal due to their periodicity. A crystalline sample is by definition periodic; a crystal is composed of many unit cells repeated indefinitely in three independent directions. Such periodic systems have a Fourier transform that is concentrated at periodically repeating points in reciprocal space known as *Bragg peaks*; the Bragg peaks correspond to the reflection spots observed in the diffraction image. Since the amplitude at these reflections grows linearly with the number N of scatterers, the observed *intensity* of these spots should grow quadratically, like N^2. In other words, using a crystal concentrates the weak scattering of the individual unit cells into a much more powerful, coherent reflection that can be observed above the noise. This is an example of constructive interference.

In a liquid, powder or amorphous sample, molecules within that sample are in random orientations. Such samples have a continuous Fourier spectrum that uniformly spreads its amplitude thereby reducing the measured signal intensity, as is observed in SAXS. More importantly, the orientational information is lost. Although theoretically possible, it is experimentally difficult to obtain atomic-resolution structures of complicated, asymmetric molecules from such rotationally averaged data. An intermediate case is fiber diffraction in which the subunits are arranged periodically in at least one dimension.

Nobel Prizes for X-ray Crystallography

Year	Laureate	Prize	Rationale
1914	Max von Laue	Physics	"For his discovery of the diffraction of X-rays by crystals", an important step in the development of X-ray spectroscopy.
1915	William Henry Bragg	Physics	"For their services in the analysis of crystal structure by means of X-rays",
1915	William Lawrence Bragg	Physics	"For their services in the analysis of crystal structure by means of X-rays",
1962	Max F. Perutz	Chemistry	"for their studies of the structures of globular proteins"
1962	John C. Kendrew	Chemistry	"for their studies of the structures of globular proteins"
1962	James Dewey Watson	Medicine	"For their discoveries concerning the molecular structure of nucleic acids and its significance for information transfer in living material"
1962	Francis Harry Compton Crick	Medicine	"For their discoveries concerning the molecular structure of nucleic acids and its significance for information transfer in living material"
1962	Maurice Hugh Frederick Wilkins	Medicine	"For their discoveries concerning the molecular structure of nucleic acids and its significance for information transfer in living material"
1964	Dorothy Hodgkin	Chemistry	"For her determinations by X-ray techniques of the structures of important biochemical substances"
1972	Stanford Moore	Chemistry	"For their contribution to the understanding of the connection between chemical structure and catalytic activity of the active centre of the ribonuclease molecule"
1972	William H. Stein	Chemistry	"For their contribution to the understanding of the connection between chemical structure and catalytic activity of the active centre of the ribonuclease molecule"
1976	William N. Lipscomb	Chemistry	"For his studies on the structure of boranes illuminating problems of chemical bonding"
1985	Jerome Karle	Chemistry	"For their outstanding achievements in developing direct methods for the determination of crystal structures"
1985	Herbert A. Hauptman	Chemistry	"For their outstanding achievements in developing direct methods for the determination of crystal structures"
1988	Johann Deisenhofer	Chemistry	"For their determination of the three-dimensional structure of a photosynthetic reaction centre"

Year	Laureate	Prize	Rationale
1988	Hartmut Michel	Chemistry	"For their determination of the three-dimensional structure of a photosynthetic reaction centre"
1988	Robert Huber	Chemistry	"For their determination of the three-dimensional structure of a photosynthetic reaction centre"
1997	John E. Walker	Chemistry	"For their elucidation of the enzymatic mechanism underlying the synthesis of adenosine triphosphate (ATP)"
2003	Roderick MacKinnon	Chemistry	"For discoveries concerning channels in cell membranes [...] for structural and mechanistic studies of ion channels"
2003	Peter Agre	Chemistry	"For discoveries concerning channels in cell membranes [...] for the discovery of water channels"
2006	Roger D. Kornberg	Chemistry	"For his studies of the molecular basis of eukaryotic transcription"
2009	Ada E. Yonath	Chemistry	"For studies of the structure and function of the ribosome"
2009	Thomas A. Steitz	Chemistry	"For studies of the structure and function of the ribosome"
2009	Venkatraman Ramakrishnan	Chemistry	"For studies of the structure and function of the ribosome"
2012	Brian Kobilka	Chemistry	"For studies of G-protein-coupled receptors"

Neutron Diffraction

Neutron diffraction or elastic neutron scattering is the application of neutron scattering to the determination of the atomic and/or magnetic structure of a material. A sample to be examined is placed in a beam of thermal or cold neutrons to obtain a diffraction pattern that provides information of the structure of the material. The technique is similar to X-ray diffraction but due to their different scattering properties, neutrons and X-rays provide complementary information: X-Rays are suited for superficial analysis, strong x-rays from synchrotron radiation are suited for shallow depths or thin specimens, while neutrons having high penetration depth are suited for bulk samples.

Instrumental and Sample Requirements

The technique requires a source of neutrons. Neutrons are usually produced in a nuclear reactor or spallation source. At a research reactor, other components are needed, including a crystal monochromators as well as filters to select the desired neutron wavelength. Some parts of the setup may also be movable. At a spallation source, the time of flight technique is used to sort the energies of the incident neutrons (higher energy neutrons are faster), so no monochromator is needed, but rather a series of aperture elements synchronized to filter neutron pulses with the desired wavelength.

The technique is most commonly performed as powder diffraction, which only requires a polycrystalline powder. Single crystal work is also possible, but the crystals must be much larger than those that are used in single-crystal X-ray crystallography. It is common to use crystals that are about 1 mm³.

Summarizing, the main disadvantage to neutron diffraction is the requirement for a nuclear reactor. For single crystal work, the technique requires relatively large crystals, which are usually challenging to grow. The advantages to the technique are many - sensitivity to light atoms, ability to distinguish isotopes, absence of radiation damage, as well as a penetration depth of several cm

Nuclear Scattering

Like all quantum particles, neutrons can exhibit wave phenomena typically associated with light or sound. Diffraction is one of these phenomena; it occurs when waves encounter obstacles whose size is comparable with the wavelength. If the wavelength of a quantum particle is short enough, atoms or their nuclei can serve as diffraction obstacles. When a beam of neutrons emanating from a reactor is slowed down and selected properly by their speed, their wavelength lies near one angstrom (0.1 nanometer), the typical separation between atoms in a solid material. Such a beam can then be used to perform a diffraction experiment. Impinging on a crystalline sample it will scatter under a limited number of well-defined angles according to the same Bragg's law that describes X-ray diffraction.

Neutrons and X-rays interact with matter differently. X-rays interact primarily with the electron cloud surrounding each atom. The contribution to the diffracted x-ray intensity is therefore larger for atoms with larger atomic number (Z). On the other hand, neutrons interact directly with the *nucleus* of the atom, and the contribution to the diffracted intensity depends on each isotope; for example, regular hydrogen and deuterium contribute differently. It is also often the case that light (low Z) atoms contribute strongly to the diffracted intensity even in the presence of large Z atoms. The scattering length varies from isotope to isotope rather than linearly with the atomic number. An element like vanadium is a strong scatterer of X-rays, but its nuclei hardly scatter neutrons, which is why it is often used as a container material. Non-magnetic neutron diffraction is directly sensitive to the positions of the nuclei of the atoms.

Unlike X-rays, neutrons scatter mostly from the nuclei of the atoms, which are tiny. Furthermore, there is no need for an atomic form factor to describe the shape of the electron cloud of the atom and the scattering power of an atom does not fall off with the scattering angle as it does for X-rays. Diffractograms therefore can show strong well defined diffraction peaks even at high angles, particularly if the experiment is done at low temperatures. Many neutron sources are equipped with liquid helium cooling systems that allow data collection at temperatures down to 4.2 K. The superb high angle (i.e. high *resolution*) information means that the atomic positions in the structure can be determined with high precision. On the other hand, Fourier maps (and to a lesser extent difference Fourier maps) derived from neutron data suffer from series termination errors, sometimes so much that the results are meaningless.

Magnetic Scattering

Although neutrons are uncharged, they carry a spin, and therefore interact with magnetic moments, including those arising from the electron cloud around an atom. Neutron diffraction can therefore reveal the microscopic magnetic structure of a material.

Magnetic scattering does require an atomic form factor as it is caused by the much larger electron cloud around the tiny nucleus. The intensity of the magnetic contribution to the diffraction peaks will therefore decrease towards higher angles.

Uses

Neutron diffraction can be used to determine the static structure factor of gases, liquids or amorphous solids. Most experiments, however, aim at the structure of crystalline solids, making neutron diffraction an important tool of crystallography.

Neutron diffraction is closely related to X-ray powder diffraction. In fact the single crystal version of the technique is less commonly used because currently available neutron sources require relatively large samples and large single crystals are hard or impossible to come by for most materials. Future developments, however, may well change this picture. Because the data is typically a 1D powder diffractogram they are usually processed using Rietveld refinement. In fact the latter found its origin in neutron diffraction (at Petten in the Netherlands) and was later extended for use in X-ray diffraction.

One practical application of elastic neutron scattering/diffraction is that the lattice constant of metals and other crystalline materials can be very accurately measured. Together with an accurately aligned micropositioner a map of the lattice constant through the metal can be derived. This can easily be converted to the stress field experienced by the material. This has been used to analyse stresses in aerospace and automotive components to give just two examples. The high penetration depth permits measuring residual stresses in bulk components as crankshafts, pistons, rails, gears. This technique has led to the development of dedicated stress diffractometers, such as the ENGIN-X instrument at the ISIS neutron source.

Neutron diffraction can also be employed to give insight into the 3D structure any material that diffracts.

Hydrogen, Null-scattering and Contrast Variation

Neutron diffraction can be used to establish the structure of low atomic number materials like proteins and surfactants much more easily with lower flux than at a synchrotron radiation source. This is because some low atomic number materials have a higher cross section for neutron interaction than higher atomic weight materials.

One major advantage of neutron diffraction over X-ray diffraction is that the latter is rather insensitive to the presence of hydrogen (H) in a structure, whereas the nuclei ^1H and ^2H (i.e. Deuterium, D) are strong scatterers for neutrons. The greater scattering power of protons and deuterons means that the position of hydrogen in a crystal and its thermal motions can be determined with greater precision by neutron diffraction. The structures of metal hydride complexes, e.g., Mg_2FeH_6 have been assessed by neutron diffraction.

The neutron scattering lengths b_H = -3.7406(11) fm and b_D = 6.671(4) fm, for H and D respectively, have opposite sign, which allows the technique to distinguish them. In fact there is a particular isotope ratio for which the contribution of the element would cancel, this is called null-scattering.

It is undesirable to work with the relatively high concentration of H in a sample. The scattering intensity by H-nuclei has a large inelastic component, which creates a large continuous background that is more or less independent of scattering angle. The elastic pattern typically consists of sharp Bragg reflections if the sample is crystalline. They tend to drown in the inelastic background. This is even more serious when the technique is used for the study of liquid structure. Nevertheless, by preparing samples with different isotope ratios it is possible to vary the scattering contrast enough to highlight one element in an otherwise complicated structure. The variation of other elements is possible but usually rather expensive. Hydrogen is inexpensive and particularly interesting because it plays an exceptionally large role in biochemical structures and is difficult to study structurally in other ways.

History

The first neutron diffraction experiments were carried out in 1945 by Ernest O. Wollan using the Graphite Reactor at Oak Ridge. He was joined shortly thereafter (June 1946) by Clifford Shull, and together they established the basic principles of the technique, and applied it successfully to many different materials, addressing problems like the structure of ice and the microscopic arrangements of magnetic moments in materials. For this achievement Shull was awarded one half of the 1994 Nobel Prize in Physics. Wollan had died in 1984. (The other half of the 1994 Nobel Prize for Physics went to Bert Brockhouse for development of the inelastic scattering technique at the Chalk River facility of AECL. This also involved the invention of the triple axis spectrometer). The delay between the achieved work (1946) and the Nobel Prize awarded to Brockhouse and Shull (1994) brings them close to the delay between the invention by Ernst Ruska of the electron microscope (1933) - also in the field of particle optics - and his own Nobel prize (1986). This in turn is near to the record of 55 years between the discoveries of Peyton Rous and his award of the Nobel Prize in 1966.

Electron Microscope

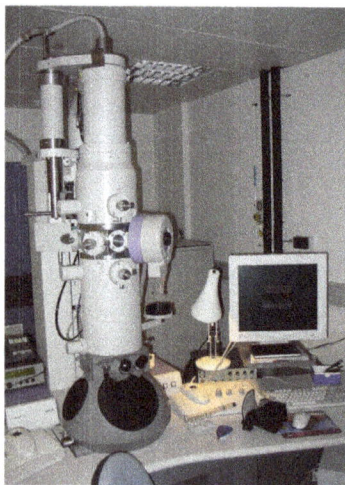

A modern transmission electron microscope

Electron microscope constructed by Ernst Ruska in 1933

An electron microscope is a microscope that uses a beam of accelerated electrons as a source of illumination. As the wavelength of an electron can be up to 100,000 times shorter than that of visible light photons, electron microscopes have a higher resolving power than light microscopes and can reveal the structure of smaller objects. A transmission electron microscope can achieve better than 50 pm resolution and magnifications of up to about 10,000,000x whereas most light microscopes are limited by diffraction to about 200 nm resolution and useful magnifications below 2000x.

Transmission electron microscopes use electrostatic and electromagnetic lenses to control the electron beam and focus it to form an image. These electron optical lenses are analogous to the glass lenses of an optical light microscope.

Electron microscopes are used to investigate the ultrastructure of a wide range of biological and inorganic specimens including microorganisms, cells, large molecules, biopsy samples, metals, and crystals. Industrially, electron microscopes are often used for quality control and failure analysis. Modern electron microscopes produce electron micrographs using specialized digital cameras and frame grabbers to capture the image.

History

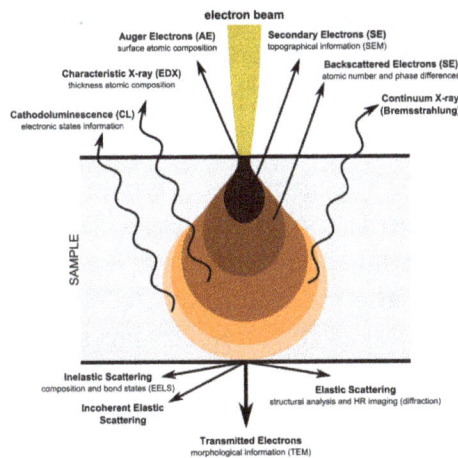

Diagram illustrating the phenomena resulting of the interaction of highly energetic electrons with matter.

The first electromagnetic lens was developed in 1926 by Hans Busch.

According to Dennis Gabor, the physicist Leó Szilárd tried in 1928 to convince Busch to build an electron microscope, for which he had filed a patent.

German physicist Ernst Ruska and the electrical engineer Max Knoll constructed the prototype electron microscope in 1931, capable of four-hundred-power magnification; the apparatus was the first demonstration of the principles of electron microscopy. Two years later, in 1933, Ruska built an electron microscope that exceeded the resolution attainable with an optical (light) microscope. Moreover, Reinhold Rudenberg, the scientific director of Siemens-Schuckertwerke, obtained the patent for the electron microscope in May 1931.

In 1932, Ernst Lubcke of Siemens & Halske built and obtained images from a prototype electron microscope, applying concepts described in the Rudenberg patent applications. Five years later (1937), the firm financed the work of Ernst Ruska and Bodo von Borries, and employed Helmut Ruska (Ernst's brother) to develop applications for the microscope, especially with biological specimens. Also in 1937, Manfred von Ardenne pioneered the scanning electron microscope. The first commercial electron microscope was produced in 1936 by L C Martin of Imperial University. The first North American electron microscope was constructed in 1938, at the University of Toronto, by

Eli Franklin Burton and students Cecil Hall, James Hillier, and Albert Prebus; and Siemens produced a transmission electron microscope (TEM) in 1939. Although contemporary transmission electron microscopes are capable of two million-power magnification, as scientific instruments, they remain based upon Ruska's prototype.

Types

Transmission Electron Microscope (Tem)

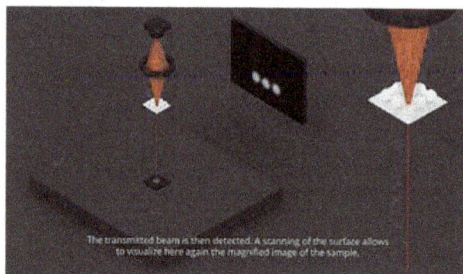

Operating principe of a Transmission Electron Microscope

The original form of electron microscope, the transmission electron microscope (TEM) uses a high voltage electron beam to illuminate the specimen and create an image. The electron beam is produced by an electron gun, commonly fitted with a tungsten filament cathode as the electron source. The electron beam is accelerated by an anode typically at +100 keV (40 to 400 keV) with respect to the cathode, focused by electrostatic and electromagnetic lenses, and transmitted through the specimen that is in part transparent to electrons and in part scatters them out of the beam. When it emerges from the specimen, the electron beam carries information about the structure of the specimen that is magnified by the objective lens system of the microscope. The spatial variation in this information (the "image") may be viewed by projecting the magnified electron image onto a fluorescent viewing screen coated with a phosphor or scintillator material such as zinc sulfide. Alternatively, the image can be photographically recorded by exposing a photographic film or plate directly to the electron beam, or a high-resolution phosphor may be coupled by means of a lens optical system or a fibre optic light-guide to the sensor of a digital camera. The image detected by the digital camera may be displayed on a monitor or computer.

The resolution of TEMs is limited primarily by spherical aberration, but a new generation of aberration correctors have been able to partially overcome spherical aberration to increase resolution. Hardware correction of spherical aberration for the high-resolution transmission electron microscopy (HRTEM) has allowed the production of images with resolution below 0.5 angstrom (50 picometres) and magnifications above 50 million times. The ability to determine the positions of atoms within materials has made the HRTEM an important tool for nano-technologies research and development.

Transmission electron microscopes are often used in electron diffraction mode. The advantages of electron diffraction over X-ray crystallography are that the specimen need not be a single crystal or even a polycrystalline powder, and also that the Fourier transform reconstruction of the object's magnified structure occurs physically and thus avoids the need for solving the phase problem faced by the X-ray crystallographers after obtaining their X-ray diffraction patterns of a single crystal or polycrystalline powder.

The major disadvantage of the transmission electron microscope is the need for extremely thin sections of the specimens, typically about 100 nanometers. Biological specimens are typically required to be chemically fixed, dehydrated and embedded in a polymer resin to stabilize them sufficiently to allow ultrathin sectioning. Sections of biological specimens, organic polymers and similar materials may require special treatment with heavy atom labels in order to achieve the required image contrast.

Scanning Electron Microscope (SEM)

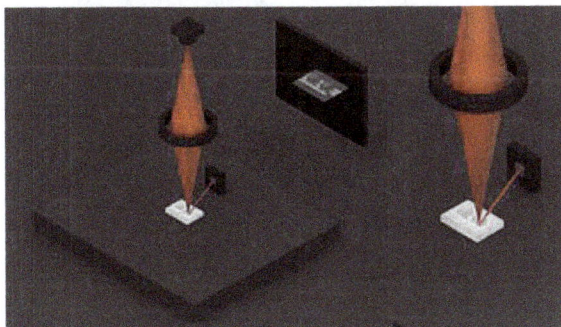

Operating principe of a Scanning Electron Microscope

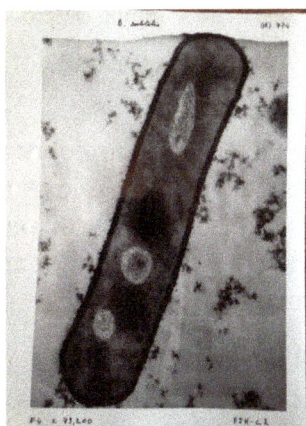

Image of bacillus subtilis taken with a 1960s electron microscope.

The SEM produces images by probing the specimen with a focused electron beam that is scanned across a rectangular area of the specimen (raster scanning). When the electron beam interacts with the specimen, it loses energy by a variety of mechanisms. The lost energy is converted into alternative forms such as heat, emission of low-energy secondary electrons and high-energy back-scattered electrons, light emission (cathodoluminescence) or X-ray emission, all of which provide signals carrying information about the properties of the specimen surface, such as its topography and composition. The image displayed by an SEM maps the varying intensity of any of these signals into the image in a position corresponding to the position of the beam on the specimen when the signal was generated. In the SEM image of an ant shown below and to the right, the image was constructed from signals produced by a secondary electron detector, the normal or conventional imaging mode in most SEMs.

Generally, the image resolution of an SEM is at least an order of magnitude poorer than that of a TEM. However, because the SEM image relies on surface processes rather than transmission, it is able to image bulk samples up to many centimeters in size and (depending on instrument design and settings) has a great depth of field, and so can produce images that are good representations of the three-dimensional shape of the sample. Another advantage of SEM is its variety called environmental scanning electron microscope (ESEM) can produce images of sufficient quality and resolution with the samples being wet or contained in low vacuum or gas. This greatly facilitates imaging biological samples that are unstable in the high vacuum of conventional electron microscopes.

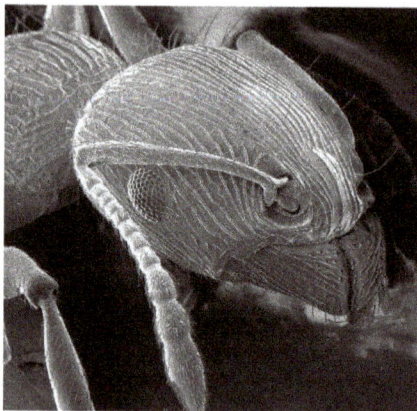

An image of an ant in a scanning electron microscope

Color

In their most common configurations, electron microscopes produce images with a single brightness value per pixel, with the results usually rendered in grayscale. However, often these images are then colorized through the use of feature-detection software, or simply by hand-editing using a graphics editor. This may be done to clarify structure or for aesthetic effect and generally does not add new information about the specimen.

In some configurations information about several specimen properties is gathered per pixel, usually by the use of multiple detectors. In SEM, the attributes of topography and material contrast can be obtained by a pair of backscattered electron detectors and such attributes can be superimposed in a single color image by assigning a different primary color to each attribute. Similarly, a combination of backscattered and secondary electron signals can be assigned to different colors and superimposed on a single color micrograph displaying simultaneously the properties of the specimen.

Some types of detectors used in SEM have analytical capabilities, and can provide several items of data at each pixel. Examples are the Energy-dispersive X-ray spectroscopy (EDS) detectors used in elemental analysis and Cathodoluminescence microscope (CL) systems that analyse the intensity and spectrum of electron-induced luminescence in (for example) geological specimens. In SEM systems using these detectors it is common to color code the signals and superimpose them in a single color image, so that differences in the distribution of the various components of the specimen can be seen clearly and compared. Optionally, the standard secondary electron image can be merged with the one or more compositional channels, so that the specimen's structure and composition can be compared. Such images can be made while maintaining the full integrity of the original signal, which is not modified in any way.

Reflection Electron Microscope (REM)

In the reflection electron microscope (REM) as in the TEM, an electron beam is incident on a surface but instead of using the transmission (TEM) or secondary electrons (SEM), the reflected beam of elastically scattered electrons is detected. This technique is typically coupled with reflection high energy electron diffraction (RHEED) and *reflection high-energy loss spectroscopy (RHELS)*. Another variation is spin-polarized low-energy electron microscopy (SPLEEM), which is used for looking at the microstructure of magnetic domains.

Scanning Transmission Electron Microscope (Stem)

The STEM rasters a focused incident probe across a specimen that (as with the TEM) has been thinned to facilitate detection of electrons scattered *through* the specimen. The high resolution of the TEM is thus possible in STEM. The focusing action (and aberrations) occur before the electrons hit the specimen in the STEM, but afterward in the TEM. The STEMs use of SEM-like beam rastering simplifies annular dark-field imaging, and other analytical techniques, but also means that image data is acquired in serial rather than in parallel fashion. Often TEM can be equipped with the scanning option and then it can function both as TEM and STEM.

Sample Preparation

An insect coated in gold for viewing with a scanning electron microscope

Materials to be viewed under an electron microscope may require processing to produce a suitable sample. The technique required varies depending on the specimen and the analysis required:

- *Chemical fixation* – for biological specimens aims to stabilize the specimen's mobile macromolecular structure by chemical crosslinking of proteins with aldehydes such as formaldehyde and glutaraldehyde, and lipids with osmium tetroxide.

- *Negative stain* – suspensions containing nanoparticles or fine biological material (such as viruses and bacteria) are briefly mixed with a dilute solution of an electron-opaque solution such as ammonium molybdate, uranyl acetate (or formate), or phosphotungstic acid. This mixture is applied to a suitably coated EM grid, blotted, then allowed to dry. Viewing of this preparation in the TEM should be carried out without delay for best results. The method is important in microbiology for fast but crude morphological identification, but can also be used as the basis for high resolution 3D reconstruction using EM tomography methodology when carbon films are used for support. Negative staining is also used for observation of nanoparticles.

- *Cryofixation* – freezing a specimen so rapidly, in liquid ethane, and maintained at liquid nitrogen or even liquid helium temperatures, so that the water forms vitreous (non-crystalline) ice. This preserves the specimen in a snapshot of its solution state. An entire field called cryo-electron microscopy has branched from this technique. With the development of cryo-electron microscopy of vitreous sections (CEMOVIS), it is now possible to observe samples from virtually any biological specimen close to its native state.

- *Dehydration* – or replacement of water with organic solvents such as ethanol or acetone, followed by critical point drying or infiltration with embedding resins. Also freeze drying.

- *Embedding, biological specimens* – after dehydration, tissue for observation in the transmission electron microscope is embedded so it can be sectioned ready for viewing. To do this the tissue is passed through a 'transition solvent' such as propylene oxide (epoxypropane) or acetone and then infiltrated with an epoxy resin such as Araldite, Epon, or Durcupan; tissues may also be embedded directly in water-miscible acrylic resin. After the resin has been polymerized (hardened) the sample is thin sectioned (ultrathin sections) and stained – it is then ready for viewing.

- *Embedding, materials* – after embedding in resin, the specimen is usually ground and polished to a mirror-like finish using ultra-fine abrasives. The polishing process must be performed carefully to minimize scratches and other polishing artifacts that reduce image quality.

- *Metal shadowing* – Metal (e.g. platinum) is evaporated from an overhead electrode and applied to the surface of a biological sample at an angle. The surface topography results in variations in the thickness of the metal that are seen as variations in brightness and contrast in the electron microscope image.

- *Replication* – A surface shadowed with metal (e.g. platinum, or a mixture of carbon and platinum) at an angle is coated with pure carbon evaporated from carbon electrodes at right angles to the surface. This is followed by removal of the specimen material (e.g. in an acid bath, using enzymes or by mechanical separation) to produce a surface replica that records the surface ultrastructure and can be examined using transmission electron microscopy.

- *Sectioning* – produces thin slices of specimen, semitransparent to electrons. These can be cut on an ultramicrotome with a diamond knife to produce ultra-thin sections about 60–90 nm thick. Disposable glass knives are also used because they can be made in the lab and are much cheaper.

- *Staining* – uses heavy metals such as lead, uranium or tungsten to scatter imaging electrons and thus give contrast between different structures, since many (especially biological) materials are nearly "transparent" to electrons (weak phase objects). In biology, specimens can be stained "en bloc" before embedding and also later after sectioning. Typically thin sections are stained for several minutes with an aqueous or alcoholic solution of uranyl acetate followed by aqueous lead citrate.

- *Freeze-fracture or freeze-etch* – a preparation method particularly useful for examining lipid membranes and their incorporated proteins in "face on" view. The fresh tissue or cell suspension is frozen rapidly (cryofixation), then fractured by breaking or by using a microtome while maintained at liquid nitrogen temperature. The cold fractured surface (sometimes "etched" by increasing the temperature to about –100 °C for several minutes to let some ice sublime) is then shadowed with evaporated platinum or gold at an average angle of 45° in a high vacuum evaporator. A second coat of carbon, evaporated perpendicular to the average surface plane is often performed to improve stability of the replica coating. The specimen is returned to room temperature and pressure, then the extremely fragile "pre-shadowed" metal replica of the fracture surface is released from the underlying biological material by careful chemical digestion with acids, hypochlorite solution or SDS detergent. The still-floating replica is thoroughly washed free from residual chemicals, carefully fished up on fine grids, dried then viewed in the TEM.

- *Ion beam milling* – thins samples until they are transparent to electrons by firing ions (typically argon) at the surface from an angle and sputtering material from the surface. A subclass of this is focused ion beam milling, where gallium ions are used to produce an electron transparent membrane in a specific region of the sample, for example through a device within a microprocessor. Ion beam milling may also be used for cross-section polishing prior to SEM analysis of materials that are difficult to prepare using mechanical polishing.

- *Conductive coating* – an ultrathin coating of electrically conducting material, deposited either by high vacuum evaporation or by low vacuum sputter coating of the sample. This is done to prevent the accumulation of static electric fields at the specimen due to the electron irradiation required during imaging. The coating materials include gold, gold/palladium, platinum, tungsten, graphite, etc.

- *Earthing* – to avoid electrical charge accumulation on a conductive coated sample, it is usually electrically connected to the metal sample holder. Often an electrically conductive adhesive is used for this purpose.

Disadvantages

Electron microscopes are expensive to build and maintain, but the capital and running costs of confocal light microscope systems now overlaps with those of basic electron microscopes. Microscopes designed to achieve high resolutions must be housed in stable buildings (sometimes underground) with special services such as magnetic field cancelling systems.

The samples largely have to be viewed in vacuum, as the molecules that make up air would scatter the electrons. An exception is the environmental scanning electron microscope, which allows hydrated samples to be viewed in a low-pressure (up to 20 Torr or 2.7 kPa) and/or wet environment.

Scanning electron microscopes operating in conventional high-vacuum mode usually image conductive specimens; therefore non-conductive materials require conductive coating (gold/palladium alloy, carbon, osmium, etc.). The low-voltage mode of modern microscopes makes possible the observation of non-conductive specimens without coating. Non-conductive materials can be imaged also by a variable pressure (or environmental) scanning electron microscope.

Small, stable specimens such as carbon nanotubes, diatom frustules and small mineral crystals (asbestos fibres, for example) require no special treatment before being examined in the electron microscope. Samples of hydrated materials, including almost all biological specimens have to be prepared in various ways to stabilize them, reduce their thickness (ultrathin sectioning) and increase their electron optical contrast (staining). These processes may result in *artifacts*, but these can usually be identified by comparing the results obtained by using radically different specimen preparation methods. It is generally believed by scientists working in the field that as results from various preparation techniques have been compared and that there is no reason that they should all produce similar artifacts, it is reasonable to believe that electron microscopy features correspond with those of living cells. Since the 1980s, analysis of cryofixed, vitrified specimens has also become increasingly used by scientists, further confirming the validity of this technique.

X-ray Absorption Spectroscopy

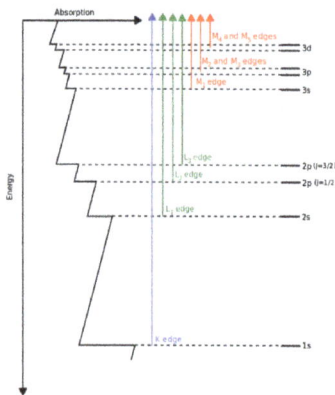

Figure 1: Transitions that contribute to XAS edges

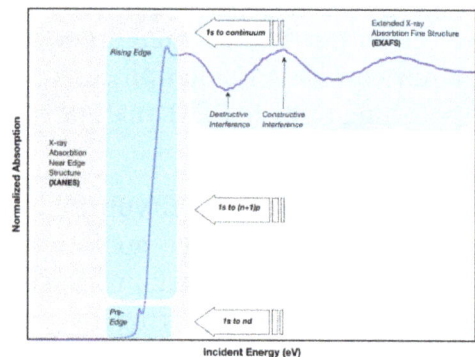

Figure 2: Three regions of XAS data

X-ray absorption spectroscopy (XAS) is a widely used technique for determining the local geometric and/or electronic structure of matter. The experiment is usually performed at synchrotron radiation sources, which provide intense and tunable X-ray beams. Samples can be in the gas-phase, solution, or condensed matter (i.e. solids).

XAS data is obtained by tuning the photon energy using a crystalline monochromator to a range where core electrons can be excited (0.1-100 keV photon energy). The "name" of the edge depends upon which core electron is excited: the principal quantum numbers n = 1, 2, and 3, correspond to the K-, L-, and M-edges, respectively. For instance, excitation of a 1s electron occurs at the K-edge, while excitation of a 2s or 2p electron occurs at an L-edge (Figure 1) .

There are three main regions found on a spectrum generated by XAS data (Figure 2):

1. The "absorption threshold" determined by the transition to the lowest unoccupied states:

   a) the states at the Fermi energy in metals giving a "rising edge" with an arc tangent shape;

   b) the bound core excitons in insulators with a Lorentzian line-shape (they occur in a pre-edge region at energies lower than the transitions to the lowest unoccupied level);

2. The X-ray Absorption Near-Edge Structure XANES introduced in 1980 and later in 1983 called also NEXAFS (Near-edge X-ray Absorption Fine Structure) which are dominated by core transitions to quasi bound states (multiple scattering resonances) for photoelectrons with kinetic energy in the range from 10 to 150 eV above the chemical potential, called "shape resonances" in molecular spectra since they are due to final states of short life-time degenerate with the continuum with the Fano line-shape. In this range multi-electron excitations and many-body final states in strongly correlated systems are relevant;

3. In the high kinetic energy range of the photoelectron the scattering cross-section with neighbor atoms is weak and the absorption spectra are dominated by EXAFS (Extended X-ray Absorption Fine Structure) where the scattering of the ejected photoelectron off neighboring atoms can be approximated by single scattering events. After it was shown in 1985 that multiple scattering theory can interpret both XANES and EXAFS the experimental analysis focusing on both regions is called XAFS .

XAS is a type of absorption spectroscopy from a core initial state with a well defined symmetry therefore the quantum mechanical selection rules select the symmetry of the final states in the continuum which usually are mixture of multiple components. The most intense features are due to electric-dipole allowed transitions (i.e. $\Delta \ell = \pm 1$) to unoccupied final states. For example, the most intense features of a K-edge are due to core transitions from 1s → p-like final states, while the most intense features of the L_3-edge are due to 2p → d-like final states.

XAS methodology can be broadly divided into four experimental categories that can give complementary results to each other: Metal K-edge, metal L-edge, ligand K-edge, and EXAFS.

Powder Diffraction

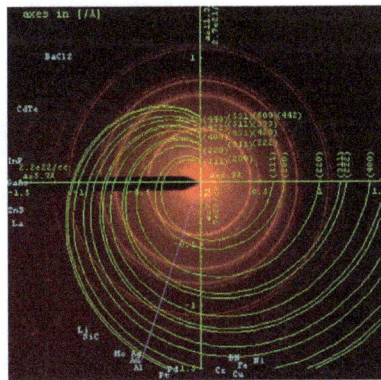

Electron powder pattern (red) of an Al film with an fcc spiral overlay (green) and a line of intersections (blue) that determines lattice parameter.

Powder diffraction is a scientific technique using X-ray, neutron, or electron diffraction on powder or microcrystalline samples for structural characterization of materials. An instrument dedicated to performing such powder measurements is called a powder diffractometer.

Powder diffraction stands in contrast to single crystal diffraction techniques, which work best with a single, well-ordered crystal.

Explanation

A diffractometer produces waves at a known frequency, which is determined by their source. The source is often x-rays, because they are the only kind of light with the correct frequency for inter-atomic-scale diffraction. However, electrons and neutrons are also common sources, with their frequency determined by their de Broglie wavelength. When these waves reach the sample, the atoms of the sample act just like a diffraction grating, producing bright spots at particular angles. By measuring the angle where these bright spots occur, the spacing of the diffraction grating can be determined by Bragg's law. Because the sample itself is the diffraction grating, this spacing is the atomic spacing.

The distinction between powder and single crystal diffraction is the degree of texturing in the sample. Single crystals have maximal texturing, and are said to be anisotropic. In contrast, in powder diffraction, every possible crystalline orientation is represented equally in a powdered sample, the isotropic case. PXRD operates under the assumption that the sample is randomly arranged. Therefore, a statistically significant number of each plane of the crystal structure will be in the proper orientation to diffract the X-rays. Therefore, each plane will be represented in the signal. In practice, it is sometimes necessary to rotate the sample orientation to eliminate the effects of texturing and achieve true randomness.

Mathematically, crystals can be described by a Bravais lattice with some regularity in the spacing between atoms. Because of this regularity, we can describe this structure in a different way using the reciprocal lattice, which is related to the original structure by a Fourier transform. This three-dimensional space can be described with reciprocal axes x*, y*, and z* or alternatively in spherical coordinates q, φ*, and χ*. In powder diffraction, intensity is homogeneous over φ* and χ*, and only q remains as an important measurable quantity. This is because orientational averaging causes the three-dimensional reciprocal space that is studied in single crystal diffraction to be projected onto a single dimension.

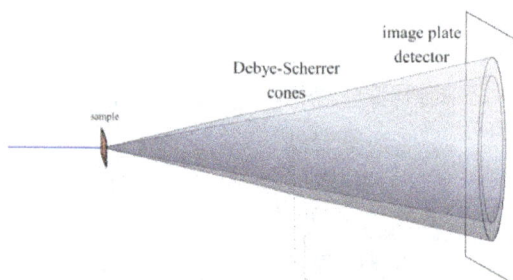

Two-dimensional powder diffraction setup with flat plate detector.

When the scattered radiation is collected on a flat plate detector, the rotational averaging leads to smooth diffraction rings around the beam axis, rather than the discrete Laue spots observed in single crystal diffraction. The angle between the beam axis and the ring is called the *scattering angle* and in X-ray crystallography always denoted as 2θ (in scattering of *visible* light the convention is usually to call it θ). In accordance with Bragg's law, each ring corresponds to a particular reciprocal lattice vector G in the sample crystal. This leads to the definition of the scattering vector as:

$$|G| = q = 2k \sin(\theta) = \frac{4\pi}{\lambda} \sin(\theta).$$

In this equation, G is the reciprocal lattice vector, q is the length of the reciprocal lattice vector, k is the momentum transfer vector, θ is half of the scattering angle, and λ is the wavelength of the radiation. Powder diffraction data are usually presented as a diffractogram in which the diffracted intensity, I, is shown as a function either of the scattering angle 2θ or as a function of the scattering vector length q. The latter variable has the advantage that the diffractogram no longer depends on the value of the wavelength λ. The advent of synchrotron sources has widened the choice of wavelength considerably. To facilitate comparability of data obtained with different wavelengths the use of q is therefore recommended and gaining acceptability.

Uses

Relative to other methods of analysis, powder diffraction allows for rapid, non-destructive analysis of multi-component mixtures without the need for extensive sample preparation. This gives laboratories around the world the ability to quickly analyze unknown materials and perform materials characterization in such fields as metallurgy, mineralogy, forensic science, archeology, condensed matter physics, and the biological and pharmaceutical sciences. Identification is performed by comparison of the diffraction pattern to a known standard or to a database such as the International Centre for Diffraction Data's Powder Diffraction File (PDF) or the Cambridge Structural Database (CSD). Advances in hardware and software, particularly improved optics and fast detectors, have dramatically improved the analytical capability of the technique, especially relative to the speed of the analysis. The fundamental physics upon which the technique is based provides high precision and accuracy in the measurement of interplanar spacings, sometimes to fractions of an Ångström, resulting in authoritative identification frequently used in patents, criminal cases and other areas of law enforcement. The ability to analyze multiphase materials also allows analysis of how materials interact in a particular matrix such as a pharmaceutical tablet, a circuit board, a mechanical weld, a geologic core sampling, cement and concrete, or a pigment found in an historic painting. The method has been historically used for the identification and classification of minerals, but it can be used for any materials, even amorphous ones, so long as a suitable reference pattern is known or can be constructed.

Phase Identification

The most widespread use of powder diffraction is in the identification and characterization of crystalline solids, each of which produces a distinctive diffraction pattern. Both the positions (corresponding to lattice spacings) and the relative intensity of the lines in a diffraction pattern are indicative of a particular phase and material, providing a "fingerprint" for comparison. A multiphase mixture, e.g. a soil sample, will show more than one pattern superposed, allowing for determination of the relative concentrations of phases in the mixture.

J.D. Hanawalt, an analytical chemist who worked for Dow Chemical in the 1930s, was the first to realize the analytical potential of creating a database. Today it is represented by the Powder Diffraction File (PDF) of the International Centre for Diffraction Data (formerly Joint Committee for Powder Diffraction Studies). This has been made searchable by computer through the work of global software developers and equipment manufacturers. There are now over 550,000 reference materials in the 2006 Powder Diffraction File Databases, and these databases are interfaced to a wide variety of diffraction analysis software and distributed globally. The Powder Diffraction File contains many subfiles, such as minerals, metals and alloys, pharmaceuticals, forensics, excipients, superconductors, semiconductors, etc., with large collections of organic, organometallic and inorganic reference materials.

Crystallinity

In contrast to a crystalline pattern consisting of a series of sharp peaks, amorphous materials (liquids, glasses etc.) produce a broad background signal. Many polymers show semicrystalline behavior, *i.e.* part of the material forms an ordered crystallite by folding of the molecule. A single polymer molecule may well be folded into two different, adjacent crystallites and thus form a tie between the two. The tie part is prevented from crystallizing. The result is that the crystallinity will never reach 100%. Powder XRD can be used to determine the crystallinity by comparing the integrated intensity of the background pattern to that of the sharp peaks. Values obtained from powder XRD are typically comparable but not quite identical to those obtained from other methods such as DSC.

Lattice Parameters

The position of a diffraction peak is 'independent' of the atomic positions within the cell and entirely determined by the size and shape of the unit cell of the crystalline phase. Each peak represents a certain lattice plane and can therefore be characterized by a Miller index. If the symmetry is high, e.g.: cubic or hexagonal it is usually not too hard to identify the index of each peak, even for an unknown phase. This is particularly important in solid-state chemistry, where one is interested in finding and identifying new materials. Once a pattern has been indexed, this characterizes the reaction product and identifies it as a new solid phase. Indexing programs exist to deal with the harder cases, but if the unit cell is very large and the symmetry low (triclinic) success is not always guaranteed.

Expansion Tensors, Bulk Modulus

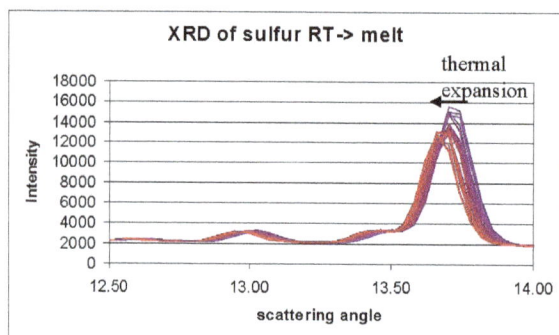

Thermal expansion of a sulfur powder

Cell parameters are somewhat temperature and pressure dependent. Powder diffraction can be combined with *in situ* temperature and pressure control. As these thermodynamic variables are changed, the observed diffraction peaks will migrate continuously to indicate higher or lower lattice spacings as the unit cell distorts. This allows for measurement of such quantities as the thermal expansion tensor and the isothermal bulk modulus, as well determination of the full equation of state of the material.

Phase Transitions

At some critical set of conditions, for example 0 °C for water at 1 atm, a new arrangement of atoms or molecules may become stable, leading to a phase transition. At this point new diffraction peaks will appear or old ones disappear according to the symmetry of the new phase. If the material melts to an isotropic liquid, all sharp lines will disappear and be replaced by a broad amorphous pattern. If the transition produces another crystalline phase, one set of lines will suddenly be replaced by another set. In some cases however lines will split or coalesce, e.g. if the material undergoes a continuous, second order phase transition. In such cases the symmetry may change because the existing structure is *distorted* rather than replaced by a completely different one. For example, the diffraction peaks for the lattice planes (100) and (001) can be found at two different values of q for a tetragonal phase, but if the symmetry becomes cubic the two peaks will come to coincide.

Crystal structure Refinement and Determination

Crystal structure determination from powder diffraction data is extremely challenging due to the overlap of reflections in a powder experiment. A number of different methods exist for structural determination, such as simulated annealing and charge flipping. The crystal structures of known materials can be refined, i.e. as a function of temperature or pressure, using the Rietveld method. The Rietveld method is a so-called full pattern analysis technique. A crystal structure, together with instrumental and microstructural information, is used to generate a theoretical diffraction pattern that can be compared to the observed data. A least squares procedure is then used to minimize the difference between the calculated pattern and each point of the observed pattern by adjusting model parameters. Techniques to determine unknown structures from powder data do exist, but are somewhat specialized. A number of programs that can be used in structure determination are TOPAS, Fox, DASH, GSAS, EXPO2004, and a few others.

Size and Strain Broadening

There are many factors that determine the width B of a diffraction peak. These include:

1. instrumental factors

2. the presence of defects to the perfect lattice

3. differences in strain in different grains

4. the size of the crystallites

It is often possible to separate the effects of size and strain. Where size broadening is independent of q (K=1/d), strain broadening increases with increasing q-values. In most cases there will be both size and strain broadening. It is possible to separate these by combining the two equations in what is known as the Hall–Williamson method:

$$B \cdot \cos(\theta) = \frac{k\lambda}{D} + \eta \cdot \sin(\theta),$$

Thus, when we plot $B \cdot \cos(\theta)$ vs. $\sin(\theta)$ we get a straight line with slope η and intercept $\frac{k\lambda}{D}$.

The expression is a combination of the Scherrer equation for size broadening and the Stokes and Wilson expression for strain broadening. The value of η is the strain in the crystallites, the value of D represents the size of the crystallites. The constant k is typically close to unity and ranges from 0.8 to 1.39.

Comparison of X-ray and Neutron Scattering

X-ray photons scatter by interaction with the electron cloud of the material, neutrons are scattered by the nuclei. This means that, in the presence of heavy atoms with many electrons, it may be difficult to detect light atoms by X-ray diffraction. In contrast, the neutron scattering lengths of most atoms are approximately equal in magnitude. Neutron diffraction techniques may therefore be used to detect light elements such as oxygen or hydrogen in combination with heavy atoms. The neutron diffraction technique therefore has obvious applications to problems such as determining oxygen displacements in materials like high temperature superconductors and ferroelectrics, or to hydrogen bonding in biological systems.

A further complication in the case of neutron scattering from hydrogenous materials is the strong incoherent scattering of hydrogen (80.27(6) barn). This leads to a very high background in neutron diffraction experiments, and may make structural investigations impossible. A common solution is deuteration, i.e., replacing the 1-H atoms in the sample with deuterium (2-H). The incoherent scattering length of deuterium is much smaller (2.05(3) barn) making structural investigations significantly easier. However, in some systems, replacing hydrogen with deuterium may alter the structural and dynamic properties of interest.

As neutrons also have a magnetic moment, they are additionally scattered by any magnetic moments in a sample. In the case of long range magnetic order, this leads to the appearance of new Bragg reflections. In most simple cases, powder diffraction may be used to determine the size of the moments and their spatial orientation.

Aperiodically Arranged Clusters

Predicting the scattered intensity in powder diffraction patterns from gases, liquids, and randomly distributed nano-clusters in the solid state is (to first order) done rather elegantly with the Debye scattering equation:

$$I(q) = \sum_{i=1}^{N} \sum_{j=1}^{N} f_i(q) f_j(q) \frac{\sin(q r_{ij})}{q r_{ij}},$$

where the magnitude of the scattering vector q is in reciprocal lattice distance units, N is the number of atoms, $f_i(q)$ is the atomic scattering factor for atom i and scattering vector q, while r_{ij} is the distance between atom i and atom j. One can also use this to predict the effect of nano-crystallite shape on detected diffraction peaks, even if in some directions the cluster is only one atom thick.

Devices

Cameras

The simplest cameras for X-ray powder diffraction consist of a small capillary and either a flat plate detector (originally a piece of X-ray film, now more and more a flat-plate detector or a CCD-camera) or a cylindrical one (originally a piece of film in a cookie-jar, but increasingly bent position sensitive detectors are used). The two types of cameras are known as the Laue and the Debye–Scherrer camera.

In order to ensure complete powder averaging, the capillary is usually spun around its axis.

For neutron diffraction vanadium cylinders are used as sample holders. Vanadium has a negligible absorption and coherent scattering cross section for neutrons and is hence nearly invisible in a powder diffraction experiment. Vanadium does however have a considerable incoherent scattering cross section which may cause problems for more sensitive techniques such as neutron inelastic scattering.

A later development in X-ray cameras is the Guinier camera. It is built around a *focusing* bent crystal monochromator. The sample is usually placed in the focusing beam, e.g. as a dusting on a piece of sticky tape. A cylindrical piece of film (or electronic multichannel detector) is put on the focusing circle, but the incident beam prevented from reaching the detector to prevent damage from its high intensity.

Diffractometers

Diffractometers can be operated both in transmission and in reflection configurations. The reflection one is more common. The powder sample is filled in a small disc-like container and its surface carefully flattened. The disc is put on one axis of the diffractometer and tilted by an angle θ while a detector (scintillation counter) rotates around it on an arm at twice this angle. This configuration is known under the name Bragg–Brentano theta-2 theta.

Another configuration is the Bragg–Brentano theta-theta configuration in which the sample is stationary while the X-ray tube and the detector are rotated around it. The angle formed between the tube and the detector is 2theta. This configuration is most convenient for loose powders.

Position-sensitive and area detectors, which allow collection from multiple angles at once, are becoming more popular on currently supplied instrumentation.

Neutron Diffraction

Sources that produce a neutron beam of suitable intensity and speed for diffraction are only available at a small number of research reactors and spallation sources in the world. Angle dispersive (fixed wavelength) instruments typically have a battery of individual detectors arranged in a cylindrical fashion around the sample holder, and can therefore collect scattered intensity simultaneously on a large 2θ range. Time of flight instruments normally have a small range of banks at different scattering angles which collect data at varying resolutions.

X-ray Tubes

Laboratory X-ray diffraction equipment relies on the use of an X-ray tube, which is used to produce the X-rays. The most commonly used laboratory X-ray tube uses a copper anode, but cobalt and molybdenum are also popular. The wavelength in nm varies for each source. The table below shows these wavelengths, determined by Bearden and quoted in the International Tables for X-ray Crystallography (all values in nm):

Element	Kα (weight average)	Kα2 (strong)	Kα1 (very strong)	Kβ (weak)
Cr	0.229100	0.229361	0.228970	0.208487
Fe	0.193736	0.193998	0.193604	0.175661
Co	0.179026	0.179285	0.178897	0.162079

Element	Kα (weight average)	Kα2 (strong)	Kα1 (very strong)	Kβ (weak)
Cu	0.154184	0.154439	0.154056	0.139222
Mo	0.071073	0.071359	0.070930	0.063229

According to the last re-examination of Holzer et al. (1997), these values are respectively:

Element	Kα2	Kα1	Kβ
Cr	0.2293663	0.2289760	0.2084920
Co	0.1792900	0.1789010	0.1620830
Cu	0.1544426	0.1540598	0.1392250
Mo	0.0713609	0.0709319	0.0632305

Other Sources

In-house applications of X-ray diffraction has always been limited to the relatively few wavelengths shown in the table above. The available choice was much needed because the combination of certain wavelengths and certain elements present in a sample can lead to strong fluorescence which increases the background in the diffraction pattern. A notorious example is the presence of iron in a sample when using copper radiation. In general elements just below the anode element in the period system need to be avoided.

Another limitation is that the intensity of traditional generators is relatively low, requiring lengthy exposure times and precluding any time dependent measurement. The advent of synchrotron sources has drastically changed this picture and caused powder diffraction methods to enter a whole new phase of development. Not only is there a much wider choice of wavelengths available, the high brilliance of the synchrotron radiation makes it possible to observe changes in the pattern during chemical reactions, temperature ramps, changes in pressure and the like.

The tunability of the wavelength also makes it possible to observe anomalous scattering effects when the wavelength is chosen close to the absorption edge of one of the elements of the sample.

Neutron diffraction has never been an in house technique because it requires the availability of an intense neutron beam only available at a nuclear reactor or spallation source. Typically the available neutron flux, and the weak interaction between neutrons and matter, require relative large samples.

Advantages and Disadvantages

Although it is possible to solve crystal structures from powder X-ray data alone, its single crystal analogue is a far more powerful technique for structure determination. This is directly related to the fact that much information is lost by the collapse of the 3D space onto a 1D axis. Nevertheless powder X-ray diffraction is a powerful and useful technique in its own right. It is mostly used to characterize and identify *phases*, and to refine details of an already known structure, rather than solving unknown structures.

Advantages of the technique are:

- simplicity of sample preparation

- rapidity of measurement

- the ability to analyze mixed phases, e.g. soil samples

- "in situ" structure determination

By contrast growth and mounting of large single crystals is notoriously difficult. In fact there are many materials for which, despite many attempts, it has not proven possible to obtain single crystals. Many materials are readily available with sufficient microcrystallinity for powder diffraction, or samples may be easily ground from larger crystals. In the field of solid-state chemistry that often aims at synthesizing *new* materials, single crystals thereof are typically not immediately available. Powder diffraction is therefore one of the most powerful methods to identify and characterize new materials in this field.

Particularly for neutron diffraction, which requires larger samples than X-ray diffraction due to a relatively weak scattering cross section, the ability to use large samples can be critical, although newer and more brilliant neutron sources are being built that may change this picture.

Since all possible crystal orientations are measured simultaneously, collection times can be quite short even for small and weakly scattering samples. This is not merely convenient, but can be essential for samples which are unstable either inherently or under X-ray or neutron bombardment, or for time-resolved studies. For the latter it is desirable to have a strong radiation source. The advent of synchrotron radiation and modern neutron sources has therefore done much to revitalize the powder diffraction field because it is now possible to study temperature dependent changes, reaction kinetics and so forth by means of time-resolved powder diffraction.

Atomic-Force Microscopy

Atomic-force microscopy (AFM) or scanning-force Microscopy (SFM) is a very-high-resolution type of scanning probe microscopy (SPM), with demonstrated resolution on the order of fractions of a nanometer, more than 1000 times better than the optical diffraction limit.

An atomic-force microscope on the left with controlling computer on the right.

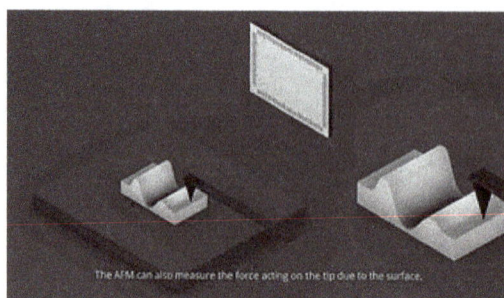

Atomic Force Microscope

Overview

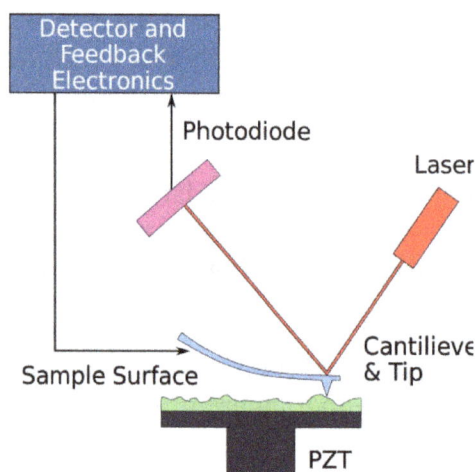

Block diagram of atomic-force microscope using beam deflection detection. As the cantilever is displaced via its interaction with the surface, so too will the reflection of the laser beam be displaced on the surface of the photodiode.

Atomic-force microscopy (AFM) or scanning-force microscopy (SFM) is a type of scanning probe microscopy (SPM), with demonstrated resolution on the order of fractions of a nanometer, more than 1000 times better than the optical diffraction limit. The information is gathered by "feeling" or "touching" the surface with a mechanical probe. Piezoelectric elements that facilitate tiny but accurate and precise movements on (electronic) command enable very precise scanning.

Abilities

The AFM has three major abilities: force measurement, imaging, and manipulation.

In force measurement, AFMs can be used to measure the forces between the probe and the sample as a function of their mutual separation. This can be applied to perform force spectroscopy.

For imaging, the reaction of the probe to the forces that the sample imposes on it can be used to form an image of the three-dimensional shape (topography) of a sample surface at a high resolution. This is achieved by raster scanning the position of the sample with respect to the tip and recording the height of the probe that corresponds to a constant probe-sample interaction. The surface topography is commonly displayed as a pseudocolor plot. An AFM image is a simulated image based on the height of each point of the surface and, in fact, each point (x, y) of the surface has a height h(x, y).

In manipulation, the forces between tip and sample can also be used to change the properties of the sample in a controlled way. Examples of this include atomic manipulation, scanning probe lithography and local stimulation of cells.

Simultaneous with the acquisition of topographical images, other properties of the sample can be measured locally and displayed as an image, often with similarly high resolution. Examples of such properties are mechanical properties like stiffness or adhesion strength and electrical properties such as conductivity or surface potential. In fact, the majority of SPM techniques are extensions of AFM that use this modality.

Other Microscopy Technologies

Compared to competitive technologies such as optical microscopy and electron microscopy, the major difference between these and the atomic-force microscope is that the latter does not use lenses or beam irradiation. Therefore, it does not suffer from a limitation of space resolution due to diffraction limit and aberration, and it is not necessary to prepare a space for guiding the beam (by creating a vacuum) or to stain the sample.

There are several types of scanning microscopy including scanning probe microscopy (which includes AFM, STM and near-field scanning optical microscope (SNOM/NSOM), STED microscopy (STED), and scanning electron microscopy). Although SNOM and STED use visible light to illuminate the sample, their resolution is not constrained by the diffraction limit.

Configuration

Fig. shows an AFM typically consisting of the following features:

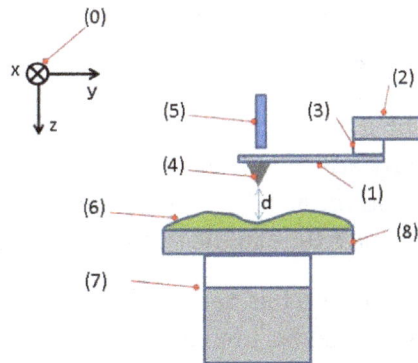

Fig. Typical configuration of an AFM. (1):: Cantilever, (2): Support for cantilever, (3): Piezoelectric element(to oscillate cantilever at its eigen frequency.), (4): Tip (Fixed to open end of a cantilever, acts as the probe), (5): Detector of deflection and motion of the cantilever, (6): Sample to be measured by AFM, (7): xyz drive, (moves sample (6) and stage (8) in x, y, and z directions with respect to a tip apex (4)), and (8): Stage.

The small spring-like cantilever (1) is carried by the support (2). Optionally, a piezoelectric element (3) oscillates the cantilever (1). The sharp tip (4) is fixed to the free end of the cantilever (1). The detector (5) records the deflection and motion of the cantilever (1). The sample (6) is mounted on the sample stage (8). An xyz drive (7) permits to displace the sample (6) and the sample stage (8) in x, y, and z directions with respect to the tip apex (4). Although Fig. 3 shows the drive attached to the sample, the drive can also be attached to the tip, or independent drives can be attached to both, since it is the relative displacement of the sample and tip that needs to be controlled. Controllers and plotter are not shown in Fig. 3. Numbers in parentheses correspond to numbered features in Fig. 3. Coordinate directions are defined by the coordinate system (0).

According to the configuration described above, the interaction between tip and sample, which can be an atomic scale phenomenon, is transduced into changes of the motion of cantilever which is a macro scale phenomenon. Several different aspects of the cantilever motion can be used to quantify the interaction between the tip and sample, most commonly the value of the deflection, the amplitude of an imposed oscillation of the cantilever, or the shift in resonance frequency of the cantilever.

Detector

The detector (5) of AFM measures the deflection (displacement with respect to the equilibrium position) of the cantilever and converts it into an electrical signal. The intensity of this signal will be proportional to the displacement of the cantilever.

Various methods of detection can be used, e.g. interferometry, optical levers, the piezoresistive method, the piezoelectric method, and STM-based detectors.

Image Formation

When using the AFM to image a sample, the tip is brought into contact with the sample, and the sample is raster scanned along an x-y grid (fig 4). Most commonly, an electronic feedback loop is employed to keep the probe-sample force constant during scanning. This feedback loop has the cantilever deflection as input, and its output controls the distance along the z axis between the probe support (2 in fig. 3) and the sample support (8 in fig 3). As long as the tip remains in contact with the sample, and the sample is scanned in the x-y plane, height variations in the sample will change the deflection of the cantilever. The feedback then adjusts the height of the probe support so that the deflection is restored to a user-defined value (the setpoint). A properly adjusted feedback loop adjusts the support-sample separation continuously during the scanning motion, such that the deflection remains approximately constant. In this situation, the feedback output equals the sample surface topography to within a small error.

Historically, a different operation method has been used, in which the sample-probe support distance is kept constant and not controlled by a feedback (servo mechanism). In this mode, usually referred to as 'constant height mode', the deflection of the cantilever is recorded as a function of the sample x-y position. As long as the tip is in contact with the sample, the deflection then corresponds to surface topography. The main reason this method is not very popular anymore, is that the forces between tip and sample are not controlled, which can lead to forces high enough to damage the tip or the sample. It is however common practice to record the deflection even when scanning in 'constant force mode', with feedback. This reveals the small tracking error of the feedback, and can sometimes reveal features that the feedback was not able to adjust for.

The AFM signals, such as sample height or cantilever deflection, are recorded on a computer during the x-y scan. They are plotted in a pseudocolor image, in which each pixel represents an x-y position on the sample, and the color represents the recorded signal.

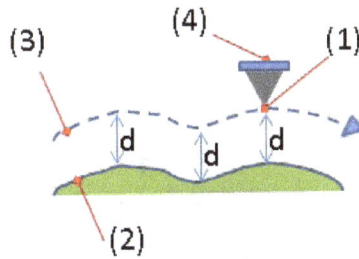

Fig. Topographic image forming by AFM. (1): Tip apex, (2): Sample surface, (3): Z-orbit of Tip apex, (4): Cantilever.

History

AFM was invented by IBM Scientists in 1982.The precursor to the AFM, the scanning tunneling microscope (STM), was developed by Gerd Binnig and Heinrich Rohrer in the early 1980s at IBM Research - Zurich, a development that earned them the Nobel Prize for Physics in 1986. Binnig invented the atomic-force microscope and the first experimental implementation was made by Binnig, Quate and Gerber in 1986.

The first commercially available atomic-force microscope was introduced in 1989. The AFM is one of the foremost tools for imaging, measuring, and manipulating matter at the nanoscale.

Applications

The AFM has been applied to problems in a wide range of disciplines of the natural sciences, including solid state physics, semiconductor science and technology, molecular engineering, polymer chemistry and physics, surface chemistry, molecular biology, cell biology and medicine.

Applications in the field of solid state physics include (a) the identification of atoms at a surface, (b) the evaluation of interactions between a specific atom and its neighboring atoms, and (c) the study of changes in physical properties arising from changes in an atomic arrangement through atomic manipulation.

In cellular biology, AFM can be used to (a) attempt to distinguish cancer cells and normal cells based on a hardness of cells, and (b) to evaluate interactions between a specific cell and its neighboring cells in a competitive culture system.

In some variations, electric potentials can also be scanned using conducting cantilevers. In more advanced versions, currents can be passed through the tip to probe the electrical conductivity or transport of the underlying surface, but this is a challenging task with few research groups reporting consistent data (as of 2004).

Principles

The AFM consists of a cantilever with a sharp tip (probe) at its end that is used to scan the specimen surface. The cantilever is typically silicon or silicon nitride with a tip radius of curvature on the order of nanometers. When the tip is brought into proximity of a sample surface, forces between the tip and the sample lead to a deflection of the cantilever according to Hooke's law. Depending on the situation, forces that are measured in AFM include mechanical contact force, van der Waals forces, capillary forces, chemical bonding, electrostatic forces, magnetic forces (see magnetic force microscope, MFM), Casimir forces, solvation forces, etc. Along with force, additional quantities may simultaneously be measured through the use of specialized types of probes.

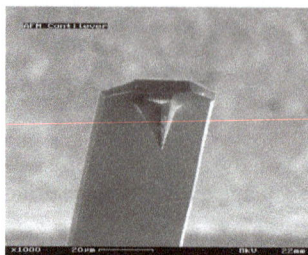

Electron micrograph of a used AFM cantilever. Image width ~100 micrometers

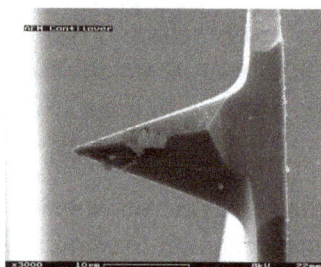

Electron micrograph of a used AFM cantilever. Image width ~30 micrometers

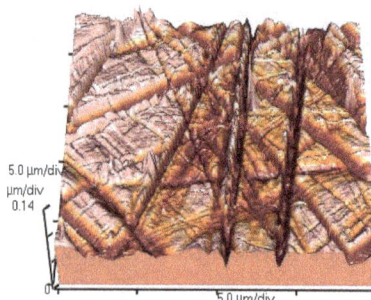

Atomic-force microscope topographical scan of a glass surface. The micro and nano-scale features of the glass can be observed, portraying the roughness of the material. The image space is (x,y,z) = (20 μm × 20 μm × 420 nm).

The AFM can be operated in a number of modes, depending on the application. In general, possible imaging modes are divided into static (also called *contact*) modes and a variety of dynamic (non-contact or "tapping") modes where the cantilever is vibrated or oscillated at a given frequency.

Imaging Modes

AFM operation is usually described as one of three modes, according to the nature of the tip motion: contact mode, also called static mode (as opposed to the other two modes, which are called dynamic modes); tapping mode, also called intermittent contact, AC mode, or vibrating mode, or, after the detection mechanism, amplitude modulation AFM; non-contact mode, or, again after the detection mechanism, frequency modulation AFM.

It should be noted that despite the nomenclature, repulsive contact can occur or be avoided both in amplitude modulation AFM and frequency modulation AFM, depending on the settings.

Contact Mode

In contact mode, the tip is "dragged" across the surface of the sample and the contours of the surface are measured either using the deflection of the cantilever directly or, more commonly, using the feedback signal required to keep the cantilever at a constant position. Because the measurement of a static signal is prone to noise and drift, low stiffness cantilevers (i.e. cantilevers with a low spring constant, k) are used to achieve a large enough deflection signal while keeping the interaction force low. Close to the surface of the sample, attractive forces can be quite strong, causing the tip to "snap-in" to the surface. Thus, contact mode AFM is almost always done at a depth where the overall force is repulsive, that is, in firm "contact" with the solid surface.

Tapping Mode

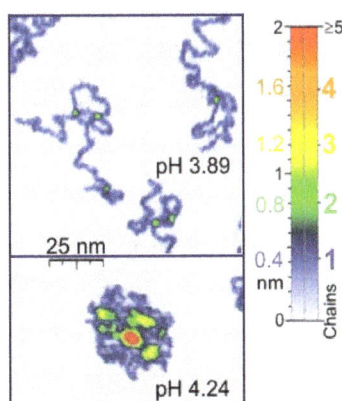

Single polymer chains (0.4 nm thick) recorded in a tapping mode under aqueous media with different pH.

In ambient conditions, most samples develop a liquid meniscus layer. Because of this, keeping the probe tip close enough to the sample for short-range forces to become detectable while preventing the tip from sticking to the surface presents a major problem for non-contact dynamic mode in ambient conditions. Dynamic contact mode (also called intermittent contact, AC mode or tapping mode) was developed to bypass this problem. Nowadays, tapping mode is the most frequently used AFM mode when operating in ambient conditions or in liquids.

In *tapping mode*, the cantilever is driven to oscillate up and down at or near its resonance frequency. This oscillation is commonly achieved with a small piezo element in the cantilever holder, but other possibilities include an AC magnetic field (with magnetic cantilevers), piezoelectric cantilevers, or periodic heating with a modulated laser beam. The amplitude of this oscillation usually varies from several nm to 200 nm. In tapping mode, the frequency and amplitude of the driving signal are kept constant, leading to a constant amplitude of the cantilever oscillation as long as there is no drift or interaction with the surface. The interaction of forces acting on the cantilever when the tip comes close to the surface, Van der Waals forces, dipole-dipole interactions, electrostatic forces, etc. cause the amplitude of the cantilever's oscillation to change (usually decrease) as the tip gets closer to the sample. This amplitude is used as the parameter that goes into the electronic servo that controls the height of the cantilever above the sample. The servo adjusts the height to maintain a set cantilever oscillation amplitude as the cantilever is scanned over the sample. A *tapping AFM* image is therefore produced by imaging the force of the intermittent contacts of the tip with the sample surface.

Although the peak forces applied during the contacting part of the oscillation can be much higher than typically used in contact mode, tapping mode generally lessens the damage done to the surface and the tip compared to the amount done in contact mode. This can be explained by the short duration of the applied force, and because the lateral forces between tip and sample are significantly lower in tapping mode over contact mode. Tapping mode imaging is gentle enough even for the visualization of supported lipid bilayers or adsorbed single polymer molecules (for instance, 0.4 nm thick chains of synthetic polyelectrolytes) under liquid medium. With proper scanning parameters, the conformation of single molecules can remain unchanged for hours, and even single molecular motors can be imaged while moving.

When operating in tapping mode, the phase of the cantilever's oscillation with respect to the driving signal can be recorded as well. This signal channel contains information about the energy dissipated by the cantilever in each oscillation cycle. Samples that contain regions of varying stiffness or with different adhesion properties can give a contrast in this channel that is not visible in the topographic image. Extracting the sample's material properties in a quantitative manner from phase images, however, is often not feasible.

Non-contact Mode

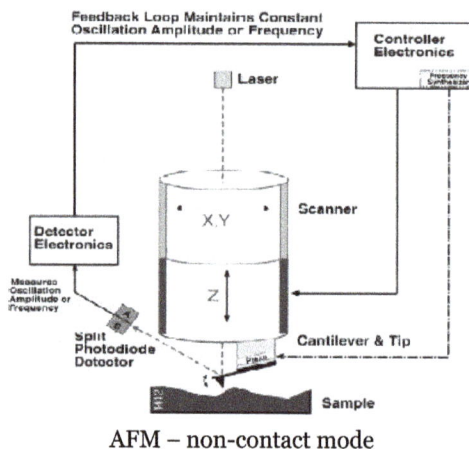

AFM – non-contact mode

In non-contact atomic force microscopy mode, the tip of the cantilever does not contact the sample surface. The cantilever is instead oscillated at either its resonant frequency (frequency modulation) or just above (amplitude modulation) where the amplitude of oscillation is typically a few nanometers (<10 nm) down to a few picometers. The van der Waals forces, which are strongest from 1 nm to 10 nm above the surface, or any other long-range force that extends above the surface acts to decrease the resonance frequency of the cantilever. This decrease in resonant frequency combined with the feedback loop system maintains a constant oscillation amplitude or frequency by adjusting the average tip-to-sample distance. Measuring the tip-to-sample distance at each (x,y) data point allows the scanning software to construct a topographic image of the sample surface.

Non-contact mode AFM does not suffer from tip or sample degradation effects that are sometimes observed after taking numerous scans with contact AFM. This makes non-contact AFM preferable to contact AFM for measuring soft samples, e.g. biological samples and organic thin film. In the case of rigid samples, contact and non-contact images may look the same. However, if a few monolayers of adsorbed fluid are lying on the surface of a rigid sample, the images may look quite different. An AFM operating in contact mode will penetrate the liquid layer to image the underlying surface, whereas in non-contact mode an AFM will oscillate above the adsorbed fluid layer to image both the liquid and surface.

Schemes for dynamic mode operation include frequency modulation where a phase-locked loop is used to track the cantilever's resonance frequency and the more common amplitude modulation with a servo loop in place to keep the cantilever excitation to a defined amplitude. In frequency modulation, changes in the oscillation frequency provide information about tip-sample interactions. Frequency can be measured with very high sensitivity and thus the frequency modulation mode allows for the use of very stiff cantilevers. Stiff cantilevers provide stability very close to the surface and, as a result, this technique was the first AFM technique to provide true atomic resolution in ultra-high vacuum conditions.

In amplitude modulation, changes in the oscillation amplitude or phase provide the feedback signal for imaging. In amplitude modulation, changes in the phase of oscillation can be used to discriminate between different types of materials on the surface. Amplitude modulation can be operated either in the non-contact or in the intermittent contact regime. In dynamic contact mode, the cantilever is oscillated such that the separation distance between the cantilever tip and the sample surface is modulated.

Amplitude modulation has also been used in the non-contact regime to image with atomic resolution by using very stiff cantilevers and small amplitudes in an ultra-high vacuum environment.

Topographic Image

Image formation is a plotting method that produces a color mapping through changing the x-y position of the tip while scanning and recording the measured variable, i.e. the intensity of control signal, to each x-y coordinate. The color mapping shows the measured value corresponding to each coordinate. The image expresses the intensity of a value as a hue. Usually, the correspondence between the intensity of a value and a hue is shown as a color scale in the explanatory notes accompanying the image.

What is the Topographic Image of Atomic-Force Microscope?

Operation mode of Image forming of the AFM are generally classified into two groups from the viewpoint whether it uses z-Feedback loop (not shown) to maintain the tip-sample distance to keep signal intensity exported by the detector. The first one (using z-Feedback loop), said to be "constant XX mode" (XX is something which kept by z-Feedback loop).

Topographic Image Formation Mode is based on abovementioned "constant XX mode", z-Feedback loop controls the relative distance between the probe and the sample through outputting control signals to keep constant one of frequency, vibration and phase which typically corresponds to the motion of cantilever (for instance, voltage is applied to the Z-piezoelectric element and it moves the sample up and down towards the Z direction.

Details will be explained in the case that especially "constant df mode"(FM-AFM) among AFM as an instance in next section.

Topographic Image of FM-AFM

When the distance between the probe and the sample is brought to the range where atomic force may be detected, while a cantilever is excited in its natural eigen frequency (f_o), a phenomenon that the resonance frequency (f) of the cantilever shifts from the original resonance frequency (natural eigen frequency) of the cantilever. In other words, in the range where atomic force may be detected, the frequency shift ($df=f-f_o$) will be observed. So, when the distance between the probe and the sample is in the non-contact region, the frequency shift increases in negative direction as the distance between the probe and the sample gets smaller.

When the sample has concavity and convexity, the distance between the tip-apex and the sample varies in accordance with the concavity and convexity accompanied with a scan of the sample along x-y direction (without height regulation in z-direction) . As a result, the frequency shift arises. The image in which the values of the frequency obtained by a raster scan along the x-y direction of the sample surface are plotted against the x-y coordination of each measurement point is called a constant-height image.

On the other hand, the df may be kept constant by moving the probe upward and downward in z-direction using a negative feedback (by using z-feedback loop) while the raster scan of the sample surface along the x-y direction . The image in which the amounts of the negative feedback (the moving distance of the probe upward and downward in z-direction) are plotted against the x-y coordination of each measurement point is a topographic image. In other words, the topographic image is a trace of the tip of the probe regulated so that the df is constant and it may also be considered to be a plot of a constant-height surface of the df.

Therefore, the topographic image of the AFM is not the exact surface morphology itself, but actually the image influenced by the bond-order between the probe and the sample, however, the topographic image of the AFM is considered to reflect the geographical shape of the surface more than the topographic image of a scanning tunnel microscope.

Force Spectroscopy

Another major application of AFM (besides imaging) is force spectroscopy, the direct measurement of tip-sample interaction forces as a function of the gap between the tip and sample (the result of this measurement is called a force-distance curve). For this method, the AFM tip is extended towards and retracted from the surface as the deflection of the cantilever is monitored as a function of piezoelectric displacement. These measurements have been used to measure nanoscale contacts, atomic bonding, Van der Waals forces, and Casimir forces, dissolution forces in liquids and single molecule stretching and rupture forces. Furthermore, AFM was used to measure, in an aqueous environment, the dispersion force due to polymer adsorbed on the substrate. Forces of the order of a few piconewtons can now be routinely measured with a vertical distance resolution of better than 0.1 nanometers. Force spectroscopy can be performed with either static or dynamic modes. In dynamic modes, information about the cantilever vibration is monitored in addition to the static deflection.

Problems with the technique include no direct measurement of the tip-sample separation and the common need for low-stiffness cantilevers, which tend to 'snap' to the surface. These problems are not insurmountable. An AFM that directly measures the tip-sample separation has been developed. The snap-in can be reduced by measuring in liquids or by using stiffer cantilevers, but in the latter case a more sensitive deflection sensor is needed. By applying a small dither to the tip, the stiffness (force gradient) of the bond can be measured as well.

Biological Applications and Other

Force spectroscopy is used in biophysics to measure the mechanical properties of living material (such as tissue or cells). Another application was to measure the interaction forces between from one hand a material stuck on the tip of the cantilever, and from another hand the surface of particles either free or occupied by the same material. From the adhesion force distribution curve, a mean value of the forces has been derived. It allowed to make a cartography of the surface of the particles, covered or not by the material.

Identification of Individual Surface Atoms

The AFM can be used to image and manipulate atoms and structures on a variety of surfaces. The atom at the apex of the tip "senses" individual atoms on the underlying surface when it forms incipient chemical bonds with each atom. Because these chemical interactions subtly alter the tip's vibration frequency, they can be detected and mapped. This principle was used to distinguish between atoms of silicon, tin and lead on an alloy surface, by comparing these 'atomic fingerprints' to values obtained from large-scale density functional theory (DFT) simulations.

The trick is to first measure these forces precisely for each type of atom expected in the sample, and then to compare with forces given by DFT simulations. The team found that the tip interacted most strongly with silicon atoms, and interacted 24% and 41% less strongly with tin and lead atoms, respectively. Thus, each different type of atom can be identified in the matrix as the tip is moved across the surface.

Probe

An AFM probe has a sharp tip on the free-swinging end of a cantilever that is protruding from a holder. The dimensions of the cantilever are in the scale of micrometers. The radius of the tip is usually on the scale of a few nanometers to a few tens of nanometers. (Specialized probes exist with much larger end radii, for example probes for indentation of soft materials.) The cantilever holder, also called holder chip – often 1.6 mm by 3.4 mm in size – allows the operator to hold the AFM cantilever/probe assembly with tweezers and fit it into the corresponding holder clips on the scanning head of the atomic-force microscope.

This device is most commonly called an "AFM probe", but other names include "AFM tip" and "cantilever" (employing the name of a single part as the name of the whole device). An AFM probe is a particular type of SPM (scanning probe microscopy) probe.

AFM probes are manufactured with MEMS technology. Most AFM probes used are made from silicon (Si), but borosilicate glass and silicon nitride are also in use. AFM probes are considered consumables as they are often replaced when the tip apex becomes dull or contaminated or when the cantilever is broken. They can cost from a couple of tens of dollars up to hundreds of dollars per cantilever for the most specialized cantilever/probe combinations.

Just the tip is brought very close to the surface of the object under investigation, the cantilever is deflected by the interaction between the tip and the surface, which is what the AFM is designed to measure. A spatial map of the interaction can be made by measuring the deflection at many points on a 2D surface.

Several types of interaction can be detected. Depending on the interaction under investigation, the surface of the tip of the AFM probe needs to be modified with a coating. Among the coatings used are gold – for covalent bonding of biological molecules and the detection of their interaction with a surface, diamond for increased wear resistance and magnetic coatings for detecting the magnetic properties of the investigated surface. Another solution exists to achieve high resolution magnetic imaging : having the probe equip with a microSQUID. The AFM tips is fabricated using silicon micro machining and the precise positioning of the microSQUID loop is done by electron beam lithography.

The surface of the cantilevers can also be modified. These coatings are mostly applied in order to increase the reflectance of the cantilever and to improve the deflection signal.

AFM Cantilever-deflection Measurement

Beam-deflection Measurement

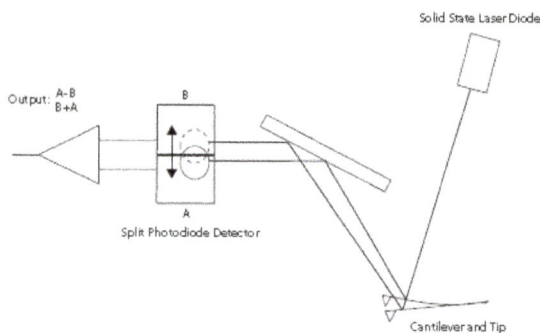

AFM beam-deflection detection

r (PSD) consisting of two closely spaced photodiodes whose output signal is collected by a differential amplifier. Angular displacement of the cantilever results in one photodiode collecting more light than the other photodiode, producing an output signal (the difference between the photodiode signals normalized by their sum), which is proportional to the deflection of the cantilever. The sensitivity of the beam-deflection method is very high, a noise floor on the order of 10 fm $Hz^{-\nu_2}$ can be obtained routinely in a well-designed system. Although this method is sometimes called the 'optical lever' method, the signal is not amplified if the beam path is made longer. A longer beam path increases the motion of the reflected spot on the photodiodes, but also widens the spot by the same amount due to diffraction, so that the same amount of optical power is moved from one photodiode to the other. The 'optical leverage' (output signal of the detector divided by deflection of the cantilever) is inversely proportional to the numerical aperture of the beam focusing optics, as long as the focused laser spot is small enough to fall completely on the cantilever. It is also inversely proportional to the length of the cantilever. The most common method for cantilever-deflection measurements is the beam-deflection method. In this method, laser light from a solid-state diode is reflected off the back of the cantilever and collected by a position-sensitive detecto

The relative popularity of the beam-deflection method can be explained by its high sensitivity and simple operation, and by the fact that cantilevers do not require electrical contacts or other special treatments, and can therefore be fabricated relatively cheaply with sharp integrated tips.

Other Deflection-measurement Methods

Many other methods for beam-deflection measurements exist.

- *Piezoelectric detection* – Cantilevers made from quartz (such as the qPlus configuration), or other piezoelectric materials can directly detect deflection as an electrical signal. Cantilever oscillations down to 10pm have been detected with this method.

- *Laser Doppler vibrometry* – A laser Doppler vibrometer can be used to produce very accurate deflection measurements for an oscillating cantilever (thus is only used in non-contact mode). This method is expensive and is only used by relatively few groups.

- *STM* — The first atomic microscope used an STM complete with its own feedback mechanism to measure deflection. This method is very difficult to implement, and is slow to react to deflection changes compared to modern methods.

- *Optical interferometry* – Optical interferometry can be used to measure cantilever deflection. Due to the nanometre scale deflections measured in AFM, the interferometer is running in the sub-fringe regime, thus, any drift in laser power or wavelength has strong effects on the measurement. For these reasons optical interferometer measurements must be done with great care (for example using index matching fluids between optical fibre junctions), with very stable lasers. For these reasons optical interferometry is rarely used.

- *Capacitive detection* – Metal coated cantilevers can form a capacitor with another contact located behind the cantilever. Deflection changes the distance between the contacts and can be measured as a change in capacitance.

- *Piezoresistive detection* – Cantilevers can be fabricated with piezoresistive elements that act as a strain gauge. Using a Wheatstone bridge, strain in the AFM cantilever due to deflection can be measured. This is not commonly used in vacuum applications, as the piezoresistive detection dissipates energy from the system affecting Q of the resonance.

Piezoelectric Scanners

AFM scanners are made from piezoelectric material, which expands and contracts proportionally to an applied voltage. Whether they elongate or contract depends upon the polarity of the voltage applied. Traditionally the tip or sample is mounted on a 'tripod' of three piezo crystals, with each responsible for scanning in the x,y and z directions. In 1986, the same year as the AFM was invented, a new piezoelectric scanner, the tube scanner, was developed for use in STM. Later tube scanners were incorporated into AFMs. The tube scanner can move the sample in the x, y, and z directions using a single tube piezo with a single interior contact and four external contacts. An advantage of the tube scanner compared to the original tripod design, is better vibrational isolation, resulting from the higher resonant frequency of the single element construction, in combination with a low resonant frequency isolation stage. A disadvantage is that the x-y motion can cause unwanted z motion resulting in distortion. Another popular design for AFM scanners is the flexure stage, which uses separate piezos for each axis, and couples them through a flexure mechanism.

Scanners are characterized by their sensitivity, which is the ratio of piezo movement to piezo voltage, i.e., by how much the piezo material extends or contracts per applied volt. Because of differences in material or size, the sensitivity varies from scanner to scanner. Sensitivity varies non-linearly with respect to scan size. Piezo scanners exhibit more sensitivity at the end than at the beginning of a scan. This causes the forward and reverse scans to behave differently and display hysteresis between the two scan directions. This can be corrected by applying a non-linear voltage to the piezo electrodes to cause linear scanner movement and calibrating the scanner accordingly. One disadvantage of this approach is that it requires re-calibration because the precise non-lin-ear voltage needed to correct non-linear movement will change as the piezo ages. This problem can be circumvented by adding a linear sensor to the sample stage or piezo stage to detect the true movement of the piezo. Deviations from ideal movement can be detected by the sensor and corrections applied to the piezo drive signal to correct for non-linear piezo movement. This design is known as a 'closed loop' AFM. Non-sensored piezo AFMs are referred to as 'open loop' AFMs.

The sensitivity of piezoelectric materials decreases exponentially with time. This causes most of the change in sensitivity to occur in the initial stages of the scanner's life. Piezoelectric scanners are run for approximately 48 hours before they are shipped from the factory so that they are past the point where they may have large changes in sensitivity. As the scanner ages, the sensitivity will change less with time and the scanner would seldom require recalibration, though various manufacturer manuals recommend monthly to semi-monthly calibration of open loop AFMs.

Advantages and Disadvantages

The first atomic-force microscope

Just like any other tool, an AFM's usefulness has limitations. When determining whether or not analyzing a sample with an AFM is appropriate, there are various advantages and disadvantages that must be considered.

Advantages

AFM has several advantages over the scanning electron microscope (SEM). Unlike the electron microscope, which provides a two-dimensional projection or a two-dimensional image of a sample, the AFM provides a three-dimensional surface profile. In addition, samples viewed by AFM do not require any special treatments (such as metal/carbon coatings) that would irreversibly change or damage the sample, and does not typically suffer from charging artifacts in the final image. While an electron microscope needs an expensive vacuum environment for proper operation, most AFM modes can work perfectly well in ambient air or even a liquid environment. This makes it possible to study biological macromolecules and even living organisms. In principle, AFM can provide higher resolution than SEM. It has been shown to give true atomic resolution in ultra-high vacuum (UHV) and, more recently, in liquid environments. High resolution AFM is comparable in resolution to scanning tunneling microscopy and transmission electron microscopy. AFM can also be combined with a variety of optical microscopy techniques such as fluorescent microscopy, further expanding its applicability. Combined AFM-optical instruments have been applied primarily in the biological sciences but have also found a niche in some materials applications, especially those involving photovoltaics research.

Disadvantages

A disadvantage of AFM compared with the scanning electron microscope (SEM) is the single scan image size. In one pass, the SEM can image an area on the order of square millimeters with a depth of field on the order of millimeters, whereas the AFM can only image a maximum scanning area of about 150×150 micrometers and a maximum height on the order of 10-20 micrometers. One method of improving the scanned area size for AFM is by using parallel probes in a fashion similar to that of millipede data storage.

The scanning speed of an AFM is also a limitation. Traditionally, an AFM cannot scan images as fast as an SEM, requiring several minutes for a typical scan, while an SEM is capable of scanning at near real-time, although at relatively low quality. The relatively slow rate of scanning during AFM imaging often leads to thermal drift in the image making the AFM less suited for measuring accurate distances between topographical features on the image. However, several fast-acting designs were suggested to increase microscope scanning productivity including what is being termed videoAFM (reasonable quality images are being obtained with videoAFM at video rate: faster than the average SEM). To eliminate image distortions induced by thermal drift, several methods have been introduced.

AFM images can also be affected by nonlinearity, hysteresis, and creep of the piezoelectric material and cross-talk between the x, y, z axes that may require software enhancement and filtering. Such filtering could "flatten" out real topographical features. However, newer AFMs utilize real-time correction software (for example, feature-oriented scanning) or closed-loop scanners, which practically eliminate these problems. Some AFMs also use separated orthogonal scanners (as opposed to a single tube), which also serve to eliminate part of the cross-talk problems.

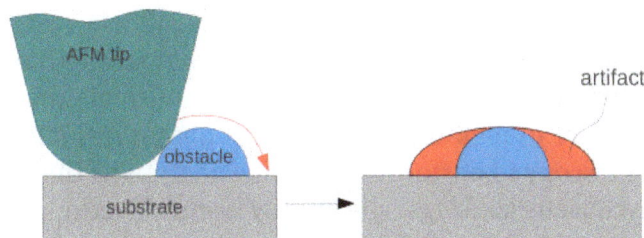

Showing an AFM artifact arising from a tip with a high radius of curvature with respect to the feature that is to be visualized.

As with any other imaging technique, there is the possibility of image artifacts, which could be induced by an unsuitable tip, a poor operating environment, or even by the sample itself, as depicted on the right. These image artifacts are unavoidable; however, their occurrence and effect on results can be reduced through various methods. Artifacts resulting from a too-coarse tip can be caused for example by inappropriate handling or de facto collisions with the sample by either scanning too fast or having an unreasonably rough surface, causing actual wearing of the tip.

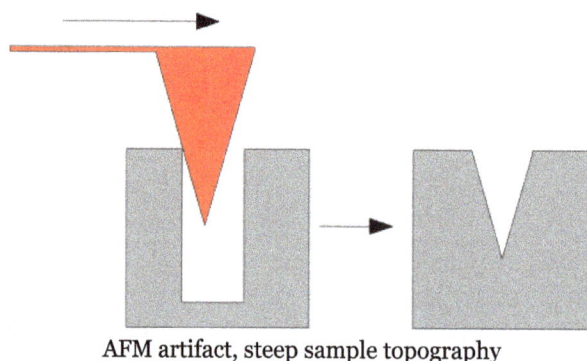

AFM artifact, steep sample topography

Due to the nature of AFM probes, they cannot normally measure steep walls or overhangs. Specially made cantilevers and AFMs can be used to modulate the probe sideways as well as up and down (as with dynamic contact and non-contact modes) to measure sidewalls, at the cost of more expensive cantilevers, lower lateral resolution and additional artifacts.

Other Applications in Various Fields of Study

The latest efforts in integrating nanotechnology and biological research have been successful and show much promise for the future. Since nanoparticles are a potential vehicle of drug delivery, the biological responses of cells to these nanoparticles are continuously being explored to optimize their efficacy and how their design could be improved. Pyrgiotakis et al. were able to study the interaction between CeO_2 and Fe_2O_3 engineered nanoparticles and cells by attaching the engineered nanoparticles to the AFM tip. Beyond the interactions with external synthetic materials, cells have been imaged with X-ray crystallography and there has been much curiosity about their behavior *in vivo*. Studies have taken advantage of AFM to obtain further information on the behavior of live cells in biological media. Real-time atomic force spectroscopy (or nanoscopy) and dynamic atomic force spectroscopy have been used to study live cells and membrane proteins and their dynamic behavior at high resolution, on the nanoscale. Imaging and obtaining information on the topography and the properties of the cells has also given insight into chemical processes and mechanisms that occur through cell-cell interaction and interactions with other signaling molecules (ex. ligands). Evans and Calderwood used single cell force microscopy to study cell adhesion forces, bond kinetics/dynamic bond strength and its role in chemical processes such as cell signaling. Scheuring, Lévy, and Rigaud reviewed studies in which AFM to explore the crystal structure of membrane proteins of photosynthetic bacteria. Alsteen et al. have used AFM-based nanoscopy to perform a real-time analysis of the interaction between live mycobacteria and antimycobacterial drugs (specifically isoniazid, ethionamide, ethambutol, and streptomycine), which serves as an example of the more in-depth analysis of pathogen-drug interactions that can be done through AFM.

References

- Shafranovskii I I & Belov N V (1962). Paul Ewald, ed. "E. S. Fedorov" (PDF). 50 Years of X-Ray Diffraction. Springer: 351. ISBN 90-277-9029-9.

- Jeruzalmi D (2006). "First analysis of macromolecular crystals: biochemistry and x-ray diffraction". Methods Mol. Biol. 364: 43–62. doi:10.1385/1-59745-266-1:43. ISBN 1-59745-266-1.

- Neutron diffraction of magnetic materials / Yu. A. Izyumov, V.E. Naish, and R.P. Ozerov ; translated from Russian by Joachim Büchner. New York : Consultants Bureau, c1991.ISBN 0-306-11030-X

- Neutron powder diffraction by Richard M. Ibberson and William I.F. David, Chapter 5 of Structure determination form powder diffraction data IUCr monographphs on crystallography, Oxford scientific publications 2002, ISBN 0-19-850091-2

- Structure determination form powder diffraction data IUCr Monographs on crystallography, Edt. W.I.F. David, K. Shankland, L.B. McCusker and Ch. Baerlocher. 2002. Oxford Science publications ISBN 0-19-850091-2

- R. V. Lapshin (2011). "Feature-oriented scanning probe microscopy". In H. S. Nalwa. Encyclopedia of Nanoscience and Nanotechnology (PDF). 14. USA: American Scientific Publishers. pp. 105–115. ISBN 1-58883-163-9.

- Perkins, Thomas. "Atomic force microscopy measures properties of proteins and protein folding". SPIE Newsroom. Retrieved 4 March 2016.

- Brown, Dwayne (October 30, 2012). "NASA Rover's First Soil Studies Help Fingerprint Martian Minerals". NASA. Retrieved October 31, 2012.

Salt: An Integrated Study

Salt is an ionic compound; there are different varieties of salts. The chapter also focuses on bresle method, acid salts, alkali salts, soap and alum. This section has been carefully written to provide an easy understanding of all the significant aspects of salt and its various applications.

Salt (chemistry)

In chemistry, a salt is an ionic compound that results from the neutralization reaction of an acid and a base. Salts are composed of related numbers of cations (positively charged ions) and anions (negative ions) so that the product is electrically neutral (without a net charge). These component ions can be inorganic, such as chloride (Cl^-), or organic, such as acetate ($CH_3CO_2^-$); and can be monatomic, such as fluoride (F^-), or polyatomic, such as sulfate (SO_4^{2-}).

The salt copper(II) sulfate as the mineral chalcanthite.

There are several varieties of salts. Salts that hydrolyze to produce hydroxide ions when dissolved in water are *basic salts*, whilst those that hydrolyze to produce hydronium ions in water are *acidic salts*. *Neutral salts* are those that are neither acid nor basic salts. Zwitterions contain an anionic centre and a cationic centre in the same molecule, but are not considered to be salts. Examples of zwitterions include amino acids, many metabolites, peptides, and proteins.

Usually, non-dissolved salts at standard temperature and pressure are solid, but there are exceptions.

Molten salts and solutions containing dissolved salts (e.g., sodium chloride in water) are called electrolytes, as they are able to conduct electricity. As observed in the cytoplasm of cells, in blood, urine, plant saps and mineral waters, mixtures of many different ions in solution usually do not form defined salts after evaporation of the water. Therefore, their salt content is given for the respective ions.

Properties

Color

Potassium dichromate, a bright orange salt used as a pigment.

Salts can appear to be clear and transparent (sodium chloride), opaque, and even metallic and lustrous (iron disulfide). In many cases, the apparent opacity or transparency are only related to the difference in size of the individual monocrystals. Since light reflects from the grain boundaries (boundaries between crystallites), larger crystals tend to be transparent, while the polycrystalline aggregates look like white powders.

Salts exist in many different colors, for example:

- yellow (sodium chromate)

- orange (potassium dichromate)

- red (cobalt nitrate)

- mauve (cobalt chloride hexahydrate)

- blue (copper sulfate pentahydrate, ferric hexacyanoferrate)

- purple (potassium permanganate)

- green (nickel chloride hexahydrate)

- colorless (sodium chloride, magnesium sulfate heptahydrate)—may appear white when powdered or in small pieces

Most minerals and inorganic pigments, as well as many synthetic organic dyes, are salts. The color of the specific salt is due to the electronic structure in the d-orbitals of transition elements or in the conjugated organic dye framework.

Taste

Different salts can elicit all five basic tastes, e.g., salty (sodium chloride), sweet (lead diacetate, which will cause lead poisoning if ingested), sour (potassium bitartrate), bitter (magnesium sulfate), and umami or savory (monosodium glutamate).

Odor

Salts of strong acids and strong bases ("strong salts") are non-volatile and odorless, whereas salts of either weak acids or weak bases ("weak salts") may smell after the conjugate acid (e.g., acetates like acetic acid (vinegar) and cyanides like hydrogen cyanide (almonds)) or the conjugate base (e.g., ammonium salts like ammonia) of the component ions. That slow, partial decomposition is usually accelerated by the presence of water, since hydrolysis is the other half of the reversible reaction equation of formation of weak salts.

Solubility

Many ionic compounds can be dissolved in water or other similar solvents. The exact combination of ions involved makes each compound have a unique solubility in any solvent. The solubility is dependent on how well each ion interacts with the solvent, so there are certain patterns. For example, all salts of sodium, potassium and ammonium are soluble in water, as are all nitrates and many sulfates – barium sulfate, calcium sulfate (sparingly soluble) and lead(II) sulfate are examples of exceptions. However, ions that bind tightly to each other and form highly stable lattices are less soluble, because it is harder for these structures to break apart for the compounds to dissolve. For example, most carbonate salts are not soluble in water, such as lead carbonate and barium carbonate. Some soluble carbonate salts are: sodium carbonate, potassium carbonate and ammonium carbonate.

Conductivity

Solid salts do not conduct electricity. However, liquid salts do. Moreover, solutions of salts also conduct electricity.

Chemical Compound

The name of a salt starts with the name of the cation (e.g., *sodium* or *ammonium*) followed by the name of the anion (e.g., *chloride* or *acetate*). Salts are often referred to only by the name of the cation (e.g., *sodium salt* or *ammonium salt*) or by the name of the anion (e.g., *chloride salt* or *acetate salt*).

Common salt-forming cations include:

- Ammonium NH_4^+

- Calcium Ca^{2+}

- Iron Fe^{2+} and Fe^{3+}

- Magnesium Mg^{2+}

- Potassium K^+

- Pyridinium $C_5H_5NH^+$

- Quaternary ammonium NR_4^+ R being an alkyl group or an aryl group

- Sodium Na^+

Common salt-forming anions (parent acids in parentheses where available) include:

- Acetate CH_3COO^- (acetic acid)

- Carbonate CO_3^{2-} (carbonic acid)

- Chloride Cl^- (hydrochloric acid)

- Citrate $HOC(COO^-)(CH_2COO^-)_2$ (citric acid)

- Cyanide $C\equiv N^-$ (hydrocyanic acid)

- Fluoride F^- (hydrofluoric acid)

- Nitrate NO_3^- (nitric acid)

- Nitrite NO_2^- (nitrous acid)

- Oxide O^{2-}

- Phosphate PO_4^{3-} (phosphoric acid)

- Sulfate SO_4^{2-} (sulfuric acid)

Formation

Solid lead(II) sulfate ($PbSO_4$)

Salts are formed by a chemical reaction between:

- A base and an acid, e.g., $NH_3 + HCl \rightarrow NH_4Cl$

- A metal and an acid, e.g., $Mg + H_2SO_4 \rightarrow MgSO_4 + H_2$

- A metal and a non-metal, e.g., $Ca + Cl_2 \rightarrow CaCl_2$

- A base and an acid anhydride, e.g., $2\,NaOH + Cl_2O \rightarrow 2\,NaClO + H_2O$

- An acid and a basic anhydride, e.g., $2\,HNO_3 + Na_2O \rightarrow 2\,NaNO_3 + H_2O$

- Salts can also form if solutions of different salts are mixed, their ions recombine, and the new salt is insoluble and precipitates, for example:

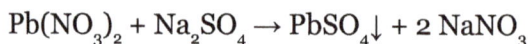

$$Pb(NO_3)_2 + Na_2SO_4 \rightarrow PbSO_4\downarrow + 2\,NaNO_3$$

Bresle Method

Bresle test kit

The Bresle method is used to determine concentration of soluble salts on metal surfaces prior to coating application, such as painting. These salts can cause serious adhesion problems after time.

Importance

Salt is ubiquitous in coastal areas. It can be tasted on the lips after walking on a beach. Salt concentration by weight is about 3.5% in sea water. With spray from waves and by other means, salt gets into the air as an aerosol, and eventually as a dust-like particle. This salt dust can be found everywhere near the coast. Salt has the property of being hygroscopic, and this property makes it harmful to coatings.

Salt contamination beneath a coating, such as paint on steel, can cause adhesion and corrosion problems due to the hygroscopic nature of salt. Its tendency to attract water through a permeable coating creates a build-up of water molecules between substrate and coating. These, together with salt and other oxidation agents entrapped during coating or migrating later through the coating, create an electrolytic cell, causing corrosion. Blast cleaning is frequently used to clean surfaces before coating; however, with salt contamination, blast cleaning may increase the problem by forcing salt into the base material. Washing a surface with deionized water before coating is a common solution.

IMO PSPC (performance standard for protective coatings) regulations set a maximum allowable concentration of soluble salts on a surface to be coated, measured as sodium chloride, of $50\,mg \cdot m^{-2}$. The maximum amount of salt allowed on a surface prior to coating application is typically determined by the coating supplier and the user, such as a shipyard. Standard values have not been established.

Origin of the Bresle Method

The Bresle method was launched in 1995 in the ISO 8502-6 and ISO 8502-9 standards. The test was developed to measure soluble salt concentration on steel surfaces prior to blasting cleaning and coating. Not only ISO, but also the US Navy, IMO, NAVSEA, and ASTM adopted this method as their standard. The method remains the primary and most flexible test method for soluble salts on metal surfaces.

Principle

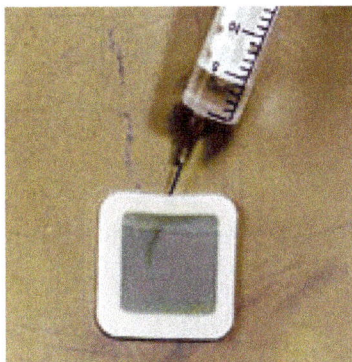

Bresle patch injected

The Bresle method uses the difference of conductivity of salts in water, each salt having a characteristic conductivity-versus-concentration relationship. The correlation between concentration and conductivity can be found in "Handbook of Chemistry and Physics". This relationship is useful only if the dissolved salt is known. Sodium chloride, the main salt in sea water, causes a big increase in conductivity with increased concentration.

A special patch is applied to the surface to be tested, and a specified volume of deionized water is injected under the patch. Any soluble salts present on the surface will dissolve in the water. The fluid is extracted and its conductivity measured.

The conductivity of the collected salt solution depends on the volume of water used and its initial conductivity, and the amount of salt in solution depends on the area of the patch. The calculation of the salt per area is based on increased conductivity but in the IMO PSPC method the salt is calculated as sodium chloride, in the ISO 8502-9 method it is calculated as a specific mixture of salts, but expressed as Sodium Chloride.

Calculation Factors

Factors are applied to the measured conductivity, depending on what is known or assumed about the salt contamination and various conditions, in order to yield meaningful measurements of contamination. Some of the variables are:

- salt type
- volume of solution (in below table factors are calculated with a volume of 15 ml solution)
- temperature
- equipment-specific scaling

A common source of error is not knowing the composition of the contamination being measured.

Conductivity µS/cm	Conductivity mS/m	As chloride µg/cm2	As chloride mg/m2	As sodium chloride µg/cm2	As sodium chloride mg/m2	As mixed salts µg/cm2	As mixed salts mg/m2
1	0.1	0.36	3.6	0.6	6	0.5	5
5	0.5	1.8	18	3	30	2.5	25
10	1	3.6	36	6	60	5	50
20	2	7.2	72	12	120	10	100

Measurement Tools

There are multiple suppliers of Bresle method test kits.

The Principle

The solubility in water depends on the type of salt. Sodium chloride can be dissolved in cold water to a concentration of 357 g·l^{-1}. Not only solubility differs between salts but also the conductivity. When performing a Bresle method test, not only sodium chloride is dissolved but also all other salts present on the surface. Because it is impossible to predict which salts are present at the surface, an assumption is made in the Bresle method. The term "measured as sodium chloride" indicates that this mixture of salts is interpreted as sodium chloride. Reporting how conductivity is factored is essential when creating a report.

In Practice

All parties involved should be clear on the impact on results of climate and the variance of the potentially different salt contents. An informed agreement should be reached between all parties as to what is an acceptable level of reading. Dependent on the size and nature of the surface to be coated, several readings may need to be taken.

Test Patches

A test patch should be as clean as possible. Contamination of a patch can influence the results significantly.

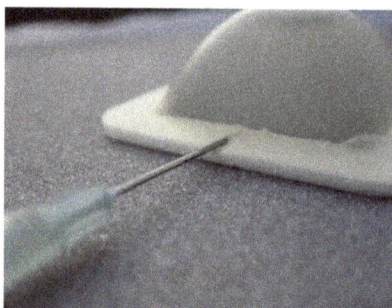

Bresle patch injected with 15ml water as in ISO 8502-6 Annex A

The ISO 8502-6 standard prescribes in annex A that certified patches shall be used. This annex describes a stress test to ensure patch adhesion and wash ability. Not being supplied with a certificate that the patches pass this test will render the results obtained by these patches useless.

Climate

A soluble salts report should include climate conditions and substrate temperature. ISO 8502-6 requires that the test is done at 23 °C and a relative humidity of 50%, with deviations reported and agreed upon by both inspector and customer. During arbitration, absence of these values in a report may render the results invalid.

Acid Salt

$$Na^+ \quad {}^-O \diagdown \underset{\underset{O}{\overset{\|}{C}}}{} \diagup OH$$

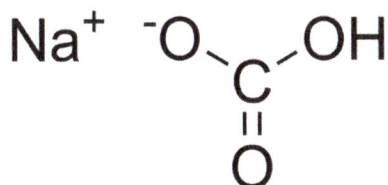

Structure of Sodium Bicarbonate

Acid salt is a term for a class of salts formed by the partial neutralization of diprotic or polyprotic acids. Because the parent acid is only partially neutralized, one or more replaceable hydrogen atoms remain. Typical acid salts have one or more alkali (alkaline) metal ions as well as one or more hydrogen atoms. Well known examples are sodium bicarbonate ($NaHCO_3$), sodium hydrosulfide ($NaHS$), sodium bisulfate ($NaHSO_4$), monosodium phosphate (NaH_2PO_4), and disodium phosphate (Na_2HPO_4). Often acid salts are used as buffers.

For example, the acid salt sodium bisulfate is the main species formed upon the *half* neutralization of sulfuric acid with sodium hydroxide:

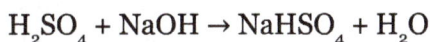

$$H_2SO_4 + NaOH \rightarrow NaHSO_4 + H_2O$$

An acid salt can act either as an acid or as a base: addition of a suitably strong acid will protonate anions, and addition of a suitably strong base will split off H^+. The pH of a solution of an acid salt will depend on the relevant equilibrium constants and the amounts of any additional base or acid. A comparison between the K_b and K_a will indicate this: if $K_b > K_a$, the solution will be basic, whereas if $K_b < K_a$, the solution will be acidic.

Use in Food

Some acid salts are used in baking. They are found in baking powders and are typically divided into low-temperature (or single-acting) and high-temperature (or double-acting) acid salts. Common low-temperature acid salts react at room temperature to produce a leavening effect. They include cream of tartar, calcium phosphate, and citrates. High-temperature acid salts produce a leavening effect during baking and are usually aluminium salts such as calcium aluminium phosphate. Some acid salts may also be found in non-dairy coffee creamers.

Baking Powder

U.S. consumer-packaged baking powder. This particular type of baking powder contains monocalcium phosphate, sodium bicarbonate, and cornstarch.

Baking powder is a dry chemical leavening agent, a mixture of a carbonate or bicarbonate and a weak acid, and is used for increasing the volume and lightening the texture of baked goods. Baking powder works by releasing carbon dioxide gas into a batter or dough through an acid-base reaction, causing bubbles in the wet mixture to expand and thus leavening the mixture. It is used instead of yeast for end-products where fermentation flavors would be undesirable or where the batter lacks the elastic structure to hold gas bubbles for more than a few minutes, or to speed the production. Because carbon dioxide is released at a faster rate through the acid-base reaction than through fermentation, breads made by chemical leavening are called quick breads.

Formulation and Mechanism

Most commercially available baking powders are made up of sodium bicarbonate (also known as baking soda or bicarbonate of soda) and one or more acid salts. Typical formulations (by weight) call for 30% sodium bicarbonate, 5-12% monocalcium phosphate, and 21-26% sodium aluminum sulfate. The last two ingredients are acidic: they combine with the sodium bicarbonate and water to produce the gaseous carbon dioxide. The use of two acidic components is the basis of the term "double acting." Another typical acid in such formulations is cream of tartar, a derivative of tartaric acid. Baking powders also include components to help with the consistency and stability of the mixture.

Commercial baking powder formulations are different from domestic ones, although the principles remain the same. Instead of sodium aluminum sulfate, commercial baking powders use sodium acid pyrophosphate as one of the two acidic components.

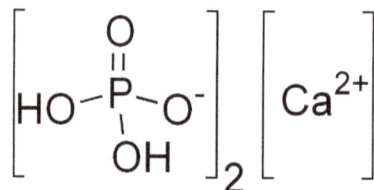

Monocalcium phosphate ("MCP") is a common acid component in domestic baking powders.

Baking soda ($NaHCO_3$) is the source of the carbon dioxide, and the acid-base reaction can be generically represented as shown:

$$NaHCO_3 + H^+ \rightarrow Na^+ + CO_2 + H_2O$$

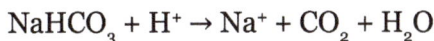

The real reactions are more complicated because the acids are complicated. For example, starting with baking soda and monocalcium phosphate the reaction produces carbon dioxide by the following stoichiometry:

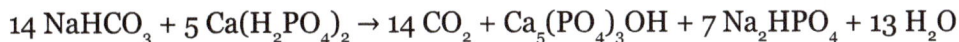

$$14\ NaHCO_3 + 5\ Ca(H_2PO_4)_2 \rightarrow 14\ CO_2 + Ca_5(PO_4)_3OH + 7\ Na_2HPO_4 + 13\ H_2O$$

Starch Component

Baking powders also include components to improve their consistency and stability. The most important additive is cornstarch, although potato starch may also be used. The inert starch serves several functions in baking powder. Primarily it is used to absorb moisture, and thus prolong shelf life by keeping the powder's additional alkaline and acidic components dry so as not to react with each other prematurely. A dry powder also flows and mixes more easily. Finally, the added bulk allows for more accurate measurements.

Single vs Double Acting Baking Powders

The acid in a baking powder can be either fast-acting or slow-acting. A fast-acting acid reacts in a wet mixture with baking soda at room temperature, and a slow-acting acid will not react until heated in an oven. Baking powders that contain both fast- and slow-acting acids are *double acting*; those that contain only one acid are *single acting*. By providing a second rise in the oven, double-acting baking powders increase the reliability of baked goods by rendering the time elapsed between mixing and baking less critical, and this is the type most widely available to consumers today. Double-acting baking powders work in two phases; once when cold, and once when hot. Common low-temperature acid salts include cream of tartar and monocalcium phosphate (also called calcium acid phosphate). High-temperature acid salts include sodium aluminium sulfate, sodium aluminum phosphate, and sodium acid pyrophosphate. Despite very low acute toxicity, exposure to aluminum salts has raised concerns regarding potential neurotoxicity.

For example, Rumford Baking Powder is a double acting consumer product that contains only monocalcium phosphate as a leavening acid. With this acid, about two-thirds of the available gas is released within about two minutes of mixing at room temperature. It then becomes dormant because an intermediate form of dicalcium phosphate is generated during the initial mixing. Further release of gas requires the batter to be heated above 140 °F (60 °C).

History

Early chemical leavening was accomplished by activating baking soda in the presence of liquid(s) and an acid such as sour milk, vinegar, lemon juice, or cream of tartar. These acidulants all react with baking soda quickly, meaning that retention of gas bubbles was dependent on batter viscosity and that it was critical for the batter to be baked before the gas escaped. The development of baking powder created a system where the gas-producing reactions could be delayed until needed.

While various baking powders were sold in the first half of the 19th century, the modern variants in use today were discovered by Alfred Bird in 1843. August Oetker, a German pharmacist, made baking powder very popular when he began selling his mixture to housewives. The recipe he created in 1891 is still sold as *Backin* in Germany. Oetker started the mass production of baking powder in 1898 and patented his technique in 1903.

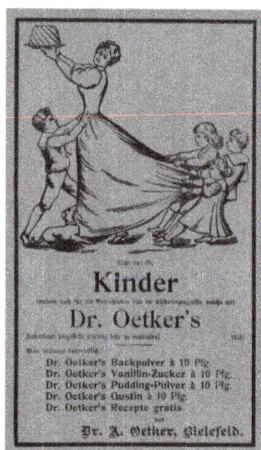

German advertisement for Dr. Oetker's baking powder in 1903.

Advertisement for DeLand & Co's Chemical Baking Powder (Established 1852, Fairport, New York.) Earliest possible date: 1877; latest possible date: 1893

In the US, Joseph and Cornelius Hoagland developed a baking powder with the help of an employee following the American Civil War. Their formula became known as Royal Baking Powder. The small company eventually moved from Fort Wayne, Indiana to New York in the 1890s and became the largest manufacturer of baking powder in the US.

Eben Norton Horsford, a student of Justus von Liebig, who began his studies on baking powder in 1856, eventually developed a variety he named in honor of Count Rumford. By the mid-1860s "Horsford's Yeast Powder" was on the market as an already-mixed leavening agent, distinct from separate packages of calcium acid phosphate and sodium bicarbonate. His product was packaged in bottles, but Horsford was interested in using metal cans for packing; this meant the mixture had to be more moisture resistant. This was accomplished by the addition of corn starch, and in 1869 Rumford began the manufacture of what can truly be considered baking powder. In 2006 Rumford Baking Powder was designated a National Historic Chemical Landmark in recognition of its significance for making baking easier, quicker, and more reliable.

During World War II, Byron H. Smith, an inventor in Bangor, Maine (U.S.), created a substitute product for American housewives, who were unable to obtain cream of tartar or baking powder due to war food shortages. Under the name "Bakewell", Smith marketed a mixture of sodium pyrophosphate mixed with corn starch to replace the acid cream of tartar component of baking powder. When mixed with baking soda, the product behaved like a single-acting baking powder, the only difference being that the acid is sodium acid pyrophosphate. A similar product is marketed today, under the name Bakewell Cream.

Use

Generally one teaspoon (5 grams (0.18 oz)) of baking powder is used to raise a mixture of one cup (125 g) of flour, one cup of liquid, and one egg. However, if the mixture is acidic, baking powder's additional acids will remain unconsumed in the chemical reaction and often lend an unpleasant taste to food. High acidity can be caused by ingredients like buttermilk, lemon juice, yogurt, citrus or honey. When excessive acid is present, some of the baking powder should be replaced with baking soda. For example, one cup of flour, one egg, and one cup of buttermilk requires only ½ teaspoon of baking powder—the remaining leavening is caused by buttermilk acids reacting with ¼ teaspoon of baking soda.

Effective baking powder foams when placed in hot water.

On the other hand, with baking powders that contain sodium acid pyrophosphate, excess alkaline substances can sometimes deprotonate the acid in two steps instead of the one that normally occurs, resulting in an offensive bitter taste to baked goods. Calcium compounds and aluminium compounds do not have that problem though, since calcium compounds that deprotonate twice are insoluble and aluminium compounds do not deprotonate in that fashion.

Moisture and heat can cause baking powder to lose its effectiveness over time, and commercial varieties have a somewhat arbitrary expiration date printed on the container. Regardless of the expiration date, the effectiveness can be tested by placing a teaspoon of the powder into a small container of hot water. If it bubbles energetically, it is still active and usable.

Different brands of baking powder can perform quite differently in the oven. In one test, six U.S. brands were used to bake white cake, cream biscuits, and chocolate cookies. Depending on the brand, the thickness of the cakes varied by up to 20 percent (from 0.89 inches to 1.24 inches). It was also found that the lower-rising products made what were judged to be better chocolate cookies.

Substituting in Recipes

Substitute Acids

As described above, baking powder is mainly just baking soda mixed with an acid. In principle, a number of kitchen acids may be combined with baking soda to simulate commercial baking powders. Vinegar (dilute acetic acid), especially white vinegar, is also a common acidifier in baking; for example, many heirloom chocolate cake recipes call for a tablespoon or two of vinegar. Where a recipe already uses buttermilk or yogurt, baking soda can be used without cream of tartar (or with less). Alternatively, lemon juice can be substituted for some of the liquid in the recipe, to provide the required acidity to activate the baking soda. The main variable with the use of these kitchen acids is the rate of leavening.

Substitutes for Baking Soda

In times past, when chemically manufactured baking soda was not available, "ash water" was used instead. Ashes from hardwood trees contain carbonates and bicarbonate salts, which can be extracted with water. This approach became obsolete with the availability of purified baking soda.

Usage of Aluminium Compounds

Baking powders are available both with and without aluminium compounds. Some people prefer not to use baking powder with aluminium because they believe it gives food a vaguely metallic taste and aluminium is not an essential mineral. Others object because of possible health concerns associated with aluminium intake. In 2015, Cook's Country, an American TV show and magazine, evaluated six baking powders marketed to consumers. They reported that 30% of their testers (n=21) noted a metallic flavor in cream biscuits made with brands containing aluminum.

Alkali Salt

Alkali salts or basic salts are salts that are the product of the neutralization of a strong base and a weak acid.

Rather than being neutral (as some other salts), alkali salts are bases as their name suggests. What makes these compounds basic is that the conjugate base from the weak acid hydrolyzes to form a basic solution. In sodium carbonate, for example, the carbonate from the carbonic acid hydrolyzes to form a basic solution. The chloride from the hydrochloric acid in sodium chloride does not hydrolyze, though, so sodium chloride is not basic.

The difference between a basic salt and an alkali is that an alkali is the soluble hydroxide compound of an alkali metal or an alkaline earth metal. A basic salt is any salt that hydrolyzes to form a basic solution. The hydroxide compounds are not salts.

Another definition of a basic salt would be a salt that contains amounts of both hydroxide and other anions. White lead is an example. It is basic lead carbonate, or lead carbonate hydroxide.

These salts are insoluble and are obtained through precipitation reactions.

Examples

- Sodium carbonate
- Sodium acetate
- Potassium cyanide
- Sodium sulfide

Alkaline Salts

'Alkaline salts' are often the major component of alkaline dishwasher detergent powders.

These salts may include:

- alkali metasilicates

- alkali metal hydroxides

- Sodium carbonate

- Sodium Bicarbonate

Sodium Carbonate

Sodium carbonate (also known as washing soda, soda ash and soda crystals), Na_2CO_3, is the water-soluble sodium salt of carbonic acid.

It most commonly occurs as a crystalline decahydrate, which readily effloresces to form a white powder, the monohydrate. Pure sodium carbonate is a white, odorless powder that is hygroscopic (absorbs moisture from the air). It has a strongly alkaline taste, and forms a moderately basic solution in water. Sodium carbonate is well known domestically for its everyday use as a water softener. It can be extracted from the ashes of many plants growing in sodium-rich soils, such as vegetation from the Middle East, kelp from Scotland and seaweed from Spain. Because the ashes of these sodium-rich plants were noticeably different from ashes of timber (used to create potash), they became known as "soda ash". It is synthetically produced in large quantities from salt (sodium chloride) and limestone by a method known as the Solvay process.

The manufacture of glass is one of the most important uses of sodium carbonate. Sodium carbonate acts as a flux for silica, lowering the melting point of the mixture to something achievable without special materials. This "soda glass" is mildly water-soluble, so some calcium carbonate is added to the melt mixture to make the glass produced insoluble. This type of glass is known as soda lime glass: "soda" for the sodium carbonate and "lime" for the calcium carbonate. Soda lime glass has been the most common form of glass for centuries.

Sodium carbonate is also used as a relatively strong base in various settings. For example, it is used as a pH regulator to maintain stable alkaline conditions necessary for the action of the majority of photographic film developing agents. It acts as an alkali because when dissolved in water, it dissociates into the weak acid: carbonic acid and the strong alkali: sodium hydroxide. This gives sodium carbonate in solution the ability to attack metals such as aluminium with the release of hydrogen gas.

It is a common additive in swimming pools used to neutralize the corrosive effects of chlorine and raise the pH.

In cooking, it is sometimes used in place of sodium hydroxide for lyeing, especially with German pretzels and lye rolls. These dishes are treated with a solution of an alkaline substance to change the pH of the surface of the food and improve browning.

In taxidermy, sodium carbonate added to boiling water will remove flesh from the skull or bones of trophies to create the "European skull mount" or for educational display in biological and historical studies.

In chemistry, it is often used as an electrolyte. Electrolytes are usually salt-based, and sodium carbonate acts as a very good conductor in the process of electrolysis. In addition, unlike chloride ions, which form chlorine gas, carbonate ions are not corrosive to the anodes. It is also used as a primary standard for acid-base titrations because it is solid and air-stable, making it easy to weigh accurately.

Domestic Use

It is used as a water softener in laundering: it competes with the magnesium and calcium ions in hard water and prevents them from bonding with the detergent being used, but doesn't prevent scaling. Sodium carbonate can be used to remove grease, oil, and wine stains.

In dyeing with fiber-reactive dyes, sodium carbonate (often under a name such as soda ash fixative or soda ash activator) is used to ensure proper chemical bonding of the dye with cellulose (plant) fibers, typically before dyeing (for tie dyes), mixed with the dye (for dye painting), or after dyeing (for immersion dyeing).

Sodium Carbonate Test

The sodium carbonate test (not to be confused with sodium carbonate extract test) is used to distinguish between some common metal ions, which are precipitated as their respective carbonates. The test can distinguish between copper (Cu), iron (Fe), and calcium (Ca), zinc (Zn) or lead (Pb). Sodium carbonate solution is added to the salt of the metal. A blue precipitate indicates Cu^{2+} ion. A dirty green precipitate indicates Fe^{2+} ion. A yellow-brown precipitate indicates Fe^{3+} ion. A white precipitate indicates Ca^{2+}, Zn^{2+}, or Pb^{2+} ion. The compounds formed are, respectively, copper(II) carbonate, iron(II) carbonate, iron(III) oxide, calcium carbonate, zinc carbonate, and lead(II) carbonate. This test is used to precipitate the ion present as almost all carbonates are insoluble. While this test is useful for telling these cations apart, it fails if other ions are present, because most metal carbonates are insoluble and will precipitate. In addition, calcium, zinc, and lead ions all produce white precipitates with carbonate, making it difficult to distinguish between them. Instead of sodium carbonate, sodium hydroxide may be added, this gives nearly the same colours, except that lead and zinc hydroxides are soluble in excess alkali, and can hence be distinguished from calcium. For the complete sequence of tests used for qualitative cation analysis, see qualitative inorganic analysis.

Other Applications

Sodium carbonate is a food additive (E500) used as an acidity regulator, anticaking agent, raising agent, and stabilizer. It is one of the components of *kansui* (かん水?), a solution of alkaline salts used to give ramen noodles their characteristic flavor and texture. It is also used in the production of *snus* (Swedish-style snuff) to stabilize the pH of the final product.

Sodium carbonate is also used in the production of sherbet powder. The cooling and fizzing sensation results from the endothermic reaction between sodium carbonate and a weak acid, commonly citric acid, releasing carbon dioxide gas, which occurs when the sherbet is moistened by saliva.

In China, it is used to replace lye-water in the crust of traditional Cantonese moon cakes, and in many other Chinese steamed buns and noodles.

Sodium carbonate is used by the brick industry as a wetting agent to reduce the amount of water needed to extrude the clay.

In casting, it is referred to as "bonding agent" and is used to allow wet alginate to adhere to gelled alginate.

Sodium carbonate is used in toothpastes, where it acts as a foaming agent and an abrasive, and to temporarily increase mouth pH.

Sodium carbonate is used by the cotton industry to neutralize the sulfuric acid needed for acid delinting of fuzzy cottonseed.

Sodium carbonate, in a solution with common salt, may be used for cleaning silver. In a nonreactive container (glass, plastic, or ceramic), aluminium foil and the silver object are immersed in the hot salt solution. The elevated pH dissolves the aluminium oxide layer on the foil and enables an electrolytic cell to be established. Hydrogen ions produced by this reaction reduce the sulfide ions on the silver restoring silver metal. The sulfide can be released as small amounts of hydrogen sulfide. Rinsing and gently polishing the silver restores a highly polished condition.

Sodium carbonate is used in some aquarium water pH buffers to maintain a desired pH and carbonate hardness (KH).

Because of its ability to absorb CO_2, sodium carbonate is being investigated as a carbon-capturing material for power plants and in other industries that produce greenhouse gases.

Physical Properties

The integral enthalpy of solution of sodium carbonate is −28.1 kJ/mol for a 10% w/w aqueous solution. The Mohs hardness of sodium carbonate monohydrate is 1.3.

Hydrates

Sodium carbonate crystallizes from water to form three different hydrates:

- sodium carbonate decahydrate (natron), $Na_2CO_3 \cdot 10H_2O$.

- sodium carbonate heptahydrate (not known in mineral form), $Na_2CO_3 \cdot 7H_2O$.

- sodium carbonate monohydrate (thermonatrite), $Na_2CO_3 \cdot H_2O$.

- anhydrous sodium carbonate also known as calcined soda is formed by heating the hydrates. It is also formed when sodium hydrogen carbonate is heated (calcined) e.g. in the final step of the Solvay process. The decahydrate is formed from water solutions crystallizing in the temperature range -2.1 to +32.0 C, the heptahydrate in the narrow range 32.0 to 35.4 C and above this temperature the monohydrate forms. In dry air the decahydrate and heptahydrate will lose water forming causing the crystals to fall apart into a white monohydrate powder. Other hydrates have been reported, e.g. with 2.5 units of water per sodium carbonate unit ("pentahemihydrate").

Occurrence as Natural Mineral

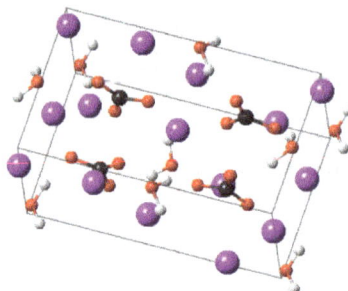

Structure of monohydrate at 346 K.

Sodium carbonate is soluble in water, and can occur naturally in arid regions, especially in mineral deposits (*evaporites*) formed when seasonal lakes evaporate. Deposits of the mineral natron have been mined from dry lake bottoms in Egypt since ancient times, when natron was used in the preparation of mummies and in the early manufacture of glass.

The anhydrous mineral form of sodium carbonate is quite rare and called natrite. Sodium carbonate also erupts from Ol Doinyo Lengai, Tanzania's unique volcano, and it is presumed to have erupted from other volcanoes in the past, but due to these minerals' instability at the earth's surface, are likely to be eroded. All three mineralogical forms of sodium carbonate, as well as trona, trisodium hydrogendicarbonate dihydrate, are also known from ultra-alkaline pegmatitic rocks, that occur for example in the Kola Peninsula in Russia.

Production

Mining

Trona, trisodium hydrogendicarbonate dihydrate ($Na_3HCO_3CO_3 \cdot 2H_2O$), is mined in several areas of the US and provides nearly all the domestic consumption of sodium carbonate. Large natural deposits found in 1938, such as the one near Green River, Wyoming, have made mining more economical than industrial production in North America. There are important reserves of trona in Turkey; two million tons of soda ash have been extracted from the reserves near Ankara. It is also mined from some alkaline lakes such as Lake Magadi in Kenya by dredging. Hot saline springs continuously replenish salt in the lake so that, provided the rate of dredging is no greater than the replenishment rate, the source is fully sustainable.

Barilla and Kelp

Several "halophyte" (salt-tolerant) plant species and seaweed species can be processed to yield an impure form of sodium carbonate, and these sources predominated in Europe and elsewhere until the early 19th century. The land plants (typically glassworts or saltworts) or the seaweed (typically *Fucus* species) were harvested, dried, and burned. The ashes were then "lixiviated" (washed with water) to form an alkali solution. This solution was boiled dry to create the final product, which was termed "soda ash"; this very old name refers to the archetypal plant source for soda ash, which was the small annual shrub *Salsola soda* ("barilla plant").

The sodium carbonate concentration in soda ash varied very widely, from 2–3 percent for the seaweed-derived form ("kelp"), to 30 percent for the best barilla produced from saltwort plants in Spain. Plant and seaweed sources for soda ash, and also for the related alkali "potash", became increasingly inadequate by the end of the 18th century, and the search for commercially viable routes to synthesizing soda ash from salt and other chemicals intensified.

Leblanc Process

In 1791, the French chemist Nicolas Leblanc patented a process for producing sodium carbonate from salt, sulfuric acid, limestone, and coal. First, sea salt (sodium chloride) was boiled in sulfuric acid to yield sodium sulfate and hydrogen chloride gas, according to the chemical equation

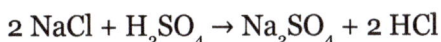

$$2\ NaCl + H_2SO_4 \rightarrow Na_2SO_4 + 2\ HCl$$

Next, the sodium sulfate was blended with crushed limestone (calcium carbonate) and coal, and the mixture was burnt, producing calcium sulfide.

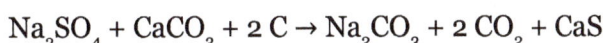

$$Na_2SO_4 + CaCO_3 + 2\ C \rightarrow Na_2CO_3 + 2\ CO_2 + CaS$$

The sodium carbonate was extracted from the ashes with water, and then collected by allowing the water to evaporate.

The hydrochloric acid produced by the Leblanc process was a major source of air pollution, and the calcium sulfide byproduct also presented waste disposal issues. However, it remained the major production method for sodium carbonate until the late 1880s.

Solvay Process

In 1861, the Belgian industrial chemist Ernest Solvay developed a method to convert sodium chloride to sodium carbonate using ammonia. The Solvay process centered around a large hollow tower. At the bottom, calcium carbonate (limestone) was heated to release carbon dioxide:

$$CaCO_3 \rightarrow CaO + CO_2$$

At the top, a concentrated solution of sodium chloride and ammonia entered the tower. As the carbon dioxide bubbled up through it, sodium bicarbonate precipitated:

$$NaCl + NH_3 + CO_2 + H_2O \rightarrow NaHCO_3 + NH_4Cl$$

The sodium bicarbonate was then converted to sodium carbonate by heating it, releasing water and carbon dioxide:

$$2\ NaHCO_3 \rightarrow Na_2CO_3 + H_2O + CO_2$$

Meanwhile, the ammonia was regenerated from the ammonium chloride byproduct by treating it with the lime (calcium hydroxide) left over from carbon dioxide generation:

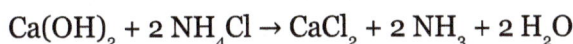

$$CaO + H_2O \rightarrow Ca(OH)_2$$

$$Ca(OH)_2 + 2\ NH_4Cl \rightarrow CaCl_2 + 2\ NH_3 + 2\ H_2O$$

Because the Solvay process recycles its ammonia, it consumes only brine and limestone, and has calcium chloride as its only waste product. This made it substantially more economical than the Leblanc process, and it soon came to dominate world sodium carbonate production. By 1900, 90% of sodium carbonate was produced by the Solvay process, and the last Leblanc process plant closed in the early 1920s.The Solvay process results in soda ash (predominantly sodium carbonate (Na_2CO_3)) from brine (as a source of sodium chloride ($NaCl$)) and from limestone (as a source of calcium carbonate ($CaCO_3$)). The overall process is:

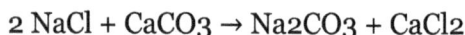

$$2\ NaCl + CaCO_3 \rightarrow Na_2CO_3 + CaCl_2$$

The actual implementation of this global, overall reaction is intricate. A simplified description can be given using the four different, interacting chemical reactions illustrated in the figure. In the first step in the process, carbon dioxide (CO_2) passes through a concentrated aqueous solution of sodium chloride (table salt, $NaCl$) and ammonia (NH_3).

$$NaCl + CO_2 + NH_3 + H_2O \rightarrow NaHCO_3 + NH_4Cl\ (I)$$

In industrial practice, the reaction is carried out by passing concentrated brine through two towers. In the first, ammonia bubbles up through the brine (salt water) and is absorbed by it. In the second, carbon dioxide bubbles up through the ammoniated brine, and sodium bicarbonate (baking soda) precipitates out of the solution. Note that, in a basic solution, $NaHCO_3$ is less water-soluble than sodium chloride. The ammonia (NH_3) buffers the solution at a basic pH; without the ammonia, a hydrochloric acid byproduct would render the solution acidic, and arrest the precipitation.

The necessary ammonia "catalyst" for reaction (I) is reclaimed in a later step, and relatively little ammonia is consumed. The carbon dioxide required for reaction (I) is produced by heating ("calcination") of the limestone at 950 - 1100 °C. The calcium carbonate ($CaCO_3$) in the limestone is partially converted to quicklime (calcium oxide (CaO)) and carbon dioxide:

$$CaCO_3 \rightarrow CO_2 + CaO\ (II)$$

The sodium bicarbonate ($NaHCO_3$) that precipitates out in reaction (I) is filtered out from the hot ammonium chloride (NH_4Cl) solution, and the solution is then reacted with the quicklime (calcium oxide (CaO)) left over from heating the limestone in step (II).

$$2\ NH_4Cl + CaO \rightarrow 2\ NH_3 + CaCl_2 + H_2O\ (III)$$

CaO makes a strong basic solution. The ammonia from reaction (III) is recycled back to the initial brine solution of reaction (I).

The sodium bicarbonate ($NaHCO_3$) precipitate from reaction (I) is then converted to the final product, sodium carbonate (washing soda: Na_2CO_3), by calcination (160 - 230 C), producing water and carbon dioxide as byproducts:

$$2\ NaHCO_3 \rightarrow Na_2CO_3 + H_2O + CO_2\ (IV)$$

The carbon dioxide from step (IV) is recovered for re-use in step (I). When properly designed and operated, a Solvay plant can reclaim almost all its ammonia, and consumes only small amounts of additional ammonia to make up for losses. The only major inputs to the Solvay process are salt, limestone and thermal energy, and its only major byproduct is calcium chloride, which is sold as road salt.

Hou's Process

This process was developed by Chinese chemist Hou Debang in the 1930s. The earlier steam reforming byproduct carbon dioxide was pumped through a saturated solution of sodium chloride and ammonia to produce sodium bicarbonate by these reactions:

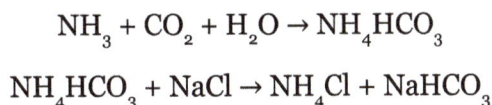

$$NH_3 + CO_2 + H_2O \rightarrow NH_4HCO_3$$

$$NH_4HCO_3 + NaCl \rightarrow NH_4Cl + NaHCO_3$$

The sodium bicarbonate was collected as a precipitate due to its low solubility and then heated to yield pure sodium carbonate similar to last step of the Solvay process. More sodium chloride is added to the remaining solution of ammonium and sodium chlorides; also, more ammonia is pumped at 30-40 °C to this solution. The solution temperature is then lowered to below 10 °C. Solubility of ammonium chloride is higher than that of sodium chloride at 30 °C and lower at 10 °C. Due to this temperature-dependent solubility difference and the common-ion effect, ammonium chloride is precipitated in a sodium chloride solution.

The Chinese name of Hou's process, *lianhe zhijian fa* (联合制碱法), means "coupled manufacturing alkali method": Hou's process is coupled to the Haber process and offers better atom economy by eliminating the production of calcium chloride, since ammonia no longer needs to be regenerated. The byproduct ammonium chloride can be sold as a fertilizer.

Soap

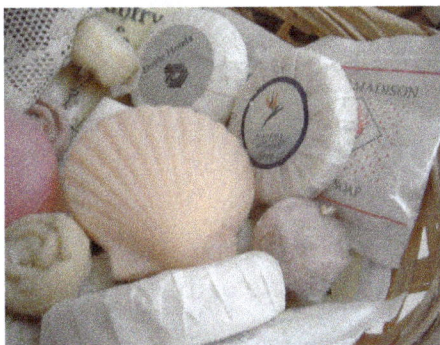

A collection of decorative soaps, as often found in hotels

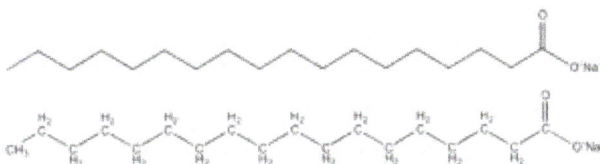

Two equivalent images of the chemical structure of sodium stearate, a typical soap.

In chemistry, a soap is a salt of a fatty acid. Household uses for soaps include washing, bathing, and other types of housekeeping, where soaps act as surfactants, emulsifying oils to enable them to be carried away by water. In industry they are also used in textile spinning and are important components of some lubricants.

Soaps for cleaning are obtained by treating vegetable or animal oils and fats with a strong base, such as sodium hydroxide or potassium hydroxide in an aqueous solution. Fats and oils are composed of triglycerides; three molecules of fatty acids attach to a single molecule of glycerol. The alkaline solution, which is often called lye (although the term "lye soap" refers almost exclusively to soaps made with sodium hydroxide), brings about a chemical reaction known as saponification.

In this reaction, the triglyceride fats first hydrolyze into free fatty acids, and then these combine with the alkali to form crude soap: an amalgam of various soap salts, excess fat or alkali, water, and liberated glycerol (glycerin). The glycerin, a useful by-product, can remain in the soap product as a softening agent, or be isolated for other uses.

Soaps are key components of most lubricating greases, which are usually emulsions of calcium soap or lithium soap and mineral oil. Many other metallic soaps are also useful, including those of aluminium, sodium, and mixtures of them. Such soaps are also used as thickeners to increase the viscosity of oils. In ancient times, lubricating greases were made by the addition of lime to olive oil.

Mechanism of Cleansing Soaps

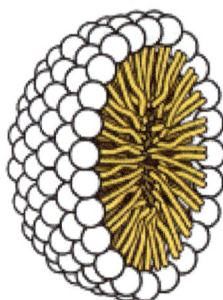

Structure of a micelle, a cell-like structure formed by the aggregation of soap subunits (such as sodium stearate): The exterior of the micelle is hydrophilic (attracted to water) and the interior is lipophilic (attracted to oils).

Action of Soap

When used for cleaning, soap allows insoluble particles to become soluble in water, so they can then be rinsed away. For example: oil/fat is insoluble in water, but when a couple of drops of dish soap are added to the mixture, the oil/fat solubilizes into the water. The insoluble oil/fat molecules become associated inside micelles, tiny spheres formed from soap molecules with polar hydrophilic (water-attracting) groups on the outside and encasing a lipophilic (fat-attracting) pocket, which shields the oil/fat molecules from the water making it soluble. Anything that is soluble will be washed away with the water.

Effect of the Alkali

The type of alkali metal used determines the kind of soap product. Sodium soaps, prepared from sodium hydroxide, are firm, whereas potassium soaps, derived from potassium hydroxide, are softer or often liquid. Historically, potassium hydroxide was extracted from the ashes of bracken or other plants. Lithium soaps also tend to be hard—these are used exclusively in greases.

Effects of Fats

Soaps are derivatives of fatty acids. Traditionally they have been made from triglycerides (oils and fats). Triglyceride is the chemical name for the triesters of fatty acids and glycerin. Tallow, *i.e.,* rendered beef fat, is the most available triglyceride from animals. Its saponified product is called sodium tallowate. Typical vegetable oils used in soap making are palm oil, coconut oil, olive oil, and laurel oil. Each species offers quite different fatty acid content and hence, results in soaps of distinct feel. The seed oils give softer but milder soaps. Soap made from pure olive oil is sometimes called Castile soap or Marseille soap, and is reputed for being extra mild. The term "Castile" is also sometimes applied to soaps from a mixture of oils, but a high percentage of olive oil.

Fatty acid content of various fats used for soapmaking							
fats	Lauric acid	Myristic acid	Palmitic acid	Stearic acid	Oleic acid	Linoleic acid	Linolenic acid
	C_{12} saturated	C_{14} saturated	C_{16} saturated	C_{18} saturated	C_{18} mono-unsaturated	C_{18} diunsaturated	C_{18} triunsaturated
Tallow	0	4	28	23	35	2	1
Coconut oil	48	18	9	3	7	2	0
Palm kernel oil	46	16	8	3	12	2	0
Laurel oil	54	0	0	0	15	17	0
Olive oil	0	0	11	2	78	10	0
Canola oil	0	1	3	2	58	9	23

History of soaps

Early history

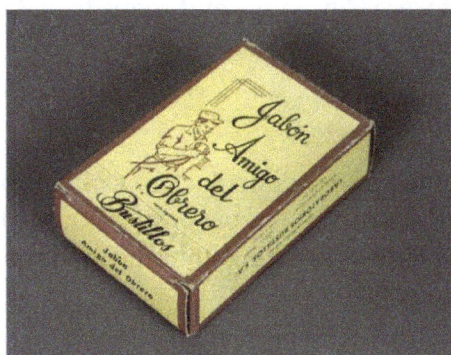

Box for Amigo del Obrero (Worker's Friend) soap from the 20th century, part of the Museo del Objeto del Objeto collection

The earliest recorded evidence of the production of soap-like materials dates back to around 2800 BC in ancient Babylon. A formula for soap consisting of water, alkali, and cassia oil was written on a Babylonian clay tablet around 2200 BC.

The Ebers papyrus (Egypt, 1550 BC) indicates the ancient Egyptians bathed regularly and combined animal and vegetable oils with alkaline salts to create a soap-like substance. Egyptian documents mention a soap-like substance was used in the preparation of wool for weaving.

In the reign of Nabonidus (556–539 BC), a recipe for soap consisted of *uhulu* [ashes], cypress [oil] and sesame [seed oil] "for washing the stones for the servant girls".

Ancient Roman Era

The word *sapo*, Latin for soap, first appears in Pliny the Elder's *Historia Naturalis*, which discusses the manufacture of soap from tallow and ashes, but the only use he mentions for it is as a pomade for hair; he mentions rather disapprovingly that the men of the Gauls and Germans were more likely to use it than their female counterparts. Aretaeus of Cappadocia, writing in the first century AD, observes among "Celts, which are men called Gauls, those alkaline substances that are made into balls [...] called *soap*". The Romans' preferred method of cleaning the body was to massage oil into the skin and then scrape away both the oil and any dirt with a strigil. The Gauls used soap made from animal fat.

A popular belief claims soap takes its name from a supposed Mount Sapo, where animal sacrifices were supposed to have taken place; tallow from these sacrifices would then have mixed with ashes from fires associated with these sacrifices and with water to produce soap, but there is no evidence of a Mount Sapo in the Roman world and no evidence for the apocryphal story. The Latin word *sapo* simply means "soap"; it was likely borrowed from an early Germanic language and is cognate with Latin *sebum*, "tallow", which appears in Pliny the Elder's account. Roman animal sacrifices usually burned only the bones and inedible entrails of the sacrificed animals; edible meat and fat from the sacrifices were taken by the humans rather than the gods.

Zosimos of Panopolis, *circa* 300 AD, describes soap and soapmaking. Galen describes soap-making using lye and prescribes washing to carry away impurities from the body and clothes. According to Galen, the best soaps were Germanic, and soaps from Gaul were second best. This is a reference to true soap in antiquity.

Ancient China

A detergent similar to soap was manufactured in ancient China from the seeds of Gleditsia sinensis. Another traditional detergent is a mixture of pig pancreas and plant ash called "Zhu yi zi". True soap, made of animal fat, did not appear in China until the modern era. Soap-like detergents were not as popular as ointments and creams.

Middle East

A 12th-century Islamic document describes the process of soap production. It mentions the key ingredient, alkali, which later becomes crucial to modern chemistry, derived from *al-qaly* or "ashes".

By the 13th century, the manufacture of soap in the Islamic world had become virtually industrialized, with sources in Nablus, Fes, Damascus, and Aleppo.

Medieval Europe

Soapmakers in Naples were members of a guild in the late sixth century (then under the control of the Eastern Roman Empire), and in the eighth century, soap-making was well known in Italy and Spain. The Carolingian capitulary *De Villis*, dating to around 800, representing the royal will of Charlemagne, mentions soap as being one of the products the stewards of royal estates are to tally. The lands of Medieval Spain were a leading soapmaker by 800, and soapmaking began in the Kingdom of England about 1200. Soapmaking is mentioned both as "women's work" and as the produce of "good workmen" alongside other necessities, such as the produce of carpenters, blacksmiths, and bakers.

15th–19th Centuries

Advertisement for Pears' Soap, 1889

A 1922 magazine advertisement for Palmolive Soap

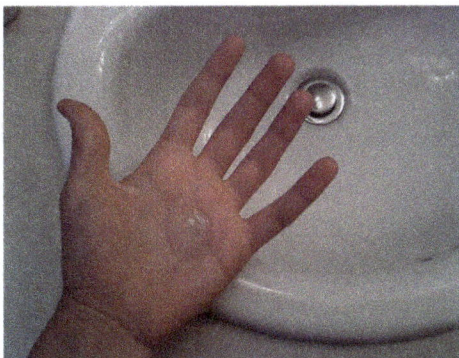

Liquid soap

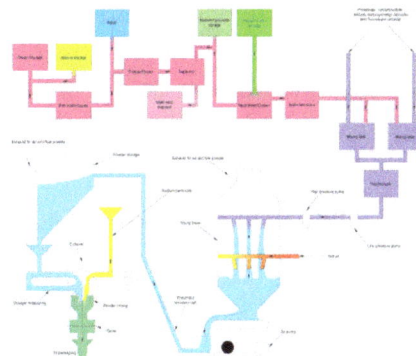

Manufacturing process of soaps/detergents

In France, by the second half of the 15th century, the semi-industrialized professional manufacture of soap was concentrated in a few centers of Provence— Toulon, Hyères, and Marseille — which supplied the rest of France. In Marseilles, by 1525, production was concentrated in at least two factories, and soap production at Marseille tended to eclipse the other Provençal centers. English manufacture tended to concentrate in London.

Finer soaps were later produced in Europe from the 16th century, using vegetable oils (such as olive oil) as opposed to animal fats. Many of these soaps are still produced, both industrially and by small-scale artisans. Castile soap is a popular example of the vegetable-only soaps derived from the oldest "white soap" of Italy.

In modern times, the use of soap has become commonplace in industrialized nations due to a better understanding of the role of hygiene in reducing the population size of pathogenic micro-organisms. Industrially manufactured bar soaps first became available in the late 18th century, as advertising campaigns in Europe and America promoted popular awareness of the relationship between cleanliness and health.

Until the Industrial Revolution, soapmaking was conducted on a small scale and the product was rough. In 1780 James Keir established a chemical works at Tipton, for the manufacture of alkali from the sulfates of potash and soda, to which he afterwards added a soap manufactory. The method of extraction proceeded on a discovery of Keir's. Andrew Pears started making a high-quality, transparent soap in 1807 in London. His son-in-law, Thomas J. Barratt, opened a factory in Isleworth in 1862.

William Gossage produced low-priced, good-quality soap from the 1850s. Robert Spear Hudson began manufacturing a soap powder in 1837, initially by grinding the soap with a mortar and pestle. American manufacturer Benjamin T. Babbitt introduced marketing innovations that included sale of bar soap and distribution of product samples. William Hesketh Lever and his brother, James, bought a small soap works in Warrington in 1886 and founded what is still one of the largest soap businesses, formerly called Lever Brothers and now called Unilever. These soap businesses were among the first to employ large-scale advertising campaigns.

Liquid Soap

Liquid soap was not invented until the nineteenth century; in 1865, William Shepphard patented a liquid version of soap. In 1898, B.J. Johnson developed a soap (made of palm and olive oils); his company (the B.J. Johnson Soap Company) introduced "Palmolive" brand soap that same year. This new brand of the new kind of soap became popular rapidly, and to such a degree that B.J. Johnson Soap Company changed its name to Palmolive.

In the early 1900s, other companies began to develop their own liquid soaps. Such products as Pine-Sol and Tide appeared on the market, making the process of cleaning things other than skin (e.g., clothing, floors, bathrooms) much easier.

Liquid soap also works better for more traditional/non-machine washing methods, such as using a washboard.

Soap-Making Processes

The industrial production of soap involves continuous processes, such as continuous addition of fat and removal of product. Smaller-scale production involves the traditional batch processes. The three variations are: the 'cold process', wherein the reaction takes place substantially at room temperature, the 'semiboiled' or 'hot process', wherein the reaction takes place near the boiling point, and the 'fully boiled process', wherein the reactants are boiled at least once and the glycerol is recovered. There are several types of 'semiboiled' hot process methods, the most common being DBHP (Double Boiler Hot Process) and CPHP (Crock Pot Hot Process). Most soapmakers, however, continue to prefer the cold process method. The cold process and hot process (semiboiled) are the simplest and typically used by small artisans and hobbyists producing handmade decorative soaps. The glycerine remains in the soap and the reaction continues for many days after the soap is poured into molds. The glycerine is left during the hot-process method, but at the high temperature employed, the reaction is practically completed in the kettle, before the soap is poured into molds. This simple and quick process is employed in small factories all over the world.

Handmade soap from the cold process also differs from industrially made soap in that an excess of fat is used, beyond that needed to consume the alkali (in a cold-pour process, this excess fat is called "superfatting"), and the glycerine left in acts as a moisturizing agent. However, the glycerine also makes the soap softer and less resistant to becoming "mushy" if left wet. Since it is better to add too much oil and have left-over fat, than to add too much lye and have left-over lye, soap produced from the hot process also contains left-over glycerine and its concomitant pros and cons. Further addition of glycerine and processing of this soap produces glycerin soap. Superfatted soap is more skin-friendly than one without extra fat. However, if too much fat is added, it can leave a "greasy" feel to the skin. Sometimes, an emollient additive, such as jojoba oil or shea butter, is added "at trace" (i.e., the point at which the saponification process is sufficiently advanced that the soap has begun to thicken in the cold process method) in the belief that nearly all the lye will be spent and it will escape saponification and remain intact. In the case of hot-process soap, an emollient may be added after the initial oils have saponified so they remain unreacted in the finished soap. Superfatting can also be accomplished through a process known as "lye discount" in which the soap maker uses less alkali than required instead of adding extra fats.

Cold Process

The lye is dissolved in water.

Even in the cold soap making process, some heat is usually required; the temperature is usually raised to a point sufficient to ensure complete melting of the fat being used. The batch may also be kept warm for some time after mixing to ensure the alkali (hydroxide) is completely used up. This soap is safe to use after about 12–48 hours, but is not at its peak quality for use for several weeks.

Cold-process soapmaking requires exact measurements of lye and fat amounts and computing their ratio, using saponification charts to ensure the finished product does not contain any excess hydroxide or too much free unreacted fat. Saponification charts should also be used in hot processes, but are not necessary for the "fully boiled hot-process" soaping.

Historically, lye used in the cold process was made from scratch using rainwater and ashes. Soapmakers deemed the lye solution ready for use when an egg would float in it. Homemade lye making for this process was unpredictable and therefore eventually led to the discovery of the sodium hydroxide by English chemist Sir Humphry Davy in the early 1800s.

A cold-process soapmaker first looks up the saponification value for each unique fat on an oil specification sheet. Oil specification sheets contain laboratory test results for each fat, including the precise saponification value of the fat. The saponification value for a specific fat will vary by season and by specimen species. This value is used to calculate the exact amount of sodium hydroxide to react with the fat to form soap. The saponification value must be converted into an equivalent sodium hydroxide value for use in cold process soapmaking. Excess unreacted lye in the soap will result in a very high pH and can burn or irritate skin; not enough lye leaves the soap greasy. Most soap makers formulate their recipes with a 2–5% deficit of lye, to account for the unknown deviation of saponification value between their oil batch and laboratory averages.

The lye is dissolved in water. Then oils are heated, or melted if they are solid at room temperature. Once the oils are liquefied and the lye is fully dissolved in water, they are combined. This lye-fat mixture is mixed until the two phases (oils and water) are fully emulsified. Emulsification is most easily identified visually when the soap exhibits some level of "trace", which is the thickening of the mixture. Many modern-day amateur soapmakers often use a stick blender to speed up this process. There are varying levels of trace. Depending on how additives will affect trace, they may be added at light trace, medium trace, or heavy trace. After much stirring, the mixture turns to the consistency of a thin pudding. "Trace" corresponds roughly to viscosity. Essential oils and fragrance oils can be added with the initial soaping oils, but solid additives such as botanicals, herbs, oatmeal, or other additives are most commonly added at light trace, just as the mixture starts to thicken.

The batch is then poured into molds, kept warm with towels or blankets, and left to continue saponification for 12 to 48 hours. (Milk soaps or other soaps with sugars added are the exception. They typically do not require insulation, as the presence of sugar increases the speed of the reaction and thus the production of heat.) During this time, it is normal for the soap to go through a "gel phase", wherein the opaque soap will turn somewhat transparent for several hours, before once again turning opaque.

After the insulation period, the soap is firm enough to be removed from the mold and cut into bars. At this time, it is safe to use the soap, since saponification is in essence complete. However, cold-process soaps are typically cured and hardened on a drying rack for 2–6 weeks before use. During this cure period, trace amounts of residual lye are consumed by saponification and excess water evaporates.

During the curing process, some molecules in the outer layer of the solid soap react with the carbon dioxide of the air and produce a dusty sheet of sodium carbonate. This reaction is more intense if the mass is exposed to wind or low temperatures.

Hot Processes

Hot-processed soaps are created by encouraging the saponification reaction by adding heat to speed up the reaction. In contrast with cold-pour soap which is poured into molds and for the most part only then saponifies, hot-process soaping for the most part saponifies the oils completely and only then is poured into molds.

In the hot process, the hydroxide and the fat are heated and mixed together at 80–100 °C, a little below boiling point, until saponification is complete, which, before modern scientific equipment, the soapmaker determined by taste (the sharp, distinctive taste of the hydroxide disappears after it is saponified) or by eye; the experienced eye can tell when gel stage and full saponification has occurred. Beginners can find this information through research and classes. Tasting soap for readiness is not recommended, as sodium and potassium hydroxides, when not saponified, are highly caustic.

An advantage of the fully boiled hot process in soapmaking is the exact amount of hydroxide required need not be known with great accuracy. They originated when the purity of the alkali hydroxides were unreliable, as these processes can use even naturally found alkalis, such as wood ashes and potash deposits. In the fully boiled process, the mix is actually boiled (100+ °C), and, after saponification has occurred, the "neat soap" is precipitated from the solution by adding common salt, and the excess liquid is drained off. This excess liquid carries away with it much of the impurities and color compounds in the fat, to leave a purer, whiter soap, and with practically all the glycerine removed. The hot, soft soap is then pumped into a mold. The spent hydroxide solution is processed for recovery of glycerine.

Molds

Logs of soap after demolding.

Many commercially available soap molds are made of silicone or various types of plastic, although many soapmaking hobbyists may use cardboard boxes lined with a plastic film. Wooden molds, unlined or lined with silicone sleeves, are also readily available to the general public. Soaps can be made in long bars that are cut into individual portions, or cast into individual molds.

Purification and Finishing

In the fully boiled process on an industrial scale, the soap is further purified to remove any excess sodium hydroxide, glycerol, and other impurities, color compounds, etc. These components are removed by boiling the crude soap curds in water and then precipitating the soap with salt.

At this stage, the soap still contains too much water, which has to be removed. This was traditionally done on chill rolls, which produced the soap flakes commonly used in the 1940s and 1950s. This process was superseded by spray dryers and then by vacuum dryers.

The dry soap (about 6–12% moisture) is then compacted into small pellets or noodles. These pellets or noodles are then ready for soap finishing, the process of converting raw soap pellets into a saleable product, usually bars.

Soap pellets are combined with fragrances and other materials and blended to homogeneity in an amalgamator (mixer). The mass is then discharged from the mixer into a refiner, which, by means of an auger, forces the soap through a fine wire screen. From the refiner, the soap passes over a roller mill (French milling or hard milling) in a manner similar to calendering paper or plastic or to making chocolate liquor. The soap is then passed through one or more additional refiners to further plasticize the soap mass. Immediately before extrusion, the mass is passed through a vacuum chamber to remove any trapped air. It is then extruded into a long log or blank, cut to convenient lengths, passed through a metal detector, and then stamped into shape in refrigerated tools. The pressed bars are packaged in many ways.

Sand or pumice may be added to produce a scouring soap. The scouring agents serve to remove dead cells from the skin surface being cleaned. This process is called exfoliation. Many newer materials that are effective, yet do not have the sharp edges and poor particle size distribution of pumice, are used for exfoliating soaps.

To make antibacterial soap, compounds such as triclosan or triclocarban can be added. There is some concern that use of antibacterial soaps and other products might encourage antibiotic resistance in microorganisms.

Azul e branco soap – a bar of blue-white soap

Handmade soaps sold at a shop in Hyères, France

Saponification

Saponification is a process that produces soap, usually from fats and lye.

Triglycerides

Vegetable oils and animal fats are the main materials that are saponified. These greasy materials, triesters called triglycerides, are mixtures derived from diverse fatty acids. Triglycerides can be converted to soap in either a one- or a two-step process. In the traditional one-step process, the triglyceride is treated with a strong base (e.g., lye), which accelerates cleavage of the ester bond and releases the fatty acid salt and glycerol. This process is the main industrial method for producing glycerol. If necessary, soaps may be precipitated by salting it out with saturated sodium chloride. The saponification value is the amount of base required to saponify a fat sample. For soap making, the triglycerides are highly purified, but saponification includes other base hydrolysis of unpurified triglycerides, for example, the conversion of the fat of a corpse into adipocere, often called "grave wax." This process is more common where the amount of fatty tissue is high, the agents of decomposition are absent or only minutely present.

Mechanism of Base Hydrolysis

The mechanism by which esters are cleaved by base involves a series of equilibria. The hydroxide anion adds to (or "attacks") the carbonyl group of the ester. The immediate product is called an orthoester:

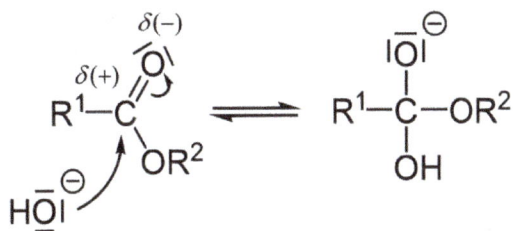

Expulsion of the alkoxide generates a carboxylic acid:

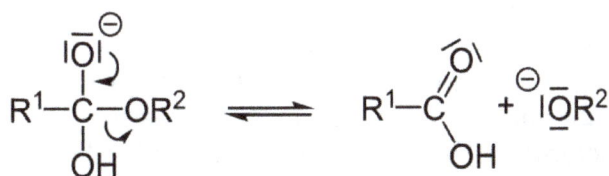

The alkoxide ion is a strong base so that the proton is transferred from the carboxylic acid to the alkoxide ion creating an alcohol:

In a classic laboratory procedure, the triglyceride trimyristin is obtained by extracting it from nutmeg with diethyl ether. Saponification to the sodium salt of myristic acid takes place with NaOH in water. The acid itself can be obtained by adding dilute hydrochloric acid.

Steam Hydrolysis

Triglycerides are also saponified in a two-step process that begins with steam hydrolysis of the triglyceride. This process gives the carboxylic acid, not its salt, as well as glycerol. Subsequently, the fatty acid is neutralized with alkali to give the soap. The advantage of the two-step process is that the fatty acids can be purified, which leads to soaps of improved quality. Steam hydrolysis proceeds via a mechanism similar to the base-catalyzed route, involving the attack of water (not hydroxide) at the carbonyl center. The process is slower, hence the requirement for steam.

Applications

Soft Versus Hard Soap

Depending on the nature of the alkali used in their production, soaps have distinct properties. Sodium hydroxide (NaOH) gives "hard soap"; hard soaps can also be used in water containing Mg, Cl, and Ca salts. By contrast, when potassium hydroxide (KOH) is used, a soft soap is formed. This form of soap cannot be used in hard water.

Lithium Soaps

Lithium derivatives of 12-hydroxystearate and several other carboxylic acids are important constituents of lubricating greases. In lithium-based greases, lithium carboxylates are thickeners. "Complex soaps" are also common, these being combinations of metallic soaps, such as lithium and calcium soaps.

Fire Extinguishers

Fires involving cooking fats and oils (classified as class K (US) or F (Australia/Europe/Asia)) burn hotter than flammable liquids, rendering a standard class B extinguisher ineffective. Flammable liquids have flash points under 37 degrees Celsius. Cooking oil is a combustible liquid, since it has a flash point over 37 degrees Celsius. Such fires should be extinguished with a wet chemical extinguisher. Extinguishers of this type are designed to extinguish cooking fats and oils through saponification. The extinguishing agent rapidly converts the burning substance to a non-combustible soap. This process is endothermic, meaning that it absorbs thermal energy from its surroundings, which decreases the temperature of the surroundings, further inhibiting the fire.

Oil Paints

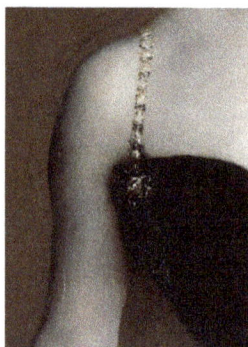

Detail of Madame X (Madame Pierre Gautreau), John Singer Sargent, 1884, showing saponification in the black dress.

Saponification can occur in oil paintings over time, causing visible damage and deformation. The ground layer or paint layers of oil paintings commonly contain heavy metals in pigments such as lead white, red lead, or zinc white. If those heavy metals react with free fatty acids in the oil medium that binds the pigments together, soaps may form in a paint layer that can then migrate outward to the painting's surface.

Saponification in oil paintings was described as early as 1912. It is believed to be widespread, having been observed in many works dating from the fifteenth through the twentieth centuries, works of different geographic origin, and works painted on various supports, such as canvas, paper, wood, and copper. Chemical analysis may reveal saponification occurring in a painting's deeper layers before any signs are visible on the surface, even in paintings centuries old.

The saponified regions may deform the painting's surface through the formation of visible lumps or protrusions that can scatter light. These soap lumps may be prominent only on certain regions of the painting rather than throughout. In John Singer Sargent's famous *Portrait of Madame X*, for example, the lumps only appear on the blackest areas, which may be because of the artist's use of more medium in those areas to compensate for the tendency of black pigments to soak it up. The process can also form chalky white deposits on a painting's surface, a deformation often described as "blooming" or "efflorescence," and may also contribute to the increased transparency of certain paint layers within an oil painting over time.

The process is still not fully understood. Saponification does not occur in all oil paintings containing the right materials. It is not yet known what triggers the process, what makes it worse, or whether it can be halted. At present, retouching is the only known restoration method.

Alum

Bulk alum

Alum /ælum/ is both a specific chemical compound and a class of chemical compounds. The specific compound is the hydrated potassium aluminium sulfate (potassium alum) with the formula $KAl(SO_4)_2 \cdot 12H_2O$. More widely, alums are double sulfate salts, with the general formula $AM(SO_4)_2 \cdot 12H_2O$, where A is a monovalent cation such as potassium or ammonium and M is a trivalent metal ion such as aluminium or chromium(III). When the trivalent ion is aluminium, the alum is named after the monovalent ion.

Chemical Properties

Alums are useful for a range of industrial processes. They are soluble in water, have a sweetish taste, react acid to litmus, and crystallize in regular octahedra. In alums each metal ion is surrounded by six water molecules . When heated, they liquefy, and if the heating is continued, the water of crystallization is driven off, the salt froths and swells, and at last an amorphous powder remains. They are astringent and acidic.

Uses

Industrial Uses

Historically, alum was used extensively in the wool industry from the Classical Times, during the Middle Ages, and well into 19th century as a mordant or dye fixative in the process of turning wool into dyed bolts of cloth.

Some alums occur as minerals. The most important members - potassium, sodium, and ammonium - are produced industrially. Typical recipes involve combining alumina, sulfuric acid, and the sulfate second cation, potassium, sodium, or ammonium.

Potassium alum has been used at least since Roman times for purification of drinking water and industrial process water. Between 30 and 40 ppm of alum for household wastewater, often more for industrial wastewater, is added to the water so that the negatively charged colloidal particles clump together into "flocs", which then float to the top of the liquid, settle to the bottom of the liquid, or can be more easily filtered from the liquid, prior to further filtration and disinfection of the water.

Alum solution has the property of dissolving steels while not affecting aluminium or base metals, and can be used to recover workpieces made in these metals with broken toolbits. lodged inside them.

Cosmetic

An alum block sold as an astringent in pharmacies in India (where it is widely known as *Fitkari*)

- Alum in block form (usually potassium alum) can be used as a blood coagulant.

- Styptic pencils containing aluminium sulfate or potassium aluminium sulfate are used as astringents to prevent bleeding from small shaving cuts.

- Alum may be used in depilatory waxes used for the removal of body hair or applied to freshly waxed skin as a soothing agent.

- In the 1950s, men sporting crewcut or flattop hairstyles sometimes applied alum to their hair as an alternative to pomade. When the hair dried, it would stay up all day.

- Alum's antiperspirant and antibacterial properties contribute to its traditional use as an underarm deodorant. It has been used for this purpose in Europe, Mexico, Thailand (where it is called *sarn-som*), throughout Asia and in the Philippines (where it is called *tawas*). Today, potassium or ammonium alum is sold commercially for this purpose as a "deodorant crystal", often in a protective plastic case.

Culinary

- Alum powder, found in the spice section of many grocery stores, may be used in pickling recipes as a preservative to maintain fruit and vegetable crispness.

- Alum is used as the acidic component of some commercial baking powders.

- Alum was used by bakers in England during the 1800s to make bread whiter. This was thought by some, without adequate scientific grounds, to cause rickets. The Sale of Food and Drugs Act 1875 prevented this and other adulterations.

Flame Retardant

- Solutions containing alum may be used to treat cloth, wood, and paper materials to increase their resistance to fire.

- Alum is also used for fireproofing wood.

Chemical Flocculant

- Alum is used to clarify water by neutralizing the electrical double layer surrounding very fine suspended particles, allowing them to flocculate (stick together). After flocculation, the particles will be large enough to settle and can be removed.

- Alum may be used to increase the viscosity of a ceramic glaze suspension; this makes the glaze more readily adherent and slows its rate of sedimentation.

Taxidermy

- Alum is used in the tanning of animal hides to remove moisture, prevent rotting, and produce a type of leather. Traditionally treating hides with alum, instead of tannic acid, is called tawing and not tanning. The product is traditionally called parchment instead of leather.

Medicine

- Alum is used in the treatment of canker sores in the mouth, as it has a significant drying effect to the area and reduces the irritation felt at the site.

- Alum is the major adjuvant used to increase the efficacy of vaccines, and has been used since the 1920s.

- Alum has been used to stop bleeding in cases of hemorrhagic cystitis.

Art

- Alum serves as a base for the majority of lake pigments

- Alum is used to fix pigments on a surface, for example in paper marbling.

- Alum is an ingredient in some recipes for homemade modeling compounds intended for use by children. These are often called "play clay" or "play dough" for their similarity to "Play-Doh".

History

In Antiquity and the Middle Ages

The word 'alumen' occurs in Pliny's *Natural History*. In the 52nd chapter of his 35th book, he gives a detailed description. By comparing this with the account of 'stupteria' given by Dioscorides in the 123rd chapter of his 5th book, it is obvious the two are identical. Pliny informs us that 'alumen' was found naturally in the earth. He calls it 'salsugoterrae'. Different substances were distinguished by the name of 'alumen', but they were all characterised by a certain degree of astringency, and were all employed in dyeing and medicine, the light-colored alumen being useful in brilliant dyes, the dark-colored only in dyeing black or very dark colors. One species was a liquid, which was apt to be adulterated; but when pure it had the property of blackening when added to pomegranate juice. This property seems to characterize a solution of iron sulfate in water; a solution of ordinary (potassium) alum would possess no such property. Pliny says that there is another kind of alum that the Greeks call 'schiston', and which "splits into filaments of a whitish colour", From the name 'schiston' and the mode of formation, it appears that this species was the salt that forms spontaneously on certain salty minerals, as alum slate and bituminous shale, and consists chiefly of sulfates of iron and aluminium. In some places the iron sulfate may have been lacking, so the salt would be white and would answer, as Pliny says it did, for dyeing bright colors. Pliny describes several other species of alumen but it is not clear as to what these minerals are.

The alumen of the ancients then, was not always the same as the alum of the moderns. They knew how to produce alum from alunite, as this process is archaeologically attested on the island Lesbos. This site was abandoned in the 7th century but dates back at least to the 2nd century CE. Native alumen from Melos appears to have been a mixture mainly of alunogen ($Al_2(SO_4)_3 \cdot 17H_2O$) with alum and other minor sulfates. The western desert of Egypt was a major source of alum substitutes in antiquity. These evaporites were mainly $FeAl_2(SO_4)_4 \cdot 22H_2O$, $MgAl_2(SO_4)_4 \cdot 22H_2O$, $NaAl(SO_4)_2 \cdot 6H_2O$, $MgSO_4 \cdot 7H_2O$ and $Al_2(SO_4)_3 \cdot 17H_2O$. Contamination with iron sulfate was greatly disliked as this darkened and dulled dye colours. They were acquainted with a variety of substances of varying degrees of purity by the names of misy, sory, and chalcanthum. As alum and green vitriol were applied to a variety of substances in common, and as both are distinguished by a sweetish and astringent taste, writers, even after the discovery of alum, do not seem to have discriminated the two salts accurately from each other. In the writings of the alchemists we find the words misy, sory, chalcanthum applied to alum as well as to iron sulfate; and the name atramentum sutorium, which one might expect to belong exclusively to green vitriol, applied indifferently to both. Various minerals are employed in the manufacture of alum, the most important being alunite, alum schist, bauxite and cryolite.

Alchemical And Later Discoveries and Uses

In the 18th century, Johann Heinrich Pott (1692–1777) and Andreas Sigismund Marggraf demonstrated that alumina was a constituent. Pott in his *Lithogeognosia* showed that the precipitate obtained when an alkali is poured into a solution of alum is quite different from lime and chalk, with which it had been confounded by G.E. Stahl. Marggraf showed that alumina is one of the constituents of alum, but that this earth possesses peculiar properties, and is one of the ingredients in common clay. He also showed that crystals of alum can be obtained by dissolving alumina in sulfuric acid and evaporating the solutions, and when a solution of potash or ammonia is dropped into this liquid, it immediately deposits perfect crystals of alum.

Torbern Bergman also observed that the addition of potash or ammonia made the solution of alumina in sulfuric acid crystallize, but that the same effect was not produced by the addition of soda or of lime, and that potassium sulfate is frequently found in alum.

After M.H. Klaproth had discovered the presence of potassium in leucite and lepidolite, it occurred to L. N. Vauquelin that it was probably an ingredient likewise in many other minerals. Knowing that alum cannot be obtained in crystals without the addition of potash, he began to suspect that this alkali constituted an essential ingredient in the salt, and in 1797 he published a dissertation demonstrating that alum is a double salt, composed of sulfuric acid, alumina, and potash. Soon after, J.A. Chaptal published the analysis of four different kinds of alum, namely, Roman alum, Levant alum, British alum and alum manufactured by himself. This analysis led to the same result as Vauquelin.

Early Uses in Industry

Egyptians reportedly used the coagulant alum as early as 1500 BC to reduce the visible cloudiness (turbidity) in the water. Alum was imported into England mainly from the Middle East, and, from the late 15th century onwards, the Papal States for hundreds of years. Its use there was as a dye-fixer (mordant) for wool (which was one of England's primary industries, the value of which increased significantly if dyed). These sources were unreliable, however, and there was a push to develop a source in England especially as imports from the Papal States were ceased following the excommunication of Henry VIII.

In the 13th and 14th centuries, alum (from alunite) was a major import from Phocaea (Gulf of Smyrna in Byzantium) by Genoans and Venetians (and was a cause of war between Genoa and Venice) and later by Florence. After the fall of Constantinople, alunite (the source of alum) was discovered at Tolfa in the Papal States (1461). The textile dyeing industry in Bruge, and many other locations in Italy, and later in England, required alum to stabilize the dyes onto the fabric (make the dyes "fast") and also to brighten the colors.

With state financing, attempts were made throughout the 16th century, but without success until early on in the 17th century. An industry was founded in Yorkshire to process the shale, which contained the key ingredient, aluminium sulfate, and made an important contribution to the Industrial Revolution. One of the oldest historic sites for the production of alum from shale and human urine are the Peak alum works in Ravenscar, North Yorkshire. By the 18th century, the landscape of northeast Yorkshire had been devastated by this process, which involved constructing 100 feet (30 m) stacks of burning shale and fuelling them with firewood continuously for months. The rest of the production process consisted of quarrying, extraction, steeping of shale ash with seaweed in urine, boiling, evaporating, crystallisation, milling and loading into sacks for export. Quarrying ate into the cliffs of the area, the forests were felled for charcoal and the land polluted by sulfuric acid and ash.

Production

From Alunite

In order to obtain alum from alunite, it is calcined and then exposed to the action of air for a considerable time. During this exposure it is kept continually moistened with water, so that it ultimately falls to a very fine powder. This powder is then lixiviated with hot water, the liquor decanted, and the alum allowed to crystallize. The alum schists employed in the manufacture of alum are mixtures of iron pyrite, aluminium silicate and various bituminous substances, and are found in upper Bavaria, Bohemia, Belgium, and Scotland. These are either roasted or exposed to the weathering action of the air. In the roasting process, sulfuric acid is formed and acts on the clay to form aluminium sulfate, a similar condition of affairs being produced during weathering. The mass is now systematically extracted with water, and a solution of aluminium sulfate of specific gravity 1.16 is prepared. This solution is allowed to stand for some time (in order that any calcium sulfate and basic ferric sulfate may separate), and is then evaporated until ferrous sulfate crystallizes on cooling; it is then drawn off and evaporated until it attains a specific gravity of 1.40. It is now allowed to stand for some time, decanted from any sediment, and finally mixed with the calculated quantity of potassium sulfate, well agitated, and the alum is thrown down as a finely divided precipitate of alum meal. If much iron should be present in the shale then it is preferable to use potassium chloride in place of potassium sulfate.

From Clays or Bauxite

In the preparation of alum from clays or from bauxite, the material is gently calcined, then mixed with sulfuric acid and heated gradually to boiling; it is allowed to stand for some time, the clear solution drawn off and mixed with acid potassium sulfate and allowed to crystallize. When cryolite is used for the preparation of alum, it is mixed with calcium carbonate and heated. By this means, sodium aluminate is formed; it is then extracted with water and precipitated either by sodium bicarbonate or by passing a current of carbon dioxide through the solution. The precipitate is then dissolved in sulfuric acid, the requisite amount of potassium sulfate added and the solution allowed to crystallize.

Types

Many trivalent metals are capable of forming alums. The general form of an alum is $AM^{III}(SO_4)_2 \cdot n\text{-}H_2O$, where A is an alkali metal or ammonium, M^{III} is a trivalent metal, and n often is 12. In general, alums are easier formed when the alkali metal atom is larger. This rule was first stated by Locke in 1902, who found that if a trivalent metal does not form a caesium alum, it neither will form an alum with any other alkali metal or with ammonium.

Double sulfates with the general formula $A_2SO_4 \cdot B_2(SO_4)_3 \cdot 24H_2O$, are known where A is a monovalent cation such as sodium, potassium, rubidium, caesium, or thallium(I), or a compound cation such as ammonium (NH_4^+), methylammonium ($CH_3NH_3^+$), hydroxylammonium ($HONH_3^+$) or hydrazinium ($N_2H_5^+$), B is a trivalent metal ion, such as aluminium, chromium, titanium, manganese, vanadium, iron(III), cobalt(III), gallium, molybdenum, indium, ruthenium, rhodium, or iridium. Analogous selenates also occur. The specific combinations of univalent cation, trivalent cation, and anion depends on the sizes of the ions. For example, unlike the other alkali metals the smallest one, lithium, does not form alums, and there is only one known sodium alum. In some cases, solid solutions of alums occur.

Alums crystallize in one of three different crystal structures. These classes are called α-, β- and γ-alums.

Potash Alum

Crystal of potassium alum

Aluminum potassium sulfate, potash alum, $KAl(SO_4)_2 \cdot 12H_2O$ is used as an astringent and antisepsis in various food preparation processes such as pickling and fermentation and as a flocculant for water purification, among other things. A common method of producing potash alum is leaching of alumina from bauxite, which is then reacted with potassium sulfate. As a naturally occurring mineral, potash alum is known as alum-(K). Other potassium aluminium sulfate minerals are alunite ($KAl(SO_4)_2 \cdot 2Al(OH)_3$) and kalinite ($KAl(SO_4)_2 \cdot 11H_2O$).

Soda Alum

Soda alum, $NaAl(SO_4)_2 \cdot 12H_2O$, mainly occurs in nature as the mineral mendozite. It is very soluble in water, and is extremely difficult to purify. In the preparation of this salt, it is preferable to mix the component solutions in the cold, and to evaporate them at a temperature not exceeding 60 °C. 100 parts of water dissolve 110 parts of sodium alum at 0 °C, and 51 parts at 16 °C. Soda alum is used in the acidulent of food as well as in the manufacture of baking powder.

Ammonium Alum

Ammonium alum, $NH_4Al(SO_4)_2 \cdot 12H_2O$, a white crystalline double sulfate of aluminium, is used in water purification, in vegetable glues, in porcelain cements, in deodorants (though potassium alum is more commonly used), in tanning, dyeing and in fireproofing textiles.

Chrome Alum

Alum crystal with small amount of chrome alum to give a slight violet color

Chrome alum, $KCr(SO_4)_2 \cdot 12H_2O$, a dark violet crystalline double sulfate of chromium and potassium, was used in tanning.

Selenate-containing Alums

Alums are also known that contain selenium in place of sulfur in the sulfate anion, making selenate (SeO_4^{2-}) instead. They are called selenium- or selenate-alums. They are strong oxidizing agents.

Aluminium Sulfate

Aluminium sulfate is referred to as papermaker's alum. Although reference to this compound as alum is quite common in industrial communication, it is not regarded as technically correct. Its properties are quite different from those of the set of alums described above. Most industrial flocculation done with alum is actually aluminium sulfate.

Solubility

The solubility of the various alums in water varies greatly, sodium alum being readily soluble in water, while caesium and rubidium alums are only sparingly soluble. The various solubilities are shown in the following table.

At temperature T, 100 parts water dissolve:				
T	Ammonium alum	Potassium alum	Rubidium alum	Cesium alum
0 °C	2.62	3.90	0.71	0.19
10 °C	4.50	9.52	1.09	0.29
50 °C	15.9	44.11	4.98	1.235
80 °C	35.20	134.47	21.60	5.29
100 °C	70.83	357.48		

Related Compounds

In addition to the alums, which are dodecahydrates, double sulfates and selenates of univalent and trivalent cations occur with other degrees of hydration. These materials may also be referred to as alums, including the undecahydrates such as mendozite and kalinite, hexahydrates such as guanidinium ($CH_6 N^{+3}$) and dimethylammonium ((CH_3) $2NH^{+2}$) "alums", tetrahydrates such as goldichite, monohydrates such as thallium plutonium sulfate and anhydrous alums (yavapaiites). These classes include differing, but overlapping, combinations of ions.

A pseudo alum is a double sulfate of the typical formula $ASO_4 \cdot B_2(SO)_3 \cdot 22H_2O$, where A is a divalent metal ion, such as cobalt (wupatkiite), manganese (apjohnite), magnesium (pickingerite) or iron (halotrichite or feather alum), and B is a trivalent metal ion.

A Tutton salt is a double sulfate of the typical formula $A_2SO_4 \cdot BSO_4 \cdot 6H_2O$, where A is a univalent cation, and B a divalent metal ion.

Double sulfates of the composition $A_2SO_4 \cdot 2BSO_4$, where A is a univalent cation and B is a divalent metal ion are referred to as langbeinites, after the prototypical potassium magnesium sulfate.

References

- Skoog, D.A; West, D.M.; Holler, J.F.; Crouch, S.R. (2004). Fundamentals of Analytical Chemistry; Chapters 14, 15 and 16 (8th ed.). Thomson Brooks/Cole. ISBN 0-03-035523-0.

- Corriher, S.O. (2008). BakeWise: The Hows and Whys of Successful Baking with Over 200 Magnificent Recipes. Scribner. ISBN 9781416560838. Retrieved 2014-10-25.

- Lide, David R., ed. (2009). CRC Handbook of Chemistry and Physics (90th ed.). Boca Raton, Florida: CRC Press. ISBN 978-1-4200-9084-0.

- Partington, James Riddick; Hall, Bert S (1999). A History of Greek Fire and Gun Powder. JHU Press. p. 307. ISBN 0-8018-5954-9.

- Jones, Geoffrey (2010). "Cleanliness and Civilization". Beauty Imagined: A History of the Global Beauty Industry. Oxford University Press. ISBN 978-0-19-160961-9.

- footnote 48, p. 104, Understanding the Middle Ages: the transformation of ideas and attitudes in the Medieval world, Harald Kleinschmidt, illustrated, revised, reprint edition, Boydell & Brewer, 2000, ISBN 0-85115-770-X.

- Anionic and Related Lime Soap Dispersants, Raymond G. Bistline, Jr., in Anionic surfactants: organic chemistry, Helmut Stache, ed., Volume 56 of Surfactant science series, CRC Press, 1996, chapter 11, p. 632, ISBN 0-8247-9394-3.

- McNeil, Ian (1990). An Encyclopaedia of the history of technology. Taylor & Francis. pp. 2003–205. ISBN 978-0-415-01306-2.

- Garzena, Patrizia, and Tadiello, Marina (2013). The Natural Soapmaking Handbook. Online information and Table of Contents. ISBN 978-0-9874995-0-9/

Electronic Band Structure: A Comprehensive Study

Electronic band structures are a spectrum of energy that determines the range of modification that electrons undergo when using quantum energy. In order to completely understand electronic band structure it is necessary to understand the process related to it. Some of the topics included in this chapter are band gap, Fermi level, tight binding and nearly free electron model.

Electronic Band Structure

In solid-state physics, the electronic band structure (or simply band structure) of a solid describes the range of energies that an electron within the solid may have (called *energy bands, allowed bands*, or simply *bands*) and ranges of energy that it may not have (called *band gaps* or *forbidden bands*).

Band theory derives these bands and band gaps by examiniwng the allowed quantum mechanical wave functions for an electron in a large, periodic lattice of atoms or molecules. Band theory has been successfully used to explain many physical properties of solids, such as electrical resistivity and optical absorption, and forms the foundation of the understanding of all solid-state devices (transistors, solar cells, etc.).

Why Bands and Band Gaps Occur

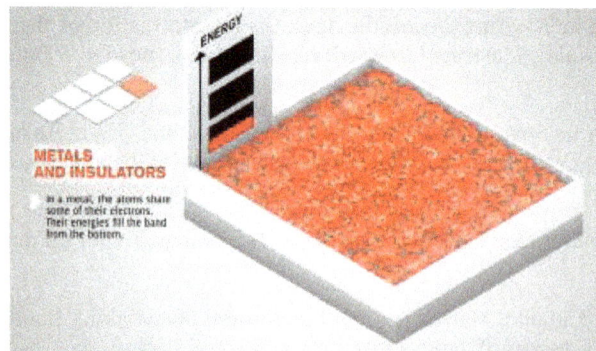

Animation of band formation and how electrons fill them in a metal and an insulator

The electrons of a single, isolated atom occupy atomic orbitals. Each orbital forms at a discrete energy level. When multiple atoms join together to form into a molecule, their atomic orbitals combine to form molecular orbitals, each of which forms at a discrete energy level. As more atoms are brought together, the molecular orbitals extend larger and larger, and the energy levels of the molecule will become increasingly dense. Eventually, the collection of atoms form a giant molecule, or in other words, a solid. For this giant molecule, the energy levels are so close that they can be considered to form a continuum.

Band gaps are essentially leftover ranges of energy not covered by any band, a result of the finite widths of the energy bands. The bands have different widths, with the widths depending upon the degree of overlap in the atomic orbitals from which they arise. Two adjacent bands may simply not be wide enough to fully cover the range of energy. For example, the bands associated with core orbitals (such as 1s electrons) are extremely narrow due to the small overlap between adjacent atoms. As a result, there tend to be large band gaps between the core bands. Higher bands involve comparatively larger orbitals with more overlap, becoming progressively wider at higher energies so that there are no band gaps at higher energies.

Basic Concepts

Assumptions and Limits of Band Structure Theory

To start out, it is important to note what has been assumed in order to gain the simplicity of the band theory:

1. *Infinite-size system*: For the bands to be continuous, we must consider a large piece of material. The concept of band structure can be extended to systems which are only "large" along reduced dimensions, such as two-dimensional electron systems.

2. *Homogeneous system*: The notion of a band structure as an intrinsic property of a material assumes that the material is homogeneous in some way. Practically, this means that band structure describes the bulk inside a uniform piece of material.

3. *Non-interactivity*: The band structure describes "single electron states". The existence of these states assumes that the electrons travel in a static potential without dynamically interacting with lattice vibrations, other electrons, photons, etc.

The above assumptions are broken in a number of important practical situations, and the use of band structure requires one to keep a close check on the limitations of band theory:

- Inhomogeneities and interfaces: Near surfaces, junctions, and other inhomogeneities, the bulk band structure is disrupted. Not only are there local small-scale disruptions (e.g., surface states or dopant states inside the band gap), but also local charge imbalances. These charge imbalances have electrostatic effects that extend deeply into semiconductors, insulators, and the vacuum.

- Along the same lines, most electronic effects (capacitance, electrical conductance, electric-field screening) involve the physics of electrons passing through surfaces and/or near interfaces. The full description of these effects, in a band structure picture, requires at least a rudimentary model of electron-electron interactions.

- Small systems: For systems which are small along every dimension (e.g., a small molecule or a quantum dot), there is no continuous band structure. The crossover between small and large dimensions is the realm of mesoscopic physics.

- Strongly correlated materials (for example, Mott insulators) simply cannot be understood in terms of single-electron states. The electronic band structures of these materials are poorly defined (or at least, not uniquely defined) and may not provide useful information about their physical state.

Crystalline Symmetry and Wavevectors

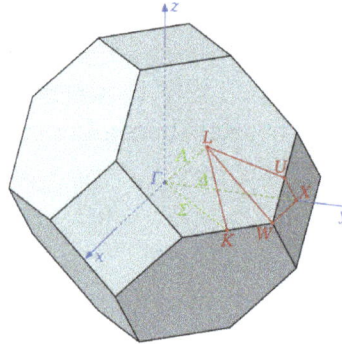

Brillouin zone of a face-centered cubic lattice showing labels for special symmetry points.

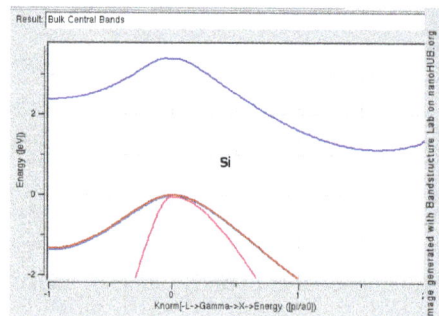

Band structure plot for Si, Ge, GaAs and InAs generated with tight binding model. Note that Si and Ge are indirect band gap materials, while GaAs and InAs are direct.

Band structure calculations take advantage of the periodic nature of a crystal lattice, exploiting its symmetry. The single-electron Schrödinger equation is solved for an electron in a lattice-periodic potential, giving Bloch waves as solutions:

$$\psi_{n\mathbf{k}}(\mathbf{r}) = e^{i\mathbf{k}\cdot\mathbf{r}} u_{n\mathbf{k}}(\mathbf{r}), ,$$

where k is called the wavevector. For each value of k, there are multiple solutions to the Schrödinger equation labelled by n, the band index, which simply numbers the energy bands. Each of these energy levels evolves smoothly with changes in k, forming a smooth band of states. For each band we can define a function $E_n(\mathbf{k})$, which is the dispersion relation for electrons in that band.

The wavevector takes on any value inside the Brillouin zone, which is a polyhedron in wavevector space that is related to the crystal's lattice. Wavevectors outside the Brillouin zone simply correspond to states that are physically identical to those states within the Brillouin zone. Special high symmetry points in the Brillouin zone are assigned labels like Γ, Δ, Λ, Σ.

It is difficult to visualize the shape of a band as a function of wavevector, as it would require a plot in four-dimensional space, E vs. k_x, k_y, k_z. In scientific literature it is common to see band structure plots which show the values of $E_n(k)$ for values of k along straight lines connecting symmetry points. Another method for visualizing band structure is to plot a constant-energy isosurface in wavevector space, showing all of the states with energy equal to a particular value. The isosurface of states with energy equal to the Fermi level is known as the Fermi surface.

Energy band gaps can be classified using the wavevectors of the states surrounding the band gap:

- Direct band gap: the lowest-energy state above the band gap has the same k as the highest-energy state beneath the band gap.

- Indirect band gap: the closest states above and beneath the band gap do not have the same k value.

Asymmetry: Band Structures in Non-crystalline Solids

Although electronic band structures are usually associated with crystalline materials, quasi-crystalline and amorphous solids may also exhibit band structures. These are somewhat more difficult to study theoretically since they lack the simple symmetry of a crystal, and it is not usually possible to determine a precise dispersion relation. As a result, virtually all of the existing theoretical work on the electronic band structure of solids has focused on crystalline materials.

Density of States

The density of states function $g(E)$ is defined as the number of electronic states per unit volume, per unit energy, for electron energies near E.

The density of states function is important for calculations of effects based on band theory. It appears in calculations for optical absorption where it provides both the number of excitable electrons and the number of final states for an electron. It appears in calculations of electrical conductivity where it provides the number of mobile states, and in computing electron scattering rates where it provides the number of final states after scattering.

For energies inside a band gap, $g(E) = 0$.

Filling of Bands

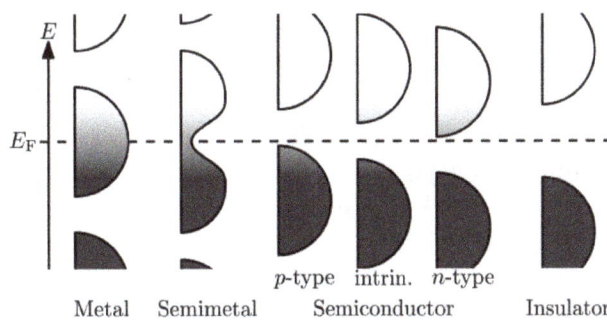

Filling of the electronic states in various types of materials at equilibrium. Here, height is energy while width is the density of available states for a certain energy in the material listed. The shade follows the Fermi–Dirac distribution (black = all states filled, white = no state filled). In metals and semimetals the Fermi level E_F lies inside at least one band. In insulators and semiconductors the Fermi level is inside a band gap; however, in semiconductors the bands are near enough to the Fermi level to be thermally populated with electrons or holes.

At thermodynamic equilibrium, the likelihood of a state of energy E being filled with an electron is given by the Fermi–Dirac distribution, a thermodynamic distribution that takes into account the Pauli exclusion principle:

$$f(E) = \frac{1}{1 + e^{(E-\mu)/k_B T}}$$

where:

- $k_B T$ is the product of Boltzmann's constant and temperature, and

- μ is the total chemical potential of electrons, or *Fermi level* (in semiconductor physics, this quantity is more often denoted E_F). The Fermi level of a solid is directly related to the voltage on that solid, as measured with a voltmeter. Conventionally, in band structure plots the Fermi level is taken to be the zero of energy (an arbitrary choice).

The density of electrons in the material is simply the integral of the Fermi–Dirac distribution times the density of states:

$$N/V = \int_{-\infty}^{\infty} g(E) f(E) dE$$

Although there are an infinite number of bands and thus an infinite number of states, there are only a finite number of electrons to place in these bands. The preferred value for the number of electrons is a consequence of electrostatics: even though the surface of a material can be charged, the internal bulk of a material prefers to be charge neutral. The condition of charge neutrality means that N/V must match the density of protons in the material. For this to occur, the material electrostatically adjusts itself, shifting its band structure up or down in energy (thereby shifting $g(E)$), until it is at the correct equilibrium with respect to the Fermi level.

Names of Bands Near the Fermi Level (Conduction Band, Valence Band)

A solid has an infinite number of allowed bands, just as an atom has infinitely many energy levels. However, most of the bands simply have too high energy, and are usually disregarded under ordinary circumstances. Conversely, there are very low energy bands associated with the core orbitals (such as 1s electrons). These low-energy *core bands* are also usually disregarded since they remain filled with electrons at all times, and are therefore inert. Likewise, materials have several band gaps throughout their band structure.

The most important bands and band gaps—those relevant for electronics and optoelectronics—are those with energies near the Fermi level. The bands and band gaps near the Fermi level are given special names, depending on the material:

- In a semiconductor or band insulator, the Fermi level is surrounded by a band gap, referred to as *the* band gap (to distinguish it from the other band gaps in the band structure). The closest band above the band gap is called *the conduction band*, and the closest band beneath the band gap is called *the valence band*. The name "valence band" was coined by analogy to chemistry, since in many semiconductors the valence band is built out of the valence orbitals.

- In a metal or semimetal, the Fermi level is inside of one or more allowed bands. In semimetals the bands are usually referred to as "conduction band" or "valence band" depending on whether the charge transport is more electron-like or hole-like, by analogy to semiconductors. In many metals, however, the bands are neither electron-like nor hole-like, and often just called "valence band" as they are made of valence orbitals. The band gaps in a metal's band structure are not important for low energy physics, since they are too far from the Fermi level.

Theory of Band Structures in Crystals

The ansatz is the special case of electron waves in a periodic crystal lattice using Bloch waves as treated generally in the dynamical theory of diffraction. Every crystal is a periodic structure which can be characterized by a Bravais lattice, and for each Bravais lattice we can determine the reciprocal lattice, which encapsulates the periodicity in a set of three reciprocal lattice vectors (b_1, b_2, b_3). Now, any periodic potential V(r) which shares the same periodicity as the direct lattice can be expanded out as a Fourier series whose only non-vanishing components are those associated with the reciprocal lattice vectors. So the expansion can be written as:

$$V(\mathbf{r}) = \sum_{\mathbf{K}} V_{\mathbf{K}} e^{i\mathbf{K} \cdot \mathbf{r}}$$

where $K = m_1 b_1 + m_2 b_2 + m_3 b_3$ for any set of integers (m_1, m_2, m_3).

From this theory, an attempt can be made to predict the band structure of a particular material, however most ab initio methods for electronic structure calculations fail to predict the observed band gap.

Nearly Free Electron Approximation

In the nearly free electron approximation, interactions between electrons are completely ignored. This approximation allows use of Bloch's Theorem which states that electrons in a periodic potential have wavefunctions and energies which are periodic in wavevector up to a constant phase shift between neighboring reciprocal lattice vectors. The consequences of periodicity are described mathematically by the Bloch wavefunction:

$$\psi_{n,\mathbf{k}}(\mathbf{r}) = e^{i\mathbf{k} \cdot \mathbf{r}} u_n(\mathbf{r})$$

where the function $u_n(\mathbf{r})$ is periodic over the crystal lattice, that is,

$$u_n(\mathbf{r}) = u_n(\mathbf{r} - \mathbf{R}).$$

Here index n refers to the n-th energy band, wavevector k is related to the direction of motion of the electron, r is the position in the crystal, and R is the location of an atomic site.

The NFE model works particularly well in materials like metals where distances between neighbouring atoms are small. In such materials the overlap of atomic orbitals and potentials on neighbouring atoms is relatively large. In that case the wave function of the electron can be approximated by a (modified) plane wave. The band structure of a metal like Aluminium even gets close to the empty lattice approximation.

Tight Binding Model

The opposite extreme to the nearly free electron approximation assumes the electrons in the crystal behave much like an assembly of constituent atoms. This tight binding model assumes the solution to the time-independent single electron Schrödinger equation Ø is well approximated by a linear combination of atomic orbitals $\psi_n(\mathbf{r})$.

$$\Psi(\mathbf{r}) = \sum_{n,\mathbf{R}} b_{n,\mathbf{R}} \psi_n(\mathbf{r} - \mathbf{R}),$$

where the coefficients $b_{n,\mathbf{R}}$ are selected to give the best approximate solution of this form. Index n refers to an atomic energy level and R refers to an atomic site. A more accurate approach using this idea employs Wannier functions, defined by:

$$a_n(\mathbf{r} - \mathbf{R}) = \frac{V_C}{(2\pi)^3} \int_{BZ} d\mathbf{k} e^{-i\mathbf{k} \cdot (\mathbf{R} - \mathbf{r})} u_{n\mathbf{k}};$$

in which $u_{n\mathbf{k}} i$ is the periodic part of the Bloch wave and the integral is over the Brillouin zone. Here index n refers to the n-th energy band in the crystal. The Wannier functions are localized near atomic sites, like atomic orbitals, but being defined in terms of Bloch functions they are accurately related to solutions based upon the crystal potential. Wannier functions on different atomic sites R are orthogonal. The Wannier functions can be used to form the Schrödinger solution for the n-th energy band as:

$$\Psi_{n,\mathbf{k}}(\mathbf{r}) = \sum_{\mathbf{R}} e^{-i\mathbf{k} \cdot (\mathbf{R} - \mathbf{r})} a_n(\mathbf{r} - \mathbf{R})$$

The TB model works well in materials with limited overlap between atomic orbitals and potentials on neighbouring atoms. Band structures of materials like Si, GaAs, SiO_2 and diamond for instance are well described by TB-Hamiltonians on the basis of atomic sp^3 orbitals. In transition metals a mixed TB-NFE model is used to describe the broad NFE conduction band and the narrow embedded TB d-bands. The radial functions of the atomic orbital part of the Wannier functions are most easily calculated by the use of pseudopotential methods. NFE, TB or combined NFE-TB band structure calculations, sometimes extended with wave function approximations based on pseudopotential methods, are often used as an economic starting point for further calculations.

KKR Model

The simplest form of this approximation centers non-overlapping spheres (referred to as *muffin tins*) on the atomic positions. Within these regions, the potential experienced by an electron is approximated to be spherically symmetric about the given nucleus. In the remaining interstitial region, the screened potential is approximated as a constant. Continuity of the potential between the atom-centered spheres and interstitial region is enforced.

A variational implementation was suggested by Korringa and by Kohn and Rostocker, and is often referred to as the *KKR model*.

Density-functional Theory

In recent physics literature, a large majority of the electronic structures and band plots are calculated using density-functional theory (DFT), which is not a model but rather a theory, i.e., a microscopic first-principles theory of condensed matter physics that tries to cope with the electron-electron many-body problem via the introduction of an exchange-correlation term in the functional of the electronic density. DFT-calculated bands are in many cases found to be in agreement with experimentally measured bands, for example by angle-resolved photoemission spectroscopy (ARPES). In particular, the band shape is typically well reproduced by DFT. But there are also systematic errors in DFT bands when compared to experiment results. In particular, DFT seems to systematically underestimate by about 30-40% the band gap in insulators and semiconductors.

It is commonly believed that DFT is a theory to predict ground state properties of a system only (e.g. the total energy, the atomic structure, etc.), and that excited state properties cannot be determined by DFT. This is a misconception. In principle, DFT can determine any property (ground state or excited state) of a system given a functional that maps the ground state density to that property. This is the essence of the Hohenburg–Kohn theorem. In practice, however, no known functional exists that maps the ground state density to excitation energies of electrons within a material. Thus, what in the literature is quoted as a DFT band plot is a representation of the DFT Kohn–Sham energies, i.e., the energies of a fictive non-interacting system, the Kohn–Sham system, which has no physical interpretation at all. The Kohn–Sham electronic structure must not be confused with the real, quasiparticle electronic structure of a system, and there is no Koopman's theorem holding for Kohn–Sham energies, as there is for Hartree–Fock energies, which can be truly considered as an approximation for quasiparticle energies. Hence, in principle, Kohn–Sham based DFT is not a band theory, i.e., not a theory suitable for calculating bands and band-plots. In principle time-dependent DFT can be used to calculate the true band structure although in practice this is often difficult. A popular approach is the use of hybrid functionals, which incorporate a portion of Hartree–Fock exact exchange; this produces a substantial improvement in predicted bandgaps of semiconductors, but is less reliable for metals and wide-bandgap materials.

Green's Function Methods and the ab Initio GW Approximation

To calculate the bands including electron-electron interaction many-body effects, one can resort to so-called Green's function methods. Indeed, knowledge of the Green's function of a system provides both ground (the total energy) and also excited state observables of the system. The poles of the Green's function are the quasiparticle energies, the bands of a solid. The Green's function can be calculated by solving the Dyson equation once the self-energy of the system is known. For real systems like solids, the self-energy is a very complex quantity and usually approximations are needed to solve the problem. One such approximation is the GW approximation, so called from the mathematical form the self-energy takes as the product $\Sigma = GW$ of the Green's function G and the dynamically screened interaction W. This approach is more pertinent when addressing the calculation of band plots (and also quantities beyond, such as the spectral function) and can also be formulated in a completely *ab initio* way. The GW approximation seems to provide band gaps of insulators and semiconductors in agreement with experiment, and hence to correct the systematic DFT underestimation.

Mott Insulators

Although the nearly free electron approximation is able to describe many properties of electron band structures, one consequence of this theory is that it predicts the same number of electrons in each unit cell. If the number of electrons is odd, we would then expect that there is an unpaired electron in each unit cell, and thus that the valence band is not fully occupied, making the material a conductor. However, materials such as CoO that have an odd number of electrons per unit cell are insulators, in direct conflict with this result. This kind of material is known as a Mott insulator, and requires inclusion of detailed electron-electron interactions (treated only as an averaged effect on the crystal potential in band theory) to explain the discrepancy. The Hubbard model is an approximate theory that can include these interactions. It can be treated non-perturbatively within the so-called dynamical mean field theory, which attempts to bridge the gap between the nearly free electron approximation and the atomic limit. Formally, however, the states are not non-interacting in this case and the concept of a band structure is not adequate to describe these cases.

Others

Calculating band structures is an important topic in theoretical solid state physics. In addition to the models mentioned above, other models include the following:

- Empty lattice approximation: the "band structure" of a region of free space that has been divided into a lattice.

- k·p perturbation theory is a technique that allows a band structure to be approximately described in terms of just a few parameters. The technique is commonly used for semiconductors, and the parameters in the model are often determined by experiment.

- The Kronig-Penney Model, a one-dimensional rectangular well model useful for illustration of band formation. While simple, it predicts many important phenomena, but is not quantitative.

- Hubbard model

The band structure has been generalised to wavevectors that are complex numbers, resulting in what is called a *complex band structure*, which is of interest at surfaces and interfaces.

Each model describes some types of solids very well, and others poorly. The nearly free electron model works well for metals, but poorly for non-metals. The tight binding model is extremely accurate for ionic insulators, such as metal halide salts (e.g. NaCl).

Band Diagrams

To understand how band structure changes relative to the Fermi level in real space, a band structure plot is often first simplified in the form of a band diagram. In a band diagram the vertical axis is energy while the horizontal axis represents real space. Horizontal lines represent energy levels, while blocks represent energy bands. When the horizontal lines in these diagram are slanted then the energy of the level or band changes with distance. Diagrammatically, this depicts the presence of an electric field within the crystal system. Band diagrams are useful in relating the general band structure properties of different materials to one another when placed in contact with each other.

Band Gap

In solid-state physics, a band gap, also called an energy gap or bandgap, is an energy range in a solid where no electron states can exist. In graphs of the electronic band structure of solids, the band gap generally refers to the energy difference (in electron volts) between the top of the valence band and the bottom of the conduction band in insulators and semiconductors. It is the energy required to promote a valence electron bound to an atom to become a conduction electron, which is free to move within the crystal lattice and serve as a charge carrier to conduct electric current. It is closely related to the HOMO/LUMO gap in chemistry. If the valence band is completely full and the conduction band is completely empty, then electrons cannot move in the solid; however, if some electrons transfer from the valence to the conduction band, then current *can* flow. Therefore, the band gap is a major factor determining the electrical conductivity of a solid. Substances with large band gaps are generally insulators, those with smaller band gaps are semiconductors, while conductors either have very small band gaps or none, because the valence and conduction bands overlap.

In Semiconductor Physics

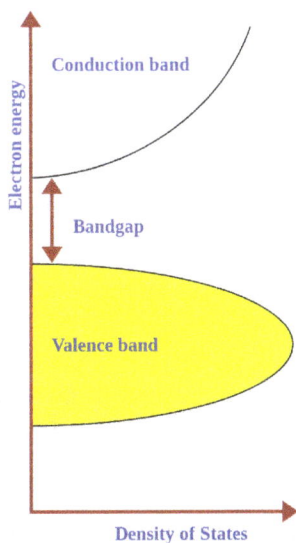

Semiconductor band structure.

Every solid has its own characteristic energy-band structure. This variation in band structure is responsible for the wide range of electrical characteristics observed in various materials. In semiconductors and insulators, electrons are confined to a number of bands of energy, and forbidden from other regions. The term "band gap" refers to the energy difference between the top of the valence band and the bottom of the conduction band. Electrons are able to jump from one band to another. However, in order for an electron to jump from a valence band to a conduction band, it requires a specific minimum amount of energy for the transition. The required energy differs with different materials. Electrons can gain enough energy to jump to the conduction band by absorbing either a phonon (heat) or a photon (light).

A semiconductor is a material with a small but non-zero band gap that behaves as an insulator at absolute zero but allows thermal excitation of electrons into its conduction band at temperatures that are below its melting point. In contrast, a material with a large band gap is an insulator. In conductors, the valence and conduction bands may overlap, so they may not have a band gap.

The conductivity of intrinsic semiconductors is strongly dependent on the band gap. The only available charge carriers for conduction are the electrons that have enough thermal energy to be excited across the band gap and the electron holes that are left off when such an excitation occurs.

Band-gap engineering is the process of controlling or altering the band gap of a material by controlling the composition of certain semiconductor alloys, such as GaAlAs, InGaAs, and InAlAs. It is also possible to construct layered materials with alternating compositions by techniques like molecular-beam epitaxy. These methods are exploited in the design of heterojunction bipolar transistors (HBTs), laser diodes and solar cells.

The distinction between semiconductors and insulators is a matter of convention. One approach is to think of semiconductors as a type of insulator with a narrow band gap. Insulators with a larger band gap, usually greater than 3 eV, are not considered semiconductors and generally do not exhibit semiconductive behaviour under practical conditions. Electron mobility also plays a role in determining a material's informal classification.

The band-gap energy of semiconductors tends to decrease with increasing temperature. When temperature increases, the amplitude of atomic vibrations increase, leading to larger interatomic spacing. The interaction between the lattice phonons and the free electrons and holes will also affect the band gap to a smaller extent. The relationship between band gap energy and temperature can be described by Varshni's empirical expression,

$$E_g(T) = E_g(0) - \frac{\alpha T^2}{T + \beta}$$, where $E_g(0)$, α and β are material constants.

In a regular semiconductor crystal, the band gap is fixed owing to continuous energy states. In a quantum dot crystal, the band gap is size dependent and can be altered to produce a range of energies between the valence band and conduction band. It is also known as quantum confinement effect.

Band gaps also depend on pressure. Band gaps can be either direct or indirect, depending on the electronic band structure.

Photovoltaic Cells

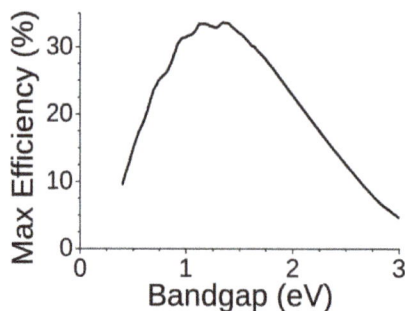

The Shockley–Queisser limit gives the maximum possible efficiency of a single-junction solar cell under un-concentrated sunlight, as a function of the semiconductor band gap. If the band gap is too high, most daylight photons cannot be absorbed; if it is too low, then most photons have much more energy than necessary to excite electrons across the band gap, and the rest is wasted. The semiconductors commonly used in commercial solar cells have band gaps near the peak of this curve, for example silicon (1.1eV) or CdTe (1.5eV). The Shockley–Queisser limit has been exceeded experimentally by combining materials with different band gap energies to make tandem solar cells.

The optical band gap determines what portion of the solar spectrum a photovoltaic cell absorbs. A semiconductor will not absorb photons of energy less than the band gap; and the energy of the electron-hole pair produced by a photon is equal to the bandgap energy. A luminescent solar converter uses a luminescent medium to downconvert photons with energies above the band gap to photon energies closer to the band gap of the semiconductor comprising the solar cell.

List of Band Gaps

Below are band gap values for some selected materials. For a comprehensive list of band gaps in semiconductors, see List of semiconductor materials.

Group	Material	Symbol	Band gap (eV) @ 302K
IV	Diamond	C	5.5
IV	Silicon	Si	1.11
IV	Germanium	Ge	0.67
III–V	Gallium nitride	GaN	3.4
III–V	Gallium phosphide	GaP	2.26
III–V	Gallium arsenide	GaAs	1.43
IV–V	Silicon nitride	Si_3N_4	5
IV–VI	Lead sulfide	PbS	0.37
IV–VI	Silicon dioxide	SiO_2	9
	Copper oxide	Cu_2O	2.1

Optical Versus Electronic Bandgap

In materials with a large exciton binding energy, it is possible for a photon to have just barely enough energy to create an exciton (bound electron–hole pair), but not enough energy to separate the electron and hole (which are electrically attracted to each other). In this situation, there is a distinction between "optical bandgap" and "electrical band gap" (or "transport gap"). The optical bandgap is the threshold for photons to be absorbed, while the transport gap is the threshold for creating an electron–hole pair that is *not* bound together. (The optical bandgap is at a lower energy than the transport gap.)

In almost all inorganic semiconductors, such as silicon, gallium arsenide, etc., there is very little interaction between electrons and holes (very small exciton binding energy), and therefore the optical and electronic bandgap are essentially identical, and the distinction between them is ignored. However, in some systems, including organic semiconductors and single-walled carbon nanotubes, the distinction may be significant.

In Photonics and Phononics

In photonics, band gaps or stop bands are ranges of photon frequencies where, if tunneling effects are neglected, no photons can be transmitted through a material. A material exhibiting this behaviour is known as a photonic crystal. The concept of hyperuniformity has broadened the range of photonic band gap materials, beyond photonic crystals. By applying the technique in supersymmetric quantum mechanics, a new class of optical disordered materials have been suggested, which support band gaps perfectly equivalent to those of crystals or quasicrystals.

Similar physics applies to phonons in a phononic crystal.

Valence and Conduction Bands

In solid-state physics, the valence band and conduction band are the bands closest to the Fermi level and thus determine the electrical conductivity of the solid. The valence band is the highest range of electron energies in which electrons are normally present at absolute zero temperature, while the conduction band is the lowest range of vacant electronic states. On a graph of the electronic band structure of a material, the valence band is located below the Fermi level, while the conduction band is located above it. This distinction is meaningless in metals as the highest band is partially filled, taking on the properties of both the valence and conduction bands.

Band Gap

In semiconductors and insulators the two bands are separated by a band gap, while in semimetals the bands overlap. A band gap is an energy range in a solid where no electron states can exist due to the quantization of energy. Electrical conductivity of non-metals is determined by the susceptibility of electrons to excitation from the valence band to the conduction band.

Electrical Conductivity

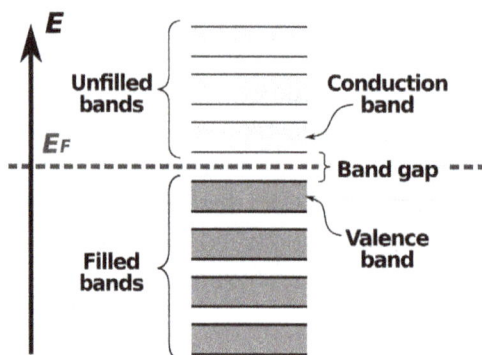

Semiconductor band structure See electrical conduction and semiconductor for a more detailed description of band structure.

In solids, the ability of electrons to act as charge carriers depends on the availability of vacant electronic states. This allows the electrons to increase their energy (i.e., accelerate) when an electric field is applied. This condition is only satisfied in the conduction band, since the valence band is full in non-metals.

As such, the electrical conductivity of a solid depends on its capability to flow electrons from the valence to the conduction band. Hence, in the case of a semimetal with an overlap region, the electrical conductivity is high. If there is a small band gap (E_g), then the flow of electrons from valence to conduction band is possible only if an external energy (thermal, etc.) is supplied; these groups with small E_g are called semiconductors. If the E_g is sufficiently high, then the flow of electrons from valence to conduction band becomes negligible under normal conditions; these groups are called insulators.

There is some conductivity in semiconductors, however. This is due to thermal excitation—some of the electrons get enough energy to jump the band gap in one go. Once they are in the conduction band, they can conduct electricity, as also can the hole they left behind in the valence band. The hole is an empty state that allows electrons in the valence band some degree of freedom.

Tight Binding

In solid-state physics, the tight-binding model (or TB model) is an approach to the calculation of electronic band structure using an approximate set of wave functions based upon superposition of wave functions for isolated atoms located at each atomic site. The method is closely related to the LCAO method (linear combination of atomic orbitals method) used in chemistry. Tight-binding models are applied to a wide variety of solids. The model gives good qualitative results in many cases and can be combined with other models that give better results where the tight-binding model fails. Though the tight-binding model is a one-electron model, the model also provides a basis for more advanced calculations like the calculation of surface states and application to various kinds of many-body problem and quasiparticle calculations.

Introduction

The name "tight binding" of this electronic band structure model suggests that this quantum mechanical model describes the properties of tightly bound electrons in solids. The electrons in this model should be tightly bound to the atom to which they belong and they should have limited interaction with states and potentials on surrounding atoms of the solid. As a result the wave function of the electron will be rather similar to the atomic orbital of the free atom to which it belongs. The energy of the electron will also be rather close to the ionization energy of the electron in the free atom or ion because the interaction with potentials and states on neighboring atoms is limited.

Though the mathematical formulation of the one-particle tight-binding Hamiltonian may look complicated at first glance, the model is not complicated at all and can be understood intuitively quite easily. There are only three kinds of matrix elements that play a significant role in the theory. Two of those three kinds of elements should be close to zero and can often be neglected. The most important elements in the model are the interatomic matrix elements, which would simply be called the bond energies by a chemist.

In general there are a number of atomic energy levels and atomic orbitals involved in the model. This can lead to complicated band structures because the orbitals belong to different point-group representations. The reciprocal lattice and the Brillouin zone often belong to a different space group than the crystal of the solid. High-symmetry points in the Brillouin zone belong to different point-group representations. When simple systems like the lattices of elements or simple compounds are studied it is often not very difficult to calculate eigenstates in high-symmetry points analytically. So the tight-binding model can provide nice examples for those who want to learn more about group theory.

The tight-binding model has a long history and has been applied in many ways and with many different purposes and different outcomes. The model doesn't stand on its own. Parts of the model can be filled in or extended by other kinds of calculations and models like the nearly-free electron model. The model itself, or parts of it, can serve as the basis for other calculations. In the study of conductive polymers, organic semiconductors and molecular electronics, for example, tight-binding-like models are applied in which the role of the atoms in the original concept is replaced by the molecular orbitals of conjugated systems and where the interatomic matrix elements are replaced by inter- or intramolecular hopping and tunneling parameters. These conductors nearly all have very anisotropic properties and sometimes are almost perfectly one-dimensional.

Historical Background

By 1928, the idea of a molecular orbital had been advanced by Robert Mulliken, who was influenced considerably by the work of Friedrich Hund. The LCAO method for approximating molecular orbitals was introduced in 1928 by B. N. Finklestein and G. E. Horowitz, while the LCAO method for solids was developed by Felix Bloch, as part of his doctoral dissertation in 1928, concurrently with and independent of the LCAO-MO approach. A much simpler interpolation scheme for approximating the electronic band structure, especially for the d-bands of transition metals, is the parameterized tight-binding method conceived in 1954 by John Clarke Slater and George Fred Koster, sometimes referred to as the SK tight-binding method. With the SK tight-binding method, electronic band structure calculations on a solid need not be carried out with full rigor as in the original Bloch's theorem but, rather, first-principles calculations are carried out only at high-symmetry points and the band structure is interpolated over the remainder of the Brillouin zone between these points.

In this approach, interactions between different atomic sites are considered as perturbations. There exist several kinds of interactions we must consider. The crystal Hamiltonian is only approximately a sum of atomic Hamiltonians located at different sites and atomic wave functions overlap adjacent atomic sites in the crystal, and so are not accurate representations of the exact wave function. There are further explanations in the next section with some mathematical expressions.

In the recent research about strongly correlated material the tight binding approach is basic approximation because highly localized electrons like 3-d transition metal electrons sometimes display strongly correlated behaviors. In this case, the role of electron-electron interaction must be considered using the many-body physics description.

The tight-binding model is typically used for calculations of electronic band structure and band gaps in the static regime. However, in combination with other methods such as the random phase approximation (RPA) model, the dynamic response of systems may also be studied.

Mathematical Formulation

We introduce the atomic orbitals $\varphi_m(r)$, which are eigenfunctions of the Hamiltonian H_{at} of a single isolated atom. When the atom is placed in a crystal, this atomic wave function overlaps adjacent atomic sites, and so are not true eigenfunctions of the crystal Hamiltonian. The overlap is less when electrons are tightly bound, which is the source of the descriptor "tight-binding". Any corrections to the atomic potential ΔUr required to obtain the true Hamiltonian H of the system, are assumed small:

$$H(\mathbf{r}) = \sum_{\mathbf{R_n}} H_{at}(\mathbf{r} - \mathbf{R_n}) + \Delta U(\mathbf{r}).$$

A solution ψ_m to the time-independent single electron Schrödinger equation is then approximated as a linear combination of atomic orbitals $\varphi_m(\mathbf{r} - \mathbf{R})$:

$$\psi_m(r) = \sum_{m,R_n} b_m(R_n)\, \varphi_m(r - R_n),$$

where m refers to the m-th atomic energy level and R_n locates an atomic site in the crystal lattice.

Translational Symmetry and Normalization

The Bloch theorem states that the wave function in crystal can change under translation only by a phase factor:

$$\psi(r + R_\ell) = e^{ik \cdot R_\ell} \psi(r),$$

where $\mathbf{k}$ is the wave vector of the wave function. Consequently, the coefficients satisfy

$$\sum_{m,R_n} b_m(R_n)\, \varphi_m(r - R_n + R_\ell) = e^{ik \cdot R_\ell} \sum_{m,R_n} b_m(R_n)\, \varphi_m(r - R_n).$$

By substituting $R_p = R_n - R_\ell$ we find

$$b_m(\mathbf{R_p} + \mathbf{R}_\ell) = e^{ik \cdot R_\ell} b_m(\mathbf{R_p}),\text{ (where in RHS we have replaced the dummy index } \mathbf{R_n} \text{ with } \mathbf{R_p})$$

or

$$b_m(\mathbf{R_l}) = e^{ik \cdot R_l} b_m(\mathbf{0}).$$

Normalizing the wave function to unity:

$$\int d^3r\, \psi_m^*(\mathbf{r}) \psi_m(\mathbf{r}) = 1$$

$$= \sum_{\mathbf{R_n}} b_m^*(\mathbf{R_n}) \sum_{\mathbf{R}_\ell} b_m(\mathbf{R}_\ell) \int d^3r\, \varphi_m^*(\mathbf{r} - \mathbf{R_n}) \varphi_m(\mathbf{r} - \mathbf{R}_\ell)$$

$$= b_m^*(0)b_m(0)\sum_{\mathbf{R_n}} e^{-i\mathbf{k}\cdot\mathbf{R_n}}\sum_{\mathbf{R_\ell}} e^{i\mathbf{k}\cdot\mathbf{R_\ell}} \int d^3r \, \varphi_m^*(\mathbf{r}-\mathbf{R_n})\varphi_m(\mathbf{r}-\mathbf{R_\ell})$$

$$= Nb_m^*(0)b_m(0)\sum_{\mathbf{R_p}} e^{-i\mathbf{k}\cdot\mathbf{R_p}} \int d^3r \, \varphi_m^*(\mathbf{r}-\mathbf{R_p})\varphi_m(\mathbf{r})$$

$$= Nb_m^*(0)b_m(0)\sum_{\mathbf{R_p}} e^{i\mathbf{k}\cdot\mathbf{R_p}} \int d^3r \, \varphi_m^*(\mathbf{r})\varphi_m(\mathbf{r}-\mathbf{R_p}),$$

so the normalization sets $b(o)$ as

$$b_m^*(0)b_m(0) = \frac{1}{N} \cdot \frac{1}{1+\sum_{\mathbf{R_p}\neq 0} e^{i\mathbf{k}\cdot\mathbf{R_p}}\alpha_m(\mathbf{R_p})},$$

where $\alpha(R_p)$ are the atomic overlap integrals, which frequently are neglected resulting in

$$b_m(0) \approx \frac{1}{\sqrt{N}},$$

and

$$\psi_m(\mathbf{r}) \approx \frac{1}{\sqrt{N}}\sum_{\mathbf{R_n}} e^{i\mathbf{k}\cdot\mathbf{R_n}}\varphi_m(\mathbf{r}-\mathbf{R_n}).$$

The Tight Binding Hamiltonian

Using the tight binding form for the wave function, and assuming only the *m-th* atomic energy level is important for the *m-th* energy band, the Bloch energies ε_m are of the form

$$\varepsilon_m = \int d^3r \, \psi^*(\mathbf{r})H(\mathbf{r})\psi(\mathbf{r})$$

$$= \sum_{\mathbf{R_n}} b^*(\mathbf{R_n}) \int d^3r \, \varphi^*(\mathbf{r}-\mathbf{R_n})H(\mathbf{r})\psi(\mathbf{r})$$

$$= \sum_{\mathbf{R_\ell}}\sum_{\mathbf{R_n}} b^*(\mathbf{R_n}) \int d^3r \, \varphi^*(\mathbf{r}-\mathbf{R_n})H_{at}(\mathbf{r}-\mathbf{R_\ell})\psi(\mathbf{r}) + \sum_{\mathbf{R_n}} b^*(\mathbf{R_n}) \int d^3r \, \varphi^*(\mathbf{r}-\mathbf{R_n})\Delta U(\mathbf{r})\psi(\mathbf{r}).$$

$$\approx E_m + b^*(0)\sum_{\mathbf{R_n}} e^{-i\mathbf{k}\cdot\mathbf{R_n}} \int d^3r \, \varphi^*(\mathbf{r}-\mathbf{R_n})\Delta U(\mathbf{r})\psi(\mathbf{r}).$$

Here terms involving the atomic Hamiltonian at sites other than where it is centered are neglected. The energy then becomes

$$\varepsilon_m(\mathbf{k}) = E_m - N\,|\,b(0)\,|^2 \left(\beta_m + \sum_{\mathbf{R_n}\neq 0}\sum_l \gamma_{m,l}(\mathbf{R_n})e^{i\mathbf{k}\cdot\mathbf{R_n}} \right),$$

$$= E_m - \frac{\beta_m + \sum_{R_n \neq 0} \sum_l e^{ik \cdot R_n} \gamma_{m,l}(R_n)}{1 + \sum_{R_n \neq 0} \sum_l e^{ik \cdot R_n} \alpha_{m,l}(R_n)},$$

where E_m is the energy of the m-th atomic level, and $\alpha_{m,l}$, β_m and $\gamma_{m,l}$ are the tight binding matrix elements.

The Tight Binding Matrix Elements

The element

$$\beta_m = -\int \varphi_m^*(\mathbf{r}) \Delta U(\mathbf{r}) \varphi_m(\mathbf{r}) d^3 r ,$$

is the atomic energy shift due to the potential on neighboring atoms. This term is relatively small in most cases. If it is large it means that potentials on neighboring atoms have a large influence on the energy of the central atom.

The next term

$$\gamma_{m,l}(\mathbf{R}_n) = -\int \varphi_m^*(\mathbf{r}) \Delta U(\mathbf{r}) \varphi_l(\mathbf{r} - \mathbf{R}_n) d^3 r ,$$

is the inter atomic matrix element between the atomic orbitals m and l on adjacent atoms. It is also called the bond energy or two center integral and it is the most important element in the tight binding model.

The last terms

$$\alpha_{m,l}(\mathbf{R}_n) = \int \varphi_m^*(\mathbf{r}) \varphi_l(\mathbf{r} - \mathbf{R}_n) d^3 r ,$$

denote the overlap integrals between the atomic orbitals m and l on adjacent atoms.

Evaluation of the Matrix Elements

As mentioned before the values of the β_m-matrix elements are not so large in comparison with the ionization energy because the potentials of neighboring atoms on the central atom are limited. If β_m is not relatively small it means that the potential of the neighboring atom on the central atom is not small either. In that case it is an indication that the tight binding model is not a very good model for the description of the band structure for some reason. The inter atomic distances can be too small or the charges on the atoms or ions in the lattice is wrong for example.

The inter atomic matrix elements $\gamma_{m,l}$ can be calculated directly if the atomic wave functions and the potentials are known in detail. Most often this is not the case. There are numerous ways to get parameters for these matrix elements. Parameters can be obtained from chemical bond energy data. Energies and eigenstates on some high symmetry points in the Brillouin zone can be evaluated and values integrals in the matrix elements can be matched with band structure data from other sources.

The inter atomic overlap matrix elements $\alpha_{m,l}$ should be rather small or neglectable. If they are large it is again an indication that the tight binding model is of limited value for some purposes. Large overlap is an indication for too short inter atomic distance for example. In metals and transition metals the broad s-band or sp-band can be fitted better to an existing band structure calculation by the introduction of next-nearest-neighbor matrix elements and overlap integrals but fits like that don't yield a very useful model for the electronic wave function of a metal. Broad bands in dense materials are better described by a nearly free electron model.

The tight binding model works particularly well in cases where the band width is small and the electrons are strongly localized, like in the case of d-bands and f-bands. The model also gives good results in the case of open crystal structures, like diamond or silicon, where the number of neighbors is small. The model can easily be combined with a nearly free electron model in a hybrid NFE-TB model.

Connection to Wannier Functions

Bloch wave functions describe the electronic states in a periodic crystal lattice. Bloch functions can be represented as a Fourier series

$$\psi_m(\mathbf{k},\mathbf{r}) = \frac{1}{\sqrt{N}}\sum_n a_m(\mathbf{R_n},\mathbf{r})e^{i\mathbf{k}\cdot\mathbf{R_n}},$$

where R_n denotes an atomic site in a periodic crystal lattice, k is the wave vector of the Bloch wave, r is the electron position, m is the band index, and the sum is over all N atomic sites. The Bloch wave is an exact eigensolution for the wave function of an electron in a periodic crystal potential corresponding to an energy $E_m(k)$, and is spread over the entire crystal volume.

Using the Fourier transform analysis, a spatially localized wave function for the m-th energy band can be constructed from multiple Bloch waves:

$$a_m(\mathbf{R_n},\mathbf{r}) = \frac{1}{\sqrt{N}}\sum_k e^{-i\mathbf{k}\cdot\mathbf{R_n}}\psi_m(\mathbf{k},\mathbf{r}) = \frac{1}{\sqrt{N}}\sum_k e^{i\mathbf{k}\cdot(\mathbf{r}-\mathbf{R_n})}u_m(\mathbf{k},\mathbf{r}).$$

These real space wave functions $a_m(\mathbf{R_n},\mathbf{r})$ are called Wannier functions, and are fairly closely localized to the atomic site R_n. Of course, if we have exact Wannier functions, the exact Bloch functions can be derived using the inverse Fourier transform.

However it is not easy to calculate directly either Bloch functions or Wannier functions. An approximate approach is necessary in the calculation of electronic structures of solids. If we consider the extreme case of isolated atoms, the Wannier function would become an isolated atomic orbital. That limit suggests the choice of an atomic wave function as an approximate form for the Wannier function, the so-called tight binding approximation.

Second Quantization

Modern explanations of electronic structure like t-J model and Hubbard model are based on tight binding model. Tight binding can be understood by working under a second quantization formalism.

Using the atomic orbital as a basis state, the second quantization Hamiltonian operator in the tight binding framework can be written as:

$$H = -t \sum_{\langle i,j \rangle, \sigma} (c_{i,\sigma}^{\dagger} c_{j,\sigma} + h.c.),$$

$c_{i\sigma}^{\dagger}, c_{j\sigma}$ - creation and annihilation operators

σ - spin polarization

t - hopping integral

$\langle i, j \rangle$ -nearest neighbor index

Here, hopping integral t corresponds to the transfer integral γ in tight binding model. Considering extreme cases of $t \to 0$, it is impossible for an electron to hop into neighboring sites. This case is the isolated atomic system. If the hopping term is turned on ($t > 0$) electrons can stay in both sites lowering their kinetic energy.

In the strongly correlated electron system, it is necessary to consider the electron-electron interaction. This term can be written in

$$H_{ee} = \frac{1}{2} \sum_{n,m,\sigma} \langle n_1 m_1, n_2 m_2 \mid \frac{e^2}{\mid r_1 - r_2 \mid} \mid n_3 m_3, n_4 m_4 \rangle c_{n_1 m_1 \sigma_1}^{\dagger} c_{n_2 m_2 \sigma_2}^{\dagger} c_{n_4 m_4 \sigma_2} c_{n_3 m_3 \sigma_1}$$

This interaction Hamiltonian includes direct Coulomb interaction energy and exchange interaction energy between electrons. There are several novel physics induced from this electron-electron interaction energy, such as metal-insulator transitions (MIT), high-temperature superconductivity, and several quantum phase transitions.

Example: one-dimensional s-band

Here the tight binding model is illustrated with a s-band model for a string of atoms with a single s-orbital in a straight line with spacing a and σ bonds between atomic sites.

To find approximate eigenstates of the Hamiltonian, we can use a linear combination of the atomic orbitals

$$\mid k \rangle = \frac{1}{\sqrt{N}} \sum_{n=1}^{N} e^{inka} \mid n \rangle$$

where N = total number of sites and k is a real parameter with $-\frac{\pi}{a} \leq k \leq \frac{\pi}{a}$. (This wave function is normalized to unity by the leading factor $1/\sqrt{N}$ provided overlap of atomic wave functions is ignored.) Assuming only nearest neighbor overlap, the only non-zero matrix elements of the Hamiltonian can be expressed as

$$\langle n \mid H \mid n \rangle = E_0 = E_i - U .$$

$$\langle n \pm 1 | H | n \rangle = -\Delta$$

$$\langle n | n \rangle = 1; \quad \langle n \pm 1 | n \rangle = S.$$

The energy E_i is the ionization energy corresponding to the chosen atomic orbital and U is the energy shift of the orbital as a result of the potential of neighboring atoms. The $\langle n \pm 1 | H | n \rangle = -\Delta$ elements, which are the Slater and Koster interatomic matrix elements, are the bond energies $E_{i,j}$. In this one dimensional s-band model we only have σ-bonds between the s-orbitals with bond energy $E_{s,s} = V_{ss\sigma}$. The overlap between states on neighboring atoms is S. We can derive the energy of the state $| k$ using the above equation:

$$H | k \rangle = \frac{1}{\sqrt{N}} \sum_n e^{inka} H | n \rangle$$

$$\langle k | H | k \rangle = \frac{1}{N} \sum_{n,m} e^{i(n-m)ka} \langle m | H | n \rangle$$

$$= \frac{1}{N} \sum_n \langle n | H | n \rangle + \frac{1}{N} \sum_n \langle n-1 | H | n \rangle e^{+ika} + \frac{1}{N} \sum_n \langle n+1 | H | n \rangle e^{-ika}$$

$$= E_0 - 2\Delta \cos(ka),$$

where, for example,

$$\frac{1}{N} \sum_n \langle n | H | n \rangle = E_0 \frac{1}{N} \sum_n 1 = E_0,$$

and

$$\frac{1}{N} \sum_n \langle n-1 | H | n \rangle e^{+ika} = -\Delta e^{ika} \frac{1}{N} \sum_n 1 = -\Delta e^{ika}.$$

$$\frac{1}{N} \sum_n \langle n-1 | n \rangle e^{+ika} = S e^{ika} \frac{1}{N} \sum_n 1 = S e^{ika}.$$

Thus the energy of this state $| k$ can be represented in the familiar form of the energy dispersion:

$$E(k) = \frac{E_0 - 2\Delta \cos(ka)}{1 + 2S \cos(ka)}.$$

- For $k = 0$ the energy is $E = (E_0 - 2\Delta)/(1 + 2S)$ and the state consists of a sum of all atomic orbitals. This state can be viewed as a chain of bonding orbitals.

- For $k = \pi/(2a)$ the energy is $E = E_0$ and the state consists of a sum of atomic orbitals which are a factor $e^{i\pi/2}$ out of phase. This state can be viewed as a chain of non-bonding orbitals.

- Finally for $k = \pi / a$ the energy is $E = E_0$ and the state consists of an alternating sum of atomic orbitals. This state can be viewed as a chain of anti-bonding orbitals.

This example is readily extended to three dimensions, for example, to a body-centered cubic or face-centered cubic lattice by introducing the nearest neighbor vector locations in place of simply $n\,a$. Likewise, the method can be extended to multiple bands using multiple different atomic orbitals at each site. The general formulation above shows how these extensions can be accomplished.

Fermi Liquid Theory

Fermi liquid theory (also known as Landau–Fermi liquid theory) is a theoretical model of interacting fermions that describes the normal state of most metals at sufficiently low temperatures. The interaction between the particles of the many-body system does not need to be small. The phenomenological theory of Fermi liquids was introduced by the Soviet physicist Lev Davidovich Landau in 1956, and later developed by Alexei Abrikosov and Isaak Khalatnikov using diagrammatic perturbation theory. The theory explains why some of the properties of an interacting fermion system are very similar to those of the Fermi gas (i.e. non-interacting fermions), and why other properties differ.

Important examples of where Fermi liquid theory has been successfully applied are most notably electrons in most metals and Liquid He-3. Liquid He-3 is a Fermi liquid at low temperatures (but not low enough to be in its superfluid phase.) He-3 is an isotope of helium, with 2 protons, 1 neutron and 2 electrons per atom. Because there is an odd number of fermions inside the atom, the atom itself is also a fermion. The electrons in a normal (non-superconducting) metal also form a Fermi liquid, as do the nucleons (protons and neutrons) in an atomic nucleus. Strontium ruthenate displays some key properties of Fermi liquids, despite being a strongly correlated material, and is compared with high temperature superconductors like cuprates.

Description

The key ideas behind Landau's theory are the notion of *adiabaticity* and the exclusion principle. Consider a non-interacting fermion system (a Fermi gas), and suppose we "turn on" the interaction slowly. Landau argued that in this situation, the ground state of the Fermi gas would adiabatically transform into the ground state of the interacting system.

By Pauli's exclusion principle, the ground state Ψ_0 of a Fermi gas consists of fermions occupying all momentum states corresponding to momentum $p < p_F$ with all higher momentum states unoccupied. As interaction is turned on, the spin, charge and momentum of the fermions corresponding to the occupied states remain unchanged, while their dynamical properties, such as their mass, magnetic moment etc. are *renormalized* to new values. Thus, there is a one-to-one correspondence between the elementary excitations of a Fermi gas system and a Fermi liquid system. In the context of Fermi liquids, these excitations are called "quasi-particles".

Landau quasiparticles are long-lived excitations with a lifetime τ that satisfies $\dfrac{\hbar}{\tau} \ll \epsilon_p$ where ϵ_p is the quasiparticle energy (measured from the Fermi energy.) At finite temperature, ϵ_p is on the order

of the thermal energy $k_B T$, and the condition for Landau quasiparticles can be reformulated as
$$\frac{\hbar}{\tau} k_B T..$$

For this system, the Green's function can be written (near its poles) in the form

$$G(\omega, p) \approx \frac{Z}{\omega + \mu - \epsilon(p)}$$

where μ is the chemical potential and $\epsilon(p)$ is the energy corresponding to the given momentum state.

The value Z is called the *quasiparticle residue* and is very characteristic of Fermi liquid theory. The spectral function for the system can be directly observed via ARPES experiment, and can be written (in the limit of low-lying excitations) in the form:

$$A(\vec{k}, \omega) = Z\delta(\omega - v_F k_{\parallel})$$

where v_F is the Fermi velocity.

Physically, we can say that a propagating fermion interacts with its surrounding in such a way that the net effect of the interactions is to make the fermion behave as a "dressed" fermion, altering its effective mass and other dynamical properties. These "dressed" fermions are what we think of as "quasiparticles".

Another important property of Fermi liquids is related to the scattering cross section for electrons. Suppose we have an electron with energy ϵ_1 above the Fermi surface, and suppose it scatters with a particle in the Fermi sea with energy ϵ_2. By Pauli's exclusion principle, both the particles after scattering have to lie above the Fermi surface, with energies $\epsilon_3, \epsilon_4 > \epsilon_F$ Now, suppose the initial electron has energy very close to the Fermi surface $\epsilon \approx \epsilon_F$ Then, we have that $\epsilon_2, \epsilon_3, \epsilon_4$ also have to be very close to the Fermi surface. This reduces the phase space volume of the possible states after scattering, and hence, by Fermi's golden rule, the scattering cross section goes to zero. Thus we can say that the lifetime of particles at the Fermi surface goes to infinity.

Similarities to Fermi Gas

The Fermi liquid is qualitatively analogous to the non-interacting Fermi gas, in the following sense: The system's dynamics and thermodynamics at low excitation energies and temperatures may be described by substituting the non-interacting fermions with interacting quasiparticles, each of which carries the same spin, charge and momentum as the original particles. Physically these may be thought of as being particles whose motion is disturbed by the surrounding particles and which themselves perturb the particles in their vicinity. Each many-particle excited state of the interacting system may be described by listing all occupied momentum states, just as in the non-interacting system. As a consequence, quantities such as the heat capacity of the Fermi liquid behave qualitatively in the same way as in the Fermi gas (e.g. the heat capacity rises linearly with temperature).

Differences from Fermi gas

The following differences to the non-interacting Fermi gas arise:

Energy

The energy of a many-particle state is not simply a sum of the single-particle energies of all occupied states. Instead, the change in energy for a given change δn_k in occupation of states k contains terms both linear and quadratic in δn_k (for the Fermi gas, it would only be linear, $\delta n_k \grave{o}_k$, where $\grave{o}_k$ denotes the single-particle energies). The linear contribution corresponds to renormalized single-particle energies, which involve, e.g., a change in the effective mass of particles. The quadratic terms correspond to a sort of "mean-field" interaction between quasiparticles, which is parameterized by so-called Landau Fermi liquid parameters and determines the behaviour of density oscillations (and spin-density oscillations) in the Fermi liquid. Still, these mean-field interactions do not lead to a scattering of quasi-particles with a transfer of particles between different momentum states.

The renormalization of the mass of a fluid of interacting fermions can be calculated from first principles using many-body computational techniques. For the two-dimensional homogeneous electron gas, *GW* calculations and quantum Monte Carlo methods have been used to calculate renormalized quasiparticle effective masses.

Specific heat and compressibility

Specific heat, compressibility, spin-susceptibility and other quantities show the same qualitative behaviour (e.g. dependence on temperature) as in the Fermi gas, but the magnitude is (sometimes strongly) changed.

Interactions

In addition to the mean-field interactions, some weak interactions between quasiparticles remain, which lead to scattering of quasiparticles off each other. Therefore, quasiparticles acquire a finite lifetime. However, at low enough energies above the Fermi surface, this lifetime becomes very long, such that the product of excitation energy (expressed in frequency) and lifetime is much larger than one. In this sense, the quasiparticle energy is still well-defined (in the opposite limit, Heisenberg's uncertainty relation would prevent an accurate definition of the energy).

Structure

The structure of the "bare" particle's (as opposed to quasiparticle) Green's function is similar to that in the Fermi gas (where, for a given momentum, the Green's function in frequency space is a delta peak at the respective single-particle energy). The delta peak in the density-of-states is broadened (with a width given by the quasiparticle lifetime). In addition (and in contrast to the quasiparticle Green's function), its weight (integral over frequency) is suppressed by a quasiparticle weight factor $0 < Z < 1$. The remainder of the total weight is in a broad "incoherent background", corresponding to the strong effects of interactions on the fermions at short time-scales.

Distribution

The distribution of particles (as opposed to quasiparticles) over momentum states at zero temperature still shows a discontinuous jump at the Fermi surface (as in the Fermi gas), but it does not drop from 1 to 0: the step is only of size Z.

Electrical Resistivity

In a metal the resistivity at low temperatures is dominated by electron-electron scattering in combination with umklapp scattering. For a Fermi liquid, the resistivity from this mechanism varies as T^2, which is often taken as an experimental check for Fermi liquid behaviour (in addition to the linear temperature-dependence of the specific heat), although it only arises in combination with the lattice. In certain cases, umklapp scattering is not required. For example, the resistivity of compensated semimetals scales as T^2 because of mutual scattering of electron and hole. This is known as the Baber mechanism.

Optical Response

Fermi liquid theory predicts that the scattering rate, which governs the optical response of metals, not only depends quadratically on temperature (thus causing the T^2 dependence of the dc resistance), but it also depends quadratically on frequency. This is in contrast to the Drude prediction for non-interacting metallic electrons, where the scattering rate is a constant as a function of frequency.One material in which optical Fermi liquid behavior was experimentally observed is the low-temperature metallic phase of Sr_2RuO_4.

Instabilities of the Fermi Liquid

The experimental observation of exotic phases in strongly correlated systems has triggered an enormous effort from the theoretical community to try to understand their microscopic origin. One possible route to detect instabilities of a Fermi liquid is precisely the analysis done by Pomeranchuk. Due to that, the Pomeranchuk instability has been studied by several authors with different techniques in the last few years and in particular, the instability of the Fermi liquid towards the nematic phase was investigated for several models.

Non-fermi Liquids

The term non-Fermi liquid, also known as "strange metal", is used to describe a system which displays breakdown of Fermi-liquid behaviour. The simplest example of such a system is the system of interacting fermions in one-dimension, called *Luttinger liquid*. Although Luttinger liquids are physically similar to Fermi liquids, the restriction to one dimension gives rise to several qualitative differences such as the absence of a *quasiparticle peak* in the momentum dependent spectral function, spin-charge separation, and the presence of spin density waves. One cannot ignore the existence of interactions in one-dimension and has to describe the problem with a non-Fermi theory, where Luttinger liquid is one of them. At small finite spin-temperatures in one-dimension the ground-state of the system is described by spin-incoherent Luttinger liquid (SILL).

Another example of such behaviour is observed at quantum critical points of certain second-order phase transitions, such as Heavy fermion criticality, Mott criticality and high-T_c cuprate phase transitions. The ground state of such transitions is characterized by the presence of a sharp Fermi surface, although there may not be well-defined quasiparticles. That is, on approaching the critical point, it is observed that the quasiparticle residue $Z \to 0$

On the other hand, specific quantum critical point represented by fermion condensation quantum phase transition (see e.g. strongly correlated quantum spin liquid) supports quasiparticles with the finite quasiparticle residue $Z > 0$. This phase transition does preserve the Pomeranchuk stability conditions, and proffers a new way to violate the stability of Fermi liquid. These unique properties of the phase transition allow one to explain both the scaling and the non-Fermi liquid behaviour observed in heavy fermion compounds.

Understanding the behaviour of non-Fermi liquids is an important problem in condensed matter physics. Approaches towards explaining these phenomena include the treatment of *marginal Fermi liquids*; attempts to understand critical points and derive scaling relations; and descriptions using *emergent* gauge theories with techniques of holographic gauge/gravity duality.

Fermi Level

The Fermi level is the total chemical potential for electrons (or electrochemical potential for electrons) and is usually denoted by μ or E_F. The Fermi level of a body is a thermodynamic quantity, and its significance is the thermodynamic work required to add one electron to the body (not counting the work required to remove the electron from wherever it came from). A precise understanding of the Fermi level—how it relates to electronic band structure in determining electronic properties, how it relates to the voltage and flow of charge in an electronic circuit—is essential to an understanding of solid-state physics.

In a band structure picture, the Fermi level can be considered to be a hypothetical energy level of an electron, such that at thermodynamic equilibrium this energy level would have a *50% probability of being occupied at any given time*. The position of the Fermi level with the relation to the band energy levels is a crucial factor in determining electrical properties. The Fermi level does not necessarily correspond to an actual energy level (in an insulator the Fermi level lies in the band gap), nor does it require the existence of a band structure. Nonetheless, the Fermi level is a precisely defined thermodynamic quantity, and differences in Fermi level can be measured simply with a voltmeter.

The Fermi Level and Voltage

Sometimes it is said that electric currents are driven by differences in electrostatic potential (Galvani potential), but this is not exactly true. As a counterexample, multi-material devices such as p–n junctions contain internal electrostatic potential differences at equilibrium, yet without any accompanying net current; if a voltmeter is attached to the junction, one simply measures zero volts. Clearly, the electrostatic potential is not the only factor influencing the flow of charge in a material—Pauli repulsion, carrier concentration gradients, electromagnetic induction, and thermal effects also play an important role.

A voltmeter measures differences in Fermi level divided by electron charge.

In fact, the quantity called *voltage* as measured in an electronic circuit has a simple relationship to the chemical potential for electrons (Fermi level). When the leads of a voltmeter are attached to two points in a circuit, the displayed voltage is a measure of the *total* work transferred when a unit charge is allowed to move from one point to the other. If a simple wire is connected between two points of differing voltage (forming a short circuit), current will flow from positive to negative voltage, converting the available work into heat.

The Fermi level of a body expresses the work required to add an electron to it, or equally the work obtained by removing an electron. Therefore, $V_A - V_B$, the observed difference in voltage between two points, A and B, in an electronic circuit is exactly related to the corresponding chemical potential difference, $\mu_A - \mu_B$, in Fermi level by the formula

$$V_A - V_B = \frac{\mu_A - \mu_B}{-e}$$

where $-e$ is the electron charge.

From the above discussion it can be seen that electrons will move from a body of high μ (low voltage) to low μ (high voltage) if a simple path is provided. This flow of electrons will cause the lower μ to increase (due to charging or other repulsion effects) and likewise cause the higher μ to decrease. Eventually, μ will settle down to the same value in both bodies. This leads to an important fact regarding the equilibrium (off) state of an electronic circuit:

> *An electronic circuit in thermodynamic equilibrium will have a constant Fermi level throughout its connected parts.*

This also means that the voltage (measured with a voltmeter) between any two points will be zero, at equilibrium. Note that thermodynamic equilibrium here requires that the circuit be internally connected and not contain any batteries or other power sources, nor any variations in temperature.

The Fermi Level and band Structure

In the band theory of solids, electrons are considered to occupy a series of bands composed of single-particle energy eigenstates each labelled by ϵ. Although this single particle picture is an approximation, it greatly simplifies the understanding of electronic behaviour and it generally provides correct results when applied correctly.

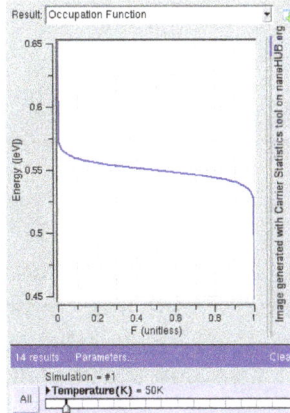

Fermi function F($\dot{O}$) vs. energy $\dot{O}$, with μ = 0.55 eV and for various temperatures in the range 50K ≤ T ≤ 375K.

The Fermi–Dirac distribution, $f(\epsilon)$,, gives the probability that (at thermodynamic equilibrium) a state having energy ϵ is occupied by an electron:

$$f(\epsilon) = \frac{1}{e^{(\epsilon-\mu)/kT}+1}$$

Here, T is the absolute temperature and k is Boltzmann's constant. If there is a state at the Fermi level ($\epsilon = \mu$), then this state will have a 50% chance of being occupied. The distribution is plotted in the left figure. The closer f is to 1, the higher chance this state is occupied. The closer f is to 0, the higher chance this state is empty.

The location of μ within a material's band structure is important in determining the electrical behaviour of the material.

- In an insulator, μ lies within a large band gap, far away from any states that are able to carry current.

- In a metal, semimetal or degenerate semiconductor, μ lies within a delocalized band. A large number of states nearby μ are thermally active and readily carry current.

- In an intrinsic or lightly doped semiconductor, μ is close enough to a band edge that there are a dilute number of thermally excited carriers residing near that band edge.

In semiconductors and semimetals the position of μ relative to the band structure can usually be controlled to a significant degree by doping or gating. These controls do not change μ which is fixed by the electrodes, but rather they cause the entire band structure to shift up and down (sometimes also changing the band structure's shape). For further information about the Fermi levels of semiconductors, see (for example) Sze.

Local Conduction Band Referencing, Internal Chemical Potential and the Parameter ζ

If the symbol ε is used to denote an electron energy level measured relative to the energy of the edge of its enclosing band, ϵ_c, then in general we have $\varepsilon = \epsilon - \epsilon_c$. We can define a parameter ζ that references the Fermi level with respect to the band edge:

$$\zeta = \mu - \epsilon_C.$$

It follows that the Fermi–Dirac distribution function can be written as

$$f(\mathcal{E}) = \frac{1}{e^{(\mathcal{E}-\zeta)/kT} + 1}.$$

The band theory of metals was initially developed by Sommerfeld, from 1927 onwards, who paid great attention to the underlying thermodynamics and statistical mechanics. Confusingly, in some contexts the band-referenced quantity ζ may be called the *Fermi level, chemical potential,* or *electrochemical potential*, leading to ambiguity with the globally-referenced Fermi level. In this article, the terms *conduction-band referenced Fermi level* or *internal chemical potential* are used to refer to ζ.

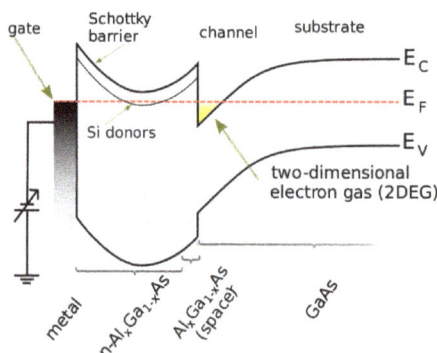

Example of variations in conduction band edge E_C in a band diagram of GaAs/AlGaAs heterojunction-based high-electron-mobility transistor.

ζ is directly related to the number of active charge carriers as well as their typical kinetic energy, and hence it is directly involved in determining the local properties of the material (such as electrical conductivity). For this reason it is common to focus on the value of ζ when concentrating on the properties of electrons in a single, homogeneous conductive material. By analogy to the energy states of a free electron, the E of a state is the kinetic energy of that state and ϵ_C is its potential energy. With this in mind, the parameter, ζ, could also be labelled the *Fermi kinetic energy*.

Unlike μ, the parameter, ζ, is not a constant at equilibrium, but rather varies from location to location in a material due to variations in ϵ_C, which is determined by factors such as material quality and impurities/dopants. Near the surface of a semiconductor or semimetal, ζ can be strongly controlled by externally applied electric fields, as is done in a field effect transistor. In a multi-band material, ζ may even take on multiple values in a single location. For example, in a piece of aluminum metal there are two conduction bands crossing the Fermi level (even more bands in other materials); each band has a different edge energy, ϵ_C, and a different ζ.

The value of ζ at zero temperature is widely known as the Fermi energy, sometimes written ζ_0. Confusingly (again), the name *Fermi energy* sometimes is used to refer to ζ at non-zero temperature.

The Fermi Level and Temperature out of Equilibrium

The Fermi level, μ, and temperature, T, are well defined constants for a solid-state device in thermodynamic equilibrium situation, such as when it is sitting on the shelf doing nothing. When the device is brought out of equilibrium and put into use, then strictly speaking the Fermi level and temperature are no longer well defined. Fortunately, it is often possible to define a quasi-Fermi level and quasi-temperature for a given location, that accurately describe the occupation of states in terms of a thermal distribution. The device is said to be in *quasi-equilibrium* when and where such a description is possible.

The quasi-equilibrium approach allows one to build a simple picture of some non-equilibrium effects as the electrical conductivity of a piece of metal (as resulting from a gradient of μ) or its thermal conductivity (as resulting from a gradient in T). The quasi-μ and quasi-T can vary (or not exist at all) in any non-equilibrium situation, such as:

- If the system contains a chemical imbalance (as in a battery).

- If the system is exposed to changing electromagnetic fields. (as in capacitors, inductors, and transformers).

- Under illumination from a light-source with a different temperature, such as the sun (as in solar cells),

- When the temperature is not constant within the device (as in thermocouples),

- When the device has been altered, but has not had enough time to re-equilibrate (as in piezoelectric or pyroelectric substances).

In some situations, such as immediately after a material experiences a high-energy laser pulse, the electron distribution cannot be described by any thermal distribution. One cannot define the quasi-Fermi level or quasi-temperature in this case; the electrons are simply said to be *non-thermalized*. In less dramatic situations, such as in a solar cell under constant illumination, a quasi-equilibrium description may be possible but requiring the assignment of distinct values of μ and T to different bands (conduction band vs. valence band). Even then, the values of μ and T may jump discontinuously across a material interface (e.g., p–n junction) when a current is being driven, and be ill-defined at the interface itself.

Technicalities

Terminology Problems

The term, *Fermi level*, is mainly used in discussing the solid state physics of electrons in semiconductors, and a precise usage of this term is necessary to describe band diagrams in devices comprising different materials with different levels of doping. In these contexts, however, one may also see Fermi level used imprecisely to refer to the *band-referenced Fermi level*, $\mu - \epsilon_c$, called ζ above. It is common to see scientists and engineers refer to "controlling", "pinning", or "tuning" the Fermi level inside a conductor, when they are in fact describing changes in ϵ_c due to doping or the field effect. In fact, thermodynamic equilibrium guarantees that the Fermi level in a conductor is *always* fixed to be exactly equal to the Fermi level of the electrodes; only the band structure (not the Fermi level) can be changed by doping or the field effect. A similar ambiguity exists between the terms, *chemical potential* and *electrochemical potential*.

It is also important to note that Fermi *level* is not necessarily the same thing as Fermi *energy*. In the wider context of quantum mechanics, the term Fermi energy usually refers to *the maximum kinetic energy of a fermion in an idealized non-interacting, disorder free, zero temperature Fermi gas*. This concept is very theoretical (there is no such thing as a non-interacting Fermi gas, and zero temperature is impossible to achieve). However, it finds some use in approximately describing white dwarfs, neutron stars, atomic nuclei, and electrons in a metal. On the other hand, in the fields of semiconductor physics and engineering, *Fermi energy* often is used to refer to the Fermi level described in this article.

Fermi Level Referencing and the Location of zero Fermi Level

Much like the choice of origin in a coordinate system, the zero point of energy can be defined arbitrarily. Observable phenomena only depend on energy differences. When comparing distinct bodies, however, it is important that they all be consistent in their choice of the location of zero energy, or else nonsensical results will be obtained. It can therefore be helpful to explicitly name a common point to ensure that different components are in agreement. On the other hand, if a reference point is inherently ambiguous (such as "the vacuum") it will instead cause more problems.

A practical and well-justified choice of common point is a bulky, physical conductor, such as the electrical ground or earth. Such a conductor can be considered to be in a good thermodynamic equilibrium and so its μ is well defined. It provides a reservoir of charge, so that large numbers of electrons may be added or removed without incurring charging effects. It also has the advantage of being accessible, so that the Fermi level of any other object can be measured simply with a voltmeter.

Why it is not Advisable to use "the Energy in Vacuum" as a Reference Zero

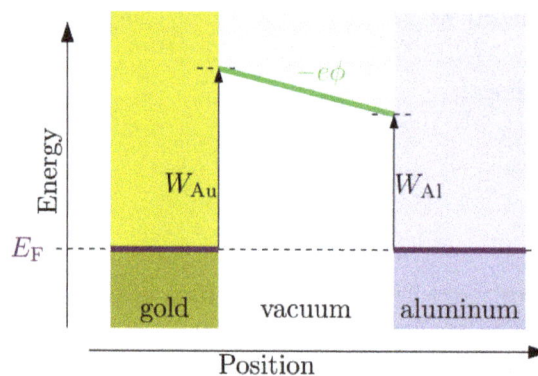

When the two metals depicted here are in thermodynamic equilibrium as shown (equal Fermi levels E_F), the vacuum electrostatic potential ⋅ is not flat due to a difference in work function.

In principle, one might consider using the state of a stationary electron in the vacuum as a reference point for energies. This approach is not advisable unless one is careful to define exactly where *the vacuum* is. The problem is that not all points in the vacuum are equivalent.

At thermodynamic equilibrium, it is typical for electrical potential differences of order 1 V to exist in the vacuum (Volta potentials). The source of this vacuum potential variation is the variation in work function between the different conducting materials exposed to vacuum. Just outside a conductor, the electrostatic potential depends sensitively on the material, as well as which surface is selected (its crystal orientation, contamination, and other details).

The parameter that gives the best approximation to universality is the Earth-referenced Fermi level suggested above. This also has the advantage that it can be measured with a voltmeter.

Discrete charging effects in small systems

In cases where the "charging effects" due to a single electron are non-negligible, the above definitions should be clarified. For example, consider a capacitor made of two identical parallel-plates. If the capacitor is uncharged, the Fermi level is the same on both sides, so one might think that it should take no energy to move an electron from one plate to the other. But when the electron has been moved, the capacitor has become (slightly) charged, so this does take a slight amount of energy. In a normal capacitor, this is negligible, but in a nano-scale capacitor it can be more important.

In this case one must be precise about the thermodynamic definition of the chemical potential as well as the state of the device: is it electrically isolated, or is it connected to an electrode?

- When the body is able to exchange electrons and energy with an electrode (reservoir), it is described by the grand canonical ensemble. The value of chemical potential μ can be said to be fixed by the electrode, and the number of electrons N on the body may fluctuate. In this case, the chemical potential of a body is the infinitesimal amount of work needed to increase the *average* number of electrons by an infinitesimal amount (even though the number of electrons at any time is an integer, the average number varies continuously.):

$$\mu(\langle N\rangle,T) = \left(\frac{\partial F}{\partial \langle N\rangle}\right)_T,$$

where $F(N, T)$ is the free energy function of the grand canonical ensemble.

- If the number of electrons in the body is fixed (but the body is still thermally connected to a heat bath), then it is in the canonical ensemble. We can define a "chemical potential" in this case literally as the work required to add one electron to a body that already has exactly N electrons,

$$\mu'(N,T) = F(N+1,T) - F(N,T),$$

where $F(N, T)$ is the free energy function of the canonical ensemble, or alternatively as the work obtained by removing an electron from that body,

$$\mu''(N,T) = F(N,T) - F(N-1,T) = \mu'(N-1,T).$$

These chemical potentials are not equivalent, $\mu \neq \mu' \neq \mu''$, except in the thermodynamic limit. The distinction is important in small systems such as those showing Coulomb blockade. The parameter, μ, (i.e., in the case where the number of electrons is allowed to fluctuate) remains exactly related to the voltmeter voltage, even in small systems. To be precise, then, the Fermi level is defined not by a deterministic charging event by one electron charge, but rather a statistical charging event by an infinitesimal fraction of an electron.

Empty Lattice Approximation

The empty lattice approximation is a theoretical electronic band structure model in which the potential is *periodic* and *weak* (close to constant). One may also consider an empty irregular lattice, in which the potential is not even periodic. The empty lattice approximation describes a number of properties of energy dispersion relations of non-interacting free electrons that move through a crystal lattice. The energy of the electrons in the "empty lattice" is the same as the energy of free electrons. The model is useful because it clearly illustrates a number of the sometimes very complex features of energy dispersion relations in solids which are fundamental to all electronic band structures.

Scattering and Periodicity

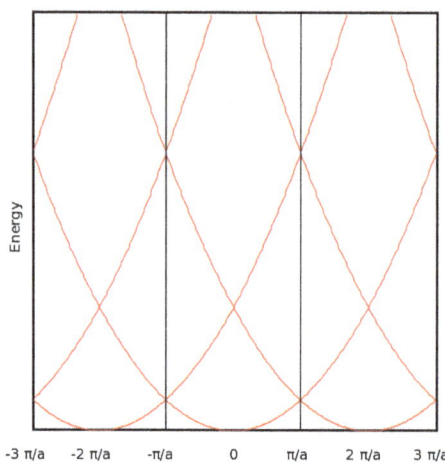

Free electron bands in a one dimensional lattice

The periodic potential of the lattice in this free electron model must be weak because otherwise the electrons wouldn't be free. The strength of the scattering mainly depends on the geometry and topology of the system. Topologically defined parameters, like scattering cross sections, depend on the magnitude of the potential and the size of the potential well. For 1-, 2- and 3-dimensional spaces potential wells do always scatter waves, no matter how small their potentials are, what their signs are or how limited their sizes are. For a particle in a one-dimensional lattice, like the Kronig-Penney model, it is possible to calculate the band structure analytically by substituting the values for the potential, the lattice spacing and the size of potential well. For two and three-dimensional problems it is more difficult to calculate a band structure based on a similar model with a few parameters accurately. Nevertheless, the properties of the band structure can easily be approximated in most regions by perturbation methods.

In theory the lattice is infinitely large, so a weak periodic scattering potential will eventually be strong enough to reflect the wave. The scattering process results in the well known Bragg reflections of electrons by the periodic potential of the crystal structure. This is the origin of the period-icity of the dispersion relation and the division of k-space in Brillouin zones. The periodic energy dispersion relation is expressed as:

$$E_n(\mathbf{k}) = \frac{\hbar^2(\mathbf{k}+\mathbf{G}_n)^2}{2m}$$

The $\mathbf{G}_n$ are the reciprocal lattice vectors to which the bands $E_n(\mathbf{k})$ belong.

The figure on the right shows the dispersion relation for three periods in reciprocal space of a one-dimensional lattice with lattice cells of length a.

The Energy Bands and the Density of States

In a one-dimensional lattice the number of reciprocal lattice vectors $\mathbf{G}_n$ that determine the bands in an energy interval is limited to two when the energy rises. In two and three dimensional lattices the number of reciprocal lattice vectors that determine the free electron bands $E_n(\mathbf{k})$ increases more rapidly when the length of the wave vector increases and the energy rises. This is because the number of reciprocal lattice vectors $\mathbf{G}_n$ that lie in an interval $[\mathbf{k}, \mathbf{k}+d\mathbf{k}]$ increases. The density of states in an energy interval $[E, E+dE]$ depends on the number of states in an interval $[\mathbf{k}, \mathbf{k}+d\mathbf{k}]$ in reciprocal space and the slope of the dispersion relation $E_n(\mathbf{k})$.

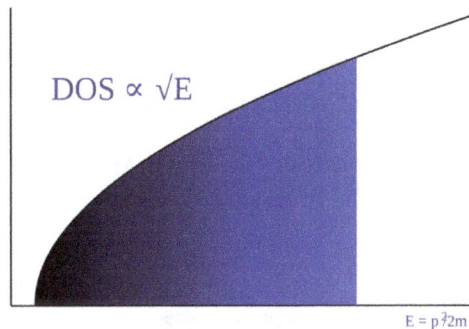

Figure 3: Free-electron DOS in 3-dimensional k-space

Though the lattice cells are not spherically symmetric, the dispersion relation still has spherical symmetry from the point of view of a fixed central point in a reciprocal lattice cell if the dispersion relation is extended outside the central Brillouin zone. The density of states in a three-dimensional lattice will be the same as in the case of the absence of a lattice. For the three-dimensional case the density of states $D_3(E)$ is;

$$D_3(E) = 2\pi\sqrt{\frac{E-E_0}{c_k^3}}.$$

In three-dimensional space the Brillouin zone boundaries are planes. The dispersion relations show conics of the free-electron energy dispersion parabolas for all possible reciprocal lattice vectors. This results in a very complicated set intersecting of curves when the dispersion relations are calculated because there is a large number of possible angles between evaluation trajectories, first and higher order Brillouin zone boundaries and dispersion parabola intersection cones.

Second, Third and Higher Brillouin Zones

"Free electrons" that move through the lattice of a solid with wave vectors **k** far outside the first Brillouin zone are still reflected back into the first Brillouin zone.

The Nearly Free Electron Model

In most simple metals, like aluminium, the screening effect strongly reduces the electric field of the ions in the solid. The electrostatic potential is expressed as

$$V(r) = \frac{Ze}{r} e^{-qr}$$

where Z is the atomic number, e is the elementary unit charge, r is the distance to the nucleus of the embedded ion and q is a screening parameter that determines the range of the potential. The Fourier transform, U_G, of the lattice potential, $V(\mathbf{r})$, is expressed as

$$U_G = \frac{4\pi Ze}{q^2 + \mathbf{G}^2}$$

When the values of the off-diagonal elements U_G between the reciprocal lattice vectors in the Hamiltonian almost go to zero. As a result, the magnitude of the band gap $2|U_G|$ collapses and the empty lattice approximation is obtained.

The electron bands of common metal crystals

Apart from a few exotic exceptions, metals crystallize in three kinds of crystal structures: the BCC and FCC cubic crystal structures and the hexagonal close-packed HCP crystal structure.

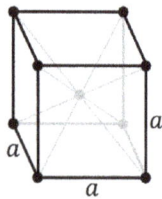

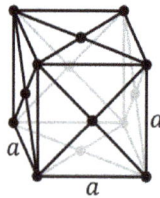

Body-centered cubic (I)   Face-centered cubic (F)   Hexagonal close-packed

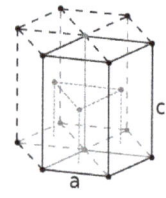

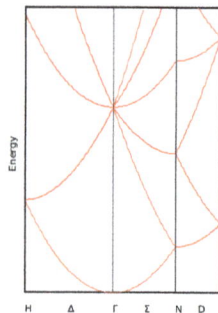

Free electron bands in a BCC crystal structure

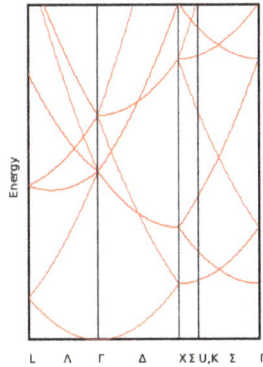

Free electron bands in a FCC crystal structure

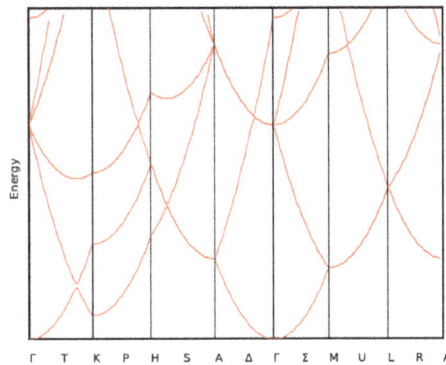

Free electron bands in a HCP crystal structure

Free Electron Model

In solid-state physics, the free electron model is a simple model for the behaviour of valence electrons in a crystal structure of a metallic solid. It was developed principally by Arnold Sommerfeld who combined the classical Drude model with quantum mechanical Fermi–Dirac statistics and hence it is also known as the Drude–Sommerfeld model.

The free electron empty lattice approximation forms the basis of the band structure model known as nearly free electron model. Given its simplicity, it is surprisingly successful in explaining many experimental phenomena, especially

- the Wiedemann–Franz law which relates electrical conductivity and thermal conductivity;

- the temperature dependence of the heat capacity;

- the shape of the electronic density of states;

- the range of binding energy values;

- electrical conductivities;

- thermal electron emission and field electron emission from bulk metals.

Ideas and Assumptions

As in the Drude model, valence electrons are assumed to be completely detached from their ions (forming an electron gas). As in an ideal gas, electron-electron interactions are completely neglected. The electrostatic fields in metals are weak because of the screening effect.

The crystal lattice is not explicitly taken into account. A quantum-mechanical justification is given by Bloch's Theorem: an unbound electron moves in a periodic potential as a free electron in vacuum, except for the electron mass m becoming an effective mass m^* which may deviate considerably from m (one can even use negative effective mass to describe conduction by electron holes). Effective masses can be derived from band structure computations. While the static lattice does not hinder the motion of the electrons, electrons can be scattered by impurities and by phonons; these two interactions determine electrical and thermal conductivity (superconductivity requires a more refined theory than the free electron model).

According to the Pauli exclusion principle, each phase space element $(\Delta k)^3 (\Delta x)^3$ can be occupied only by two electrons (one per spin quantum number). This restriction of available electron states is taken into account by Fermi–Dirac statistics. Main predictions of the free-electron model are derived by the Sommerfeld expansion of the Fermi–Dirac occupancy for energies around the Fermi level.

Energy and Wave Function of a Free Electron

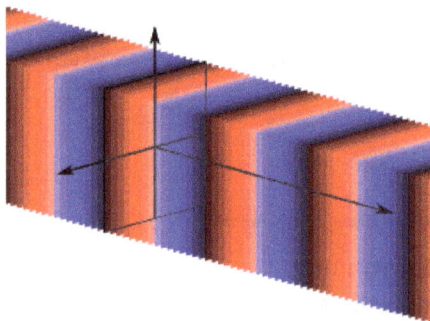

Plane wave traveling in the x-direction

For a free particle the potential is $V(\mathbf{r}) = 0$. The Schrödinger equation for such a particle, like the free electron, is

$$-\frac{\hbar^2}{2m}\nabla^2\Psi(\mathbf{r},t) = i\hbar\frac{\partial}{\partial t}\Psi(\mathbf{r},t)$$

The wave function $\Psi(\mathbf{r},t)$ can be split into a solution of a time dependent and a solution of a time independent equation. The solution of the time dependent equation is

$$\Psi(\mathbf{r},t) = \psi(\mathbf{r})e^{-i\omega t}$$

with energy

$$E = \hbar\omega$$

The solution of the time independent equation is

$$\psi_{\mathbf{k}}(\mathbf{r}) = \frac{1}{\sqrt{\Omega_r}} e^{i\mathbf{k}\cdot\mathbf{r}}$$

with a wave vector $\mathbf{k}$. Ω_r is the volume of space where the electron can be found. The electron has a kinetic energy

$$E = \frac{\hbar^2 k^2}{2m}$$

The plane wave solution of this Schrödinger equation is

$$\Psi(\mathbf{r},t) = \frac{1}{\sqrt{\Omega_r}} e^{i\mathbf{k}\cdot\mathbf{r}-i\omega t}$$

For solid state and condensed matter physics the time independent solution $\psi_{\mathbf{k}}(\mathbf{r})$ is of major interest. It is the basis of electronic band structure models that are widely used in solid-state physics for model Hamiltonians like the nearly free electron model and the Tight binding model and different models that use a Muffin-tin approximation. The eigenfunctions of these Hamiltonians are Bloch waves which are modulated plane waves.

Dielectric Function of the Electron Gas

On a scale much larger than the inter atomic distance a solid can be viewed as an aggregate of a negatively charged plasma of the free electron gas and a positively charged background of atomic cores. The background is the rather stiff and massive background of atomic nuclei and core electrons which we will consider to be infinitely massive and fixed in space. The negatively charged plasma is formed by the valence electrons of the free electron model that are uniformly distributed over the interior of the solid. If an oscillating electric field is applied to the solid, the negatively charged plasma tends to move a distance x apart from the positively charged background. As a result, the sample is polarized and there will be an excess charge at the opposite surfaces of the sample. The surface charge density is

$$\rho_s = -nex$$

which produces a restoring electric field in the sample

$$E = \frac{nex}{\epsilon_0}$$

The dielectric function of the sample is expressed as

$$\epsilon(\omega) = \frac{D(\omega)}{\epsilon_0 E(\omega)} = 1 + \frac{P(\omega)}{\epsilon_0 E(\omega)}$$

where $D(\omega)$ is the electric displacement and $P(\omega)$ is the polarization density.

The electric field and polarization densities are

$$E(\omega) = E_0 e^{-i\omega t}, \quad P(\omega) = P_0 e^{-i\omega t}$$

and the polarization density with n electron density is

$$P = -nex$$

The force F of the oscillating electric field causes the electrons with charge e and mass m to accelerate with an acceleration a

$$F = -eE = ma = m\frac{d^2 x}{dt^2}$$

which, after substitution of E, P and x, yields an harmonic oscillator equation.

After a little algebra the relation between polarization density and electric field can be expressed as

$$P(\omega) = -\frac{ne^2}{m\omega^2} E(\omega)$$

The frequency dependent dielectric function of the solid is

$$\epsilon(\omega) = 1 - \frac{ne^2}{\epsilon_0 m\omega^2}$$

At a resonance frequency ω_p, called the plasma frequency, the dielectric function changes sign from negative to positive and real part of the dielectric function drops to zero.

$$\omega_p = \sqrt{\frac{ne^2}{\epsilon_0 m}}$$

This is a plasma oscillation resonance or plasmon. The plasma frequency is a direct measure of the square root of the density of valence electrons in a solid. Observed values are in reasonable agreement with this theoretical prediction for a large number of materials. Below the plasma frequency, the dielectric function is negative and the field cannot penetrate the sample. Light with angular frequency below the plasma frequency will be totally reflected. Above the plasma frequency the light waves can penetrate the sample.

Solution of the Schrödinger Equation

The Schrödinger Equation

For a free particle the potential is $V(\mathbf{r}) = 0$, so the Schrödinger equation for the free electron is

$$-\frac{\hbar^2}{2m}\nabla^2\Psi(\mathbf{r},t)=i\hbar\frac{\partial}{\partial t}\Psi(\mathbf{r},t)$$

This is a type of wave equation that has numerous kinds of solutions. One way of solving the equation is splitting it in a time-dependent oscillator equation and a space-dependent wave equation like

$$i\hbar\frac{\partial}{\partial t}\Psi(\mathbf{r},t)=E\Psi(\mathbf{r},t)$$

and

$$-\frac{\hbar^2}{2m}\nabla^2\Psi(\mathbf{r},t)=E\Psi(\mathbf{r},t)$$

and substituting a product of solutions like

$$\Psi(\mathbf{r},t)=\psi(\mathbf{r})e^{\alpha t}$$

The Schrödinger equation can be split in a time dependent part and a time independent part. Which is derived.

Solution of the Time Dependent Equation

The peculiar time dependent part of the Schrödinger equation is, unlike the Klein–Gordon equation for pions and most of the other well known wave equations, a first order in time differential equation with a 90° out of phase driving mechanism, while most oscillator equations are second order in time differential equations with 180° out of phase driving mechanisms.

The equation that has to be solved is

$$i\hbar\frac{\partial}{\partial t}e^{\alpha t}=Ee^{\alpha t}.$$

The complex (imaginary) exponent is proportional to the energy

$$\alpha=-\frac{iE}{\hbar}$$

The imaginary exponent can be transformed to an angular frequency

$$E=i\hbar\alpha=\hbar\omega$$

The wave function now has a stationary and an oscillating part

$$\Psi(\mathbf{r},t)=\psi(\mathbf{r})e^{-i\omega t}$$

The stationary part is of major importance to the physical properties of the electronic structure of matter.

Solution of the Time Independent Equation

The wave function of free electrons is in general described as the solution of the time independent Schrödinger equation for free electrons

$$-\frac{\hbar^2}{2m}\nabla^2\psi(\mathbf{r}) = E\psi(\mathbf{r})$$

The Laplace operator in Cartesian coordinates is

$$\nabla^2 = \frac{\partial^2}{\partial x^2} + \frac{\partial^2}{\partial y^2} + \frac{\partial^2}{\partial z^2}$$

The wave function can be factorized for the three Cartesian directions

$$\psi(\mathbf{r}) = \phi_x(x)\phi_y(y)\phi_z(z)$$

Now the time independent Schrödinger equation can be split in three independent parts for the three different Cartesian directions

$$-\frac{\hbar^2}{2m}\frac{\partial^2}{\partial x^2}\phi_x(x) = E_x\phi_x(x)$$

As a solution an exponential function is substituted in the time independent Schrödinger equation

$$\phi_x(x) = N_x e^{\kappa x}$$

The solution of

$$\frac{\partial^2}{\partial x^2}\phi_x(x) = \kappa^2 N_x e^{\kappa x} = -\frac{2m}{\hbar^2}E_x N_x e^{\kappa x}$$

gives the exponent

$$\kappa = ik_x = i\sqrt{\frac{2mE_x}{\hbar^2}}$$

which yields the wave equation

$$\psi(\mathbf{r}) = N_x N_y N_z e^{i(k_x x + k_y y + k_z z)} = N_{\mathbf{k}} e^{i\mathbf{k}\cdot\mathbf{r}}$$

and the energy

$$E = \frac{\hbar^2}{2m}(k_x^2 + k_y^2 + k_z^2)$$

With the normalization

$$1 = \int_{\Omega_r} |\psi_{\mathbf{k}}(\mathbf{r})|^2 \, d\mathbf{r} = \int_{\Omega_r} \psi_{\mathbf{k}}^*(\mathbf{r})\psi_{\mathbf{k'}}(\mathbf{r})d\mathbf{r} = \int_{\Omega_r} (N_{\mathbf{k}}^* e^{-i\mathbf{k}\cdot\mathbf{r}})(N_{\mathbf{k}} e^{i\mathbf{k}\cdot\mathbf{r}})d\mathbf{r} = N_{\mathbf{k}}^2 \, |\Omega_r \, |_{vol}$$

and the wave vector magnitude

$$k = \sqrt{k_x^2 + k_y^2 + k_z^2}$$

we arrive at the plane wave solution with a wave function

$$\psi_{\mathbf{k}}(\mathbf{r}) = \frac{1}{\sqrt{\Omega_r}} e^{i\mathbf{k}\cdot\mathbf{r}}$$

for free electrons with a wave vector $\mathbf{k}$ and a kinetic energy

$$E = \frac{\hbar^2 k^2}{2m}$$

in which Ω_r is the volume of space occupied by the electron.

The traveling Plane Wave Solution

The product of the time independent stationary wave solution and time dependent oscillator solution

$$\Psi(\mathbf{r},t) = \psi(\mathbf{r})e^{-i\omega t}$$

gives the traveling plane wave solution

$$\Psi(\mathbf{r},t) = \frac{1}{\sqrt{\Omega_r}} e^{i\mathbf{k}\cdot\mathbf{r}-i\omega t}$$

which is the final solution for the free electron wave function.

Fermi Energy

According to the Pauli principle, the electrons in the ground state occupy all the lowest-energy states, up to some Fermi energy E_F. Since the energy is given by

$$E(\mathbf{k}) = \frac{\hbar^2 k^2}{2m},$$

this corresponds to occupying all the states with wave vectors $|\mathbf{k}| < k_F$, where k_F is so-called Fermi wave vector, given by

$$k_F = (3\pi^2 N_e / V)^{1/3},$$

where N_e is the total number of electrons in the system, and V is the total volume. The Fermi energy is then

$$E_F = \frac{\hbar^2}{2m}\left(\frac{3\pi^2 N_e}{V}\right)^{2/3}$$

In a nearly-free-electron model of a Z-valent metal, one can replace N_e with NZ, where N is the total number of metal ions.

Density of States

The density of states (DOS) corresponds to electrons with a spherically-symmetric parabolic dispersion

$$E(\mathbf{k}) = \frac{\hbar^2 k^2}{2m},$$

with two electrons (one of each spin) per each "quantum" of the phase space, $\Delta V \Delta \mathbf{k} = (2\pi)^3$. In 3D, this corresponds to

$$N(E) = \frac{V}{2\pi^2}\left(\frac{2m}{\hbar^2}\right)^{3/2}\sqrt{E},,$$

where V is the total volume.

Combining these expressions for the Fermi energy and the DOS, one can show that the following relationship holds at the Fermi level:

$$N(E_F) = \frac{3ZN}{2}\frac{1}{E_F},$$

where Z is the charge of each of the N metal ions in the crystal.

Nearly Free Electron Model

In solid-state physics, the nearly free electron model (or NFE model) is a quantum mechanical model of physical properties of electrons that can move almost freely through the crystal lattice of a solid. The model is closely related to the more conceptual Empty Lattice Approximation. The model enables understanding and calculating the electronic band structure of especially metals.

Introduction - a Heuristic Argument

Free electrons are traveling plane waves. Generally the time independent part of their wave function is expressed as

$$\psi_{\mathbf{k}}(\mathbf{r}) = \frac{1}{\sqrt{\Omega_r}} e^{i\mathbf{k}\cdot\mathbf{r}}$$

These plane wave solutions have an energy of

$$E_k = \frac{\hbar^2 k^2}{2m}$$

The expression of the plane wave as a complex exponential function can also be written as the sum of two periodic functions which are mutually shifted a quarter of a period.

$$\psi_{\mathbf{k}}(\mathbf{r}) = \frac{1}{\sqrt{\Omega_r}} \big[\cos(\mathbf{k}\cdot\mathbf{r}) + i\sin(\mathbf{k}\cdot\mathbf{r})\big]$$

In this light the wave function of a free electron can be viewed as the sum of two plane waves. Sine and cosine functions can also be expressed as sums or differences of plane waves moving in opposite directions

$$\cos(\mathbf{k}\cdot\mathbf{r}) = \frac{1}{2}[e^{i\mathbf{k}\cdot\mathbf{r}} + e^{-i\mathbf{k}\cdot\mathbf{r}}]$$

Assume that there is only one kind of atom present in the lattice and that the atoms are located at the lattice points. The potential of the atoms is attractive (negative) and concentrated near the lattice points. In the remainder of the cell the potential is close to zero.

The Hamiltonian is expressed as

$$H = T + V = -\frac{\hbar^2}{2m}\nabla^2 + V(\mathbf{r})$$

in which T is the kinetic and V is the potential energy. From this expression the energy expectation value, or the statistical average, of the energy of the electron can be calculated with

$$E = \langle H \rangle = \int_{\Omega_r} \psi_{\mathbf{k}}^*(\mathbf{r})[T + V]\psi_{\mathbf{k}}(\mathbf{r})d\mathbf{r}$$

where we integrate in $\mathbf{r}$ over a single lattice cell. If we assume that the electron is given by a plane wave of wave number $\mathbf{k}$ (despite the nonconstant potential V), the energy of the electron is:

$$E_k = \frac{1}{\Omega_r}\int_{\Omega_r} e^{-i\mathbf{k}\cdot\mathbf{r}}\left[\frac{\hbar^2 k^2}{2m} + V(\mathbf{r})\right]e^{i\mathbf{k}\cdot\mathbf{r}}d\mathbf{r} = \frac{\hbar^2 k^2}{2m} + \langle V \rangle$$

This means that at each $\mathbf{k}$ the energy is lowered below the free space value by the average V of the attractive potential of the atom. If the potential is very small we get the Empty Lattice Approximation. This isn't a very sensational result and it doesn't say anything about what happens when we get close to the Brillouin zone boundary. We will look at those regions in $\mathbf{k}$-space now.

Let's assume that we look at the problem from the origin, at position $\mathbf{r} = 0$. If $\mathbf{k} = 0$ only the cosine part is present and the sine part is moved to ∞. If we let the length of the wave vector $\mathbf{k}$ grow, then the central maximum of the cosine part stays at $\mathbf{r} = 0$. The first maximum and minimum of the sine part are at $\mathbf{r} = \pm\pi / (2\mathbf{k})$. They come nearer as $\mathbf{k}$ grows. Let's assume that $\mathbf{k}$ is close to the Brillouin zone boundary for the analysis in the next part of this introduction.

The atomic positions coincide with the maximum of the $\cos(\mathbf{k} \cdot \mathbf{r})$-component of the wave function. The interaction of the $\cos(\mathbf{k} \cdot \mathbf{r})$-component of the wave function with the potential will be different from the interaction of the $\cos(\mathbf{k} \cdot \mathbf{r})$-component of the wave function with the potential because their phases are shifted. The charge density $\rho_\mathbf{k}$ is proportional to the absolute value squared, $|\psi_\mathbf{k}(\mathbf{r})|^2$, of the wave function. It is useful to split it into two parts, $\rho_\mathbf{k} = \rho_\mathbf{k}^c(\mathbf{r}) + \rho_\mathbf{k}^s(\mathbf{r}),$, coming from the $\cos(\mathbf{k} \cdot \mathbf{r})$ and -components. For the former component it is

$$\rho_\mathbf{k}^c(\mathbf{r}) = \frac{1}{2\Omega}\left[1 + \cos(2\mathbf{k} \cdot \mathbf{r})\right]$$

and for the $\sin(\mathbf{k} \cdot \mathbf{r})$ – component it is

$$\rho_\mathbf{k}^s(\mathbf{r}) = \frac{1}{2\Omega}\left[1 - \cos(2\mathbf{k} \cdot \mathbf{r})\right]$$

For values of $\mathbf{k}$ close to the Brillouin zone boundary, the length of the two waves and the period of the two different charge density distributions almost coincide with the periodic potential of the lattice. As a result the charge densities of the two components have a different energy because the maximum of the charge density of the $\cos(\mathbf{k} \cdot \mathbf{r})$-component coincides with the attractive potential of the atoms while the maximum of the charge density of the $\cos(\mathbf{k} \cdot \mathbf{r})$-component lies in the regions with a higher electrostatic potential between the atoms.

As a result the aggregate will be split in high and low energy components when the kinetic energy increases and the wave vector approaches the length of the reciprocal lattice vectors. The potentials of the atomic cores can be decomposed into Fourier components to meet the requirements of a description in terms of reciprocal space parameters.

Mathematical Formulation

The nearly free electron model is a modification of the free-electron gas model which includes a *weak* periodic perturbation meant to model the interaction between the conduction electrons and the ions in a crystalline solid. This model, like the free-electron model, does not take into account electron-electron interactions; that is, the independent-electron approximation is still in effect.

As shown by Bloch's theorem, introducing a periodic potential into the Schrödinger equation results in a wave function of the form

$$\psi_\mathbf{k}(\mathbf{r}) = u_\mathbf{k}(\mathbf{r})e^{i\mathbf{k}\cdot\mathbf{r}}$$

where the function u has the same periodicity as the lattice:

$$u_\mathbf{k}(\mathbf{r}) = u_\mathbf{k}(\mathbf{r} + \mathbf{T})$$

(where T is a lattice translation vector.)

Because it is a *nearly* free electron approximation we can assume that

$$u_{\mathbf{k}}(\mathbf{r}) \approx \frac{1}{\sqrt{\Omega_r}}$$

A solution of this form can be plugged into the Schrödinger equation, resulting in the central equation:

$$(\lambda_{\mathbf{k}} - \epsilon)C_{\mathbf{k}} + \sum_{\mathbf{G}} U_{\mathbf{G}} C_{\mathbf{k}-\mathbf{G}} = 0$$

where the kinetic energy $\lambda_{\mathbf{k}}$ is given by

$$\lambda_{\mathbf{k}}\psi_{\mathbf{k}}(\mathbf{r}) = -\frac{\hbar^2}{2m}\nabla^2\psi_{\mathbf{k}}(\mathbf{r}) = -\frac{\hbar^2}{2m}\nabla^2(u_{\mathbf{k}}(\mathbf{r})e^{i\mathbf{k}\cdot\mathbf{r}})$$

which, after dividing by $\psi_{\mathbf{k}}(\mathbf{r})$, reduces to

$$\lambda_{\mathbf{k}} = \frac{\hbar^2 k^2}{2m}$$

if we assume that $u_{\mathbf{k}}(\mathbf{r})$ is almost constant and

The reciprocal parameters C_k and U_G are the Fourier coefficients of the wave function $\psi(r)$ and the screened potential energy $U(r)$, respectively:

$$U(\mathbf{r}) = \sum_{\mathbf{G}} U_{\mathbf{G}} e^{i\mathbf{G}\cdot\mathbf{r}}$$

$$\psi(\mathbf{r}) = \sum_{\mathbf{k}} C_{\mathbf{k}} e^{i\mathbf{k}\cdot\mathbf{r}}$$

The vectors G are the reciprocal lattice vectors, and the discrete values of k are determined by the boundary conditions of the lattice under consideration.

In any perturbation analysis, one must consider the base case to which the perturbation is applied. Here, the base case is with $U(x) = o$, and therefore all the Fourier coefficients of the potential are also zero. In this case the central equation reduces to the form

$$(\lambda_{\mathbf{k}} - \epsilon)C_{\mathbf{k}} = 0$$

This identity means that for each k, one of the two following cases must hold:

$$C_{\mathbf{k}} = 0,$$

$$\lambda_{\mathbf{k}} = \epsilon$$

If the values of λ_k are non-degenerate, then the second case occurs for only one value of k, while for the rest, the Fourier expansion coefficient C_k must be zero. In this non-degenerate case, the standard free electron gas result is retrieved:

$$\psi_k \propto e^{i\mathbf{k}\cdot\mathbf{r}}$$

In the degenerate case, however, there will be a set of lattice vectors $k_1, ..., k_m$ with $\lambda_1 = ... = \lambda_m$. When the energy $\dot{o}$ is equal to this value of λ, there will be m independent plane wave solutions of which any linear combination is also a solution:

$$\psi \propto \sum_{i=1}^{m} A_i e^{i\mathbf{k}_i \cdot \mathbf{r}}$$

Non-degenerate and degenerate perturbation theory can be applied in these two cases to solve for the Fourier coefficients C_k of the wavefunction (correct to first order in U) and the energy eigenvalue (correct to second order in U). An important result of this derivation is that there is no first-order shift in the energy ε in the case of no degeneracy, while there is in the case of near-degeneracy, implying that the latter case is more important in this analysis. Particularly, at the Brillouin zone boundary (or, equivalently, at any point on a Bragg plane), one finds a twofold energy degeneracy that results in a shift in energy given by:

$$\epsilon = \lambda_{\mathbf{k}} \pm |U_{\mathbf{k}}|$$

This energy gap between Brillouin zones is known as the band gap, with a magnitude of $2|U_K|$..

Results

Introducing this weak perturbation has significant effects on the solution to the Schrödinger equation, most significantly resulting in a band gap between wave vectors in different Brillouin zones.

Justifications

In this model, the assumption is made that the interaction between the conduction electrons and the ion cores can be modeled through the use of a "weak" perturbing potential. This may seem like a severe approximation, for the Coulomb attraction between these two particles of opposite charge can be quite significant at short distances. It can be partially justified, however, by noting two important properties of the quantum mechanical system:

1. The force between the ions and the electrons is greatest at very small distances. However, the conduction electrons are not "allowed" to get this close to the ion cores due to the Pauli exclusion principle: the orbitals closest to the ion core are already occupied by the core electrons. Therefore, the conduction electrons never get close enough to the ion cores to feel their full force.

2. Furthermore, the core electrons shield the ion charge magnitude "seen" by the conduction electrons. The result is an *effective nuclear charge* experienced by the conduction electrons which is significantly reduced from the actual nuclear charge.

References

- Daniel Charles Mattis (1994). The Many-Body Problem: Encyclopaedia of Exactly Solved Models in One Dimension. World Scientific. p. 340. ISBN 981-02-1476-6.

- Streetman, Ben G.; Sanjay Banerjee (2000). Solid State electronic Devices (5th ed.). New Jersey: Prentice Hall. p. 524. ISBN 0-13-025538-6.

- Alexander Altland and Ben Simons (2006). "Interaction effects in the tight-binding system". Condensed Matter Field Theory. Cambridge University Press. pp. 58 ff. ISBN 978-0-521-84508-3.

- Sir Nevill F Mott & H Jones (1958). "II §4 Motion of electrons in a periodic field". The theory of the properties of metals and alloys (Reprint of Clarendon Press (1936) ed.). Courier Dover Publications. pp. 56 ff. ISBN 0-486-60456-X.

- Ong, edited by N. Phuan; Bhatt, Ravin N. (2001). More is different : fifty years of condensed matter physics. Princeton (N.J.): Princeton university press. p. 65. ISBN 0691088667. Retrieved 2 February 2015.

- Amusia, M.; Popov, K.; Shaginyan, V.; Stephanovich, V. (2014). "Theory of Heavy-Fermion Compounds - Theory of Strongly Correlated Fermi-Systems". Springer. ISBN 978-3-319-10825-4.

- For example: D. Chattopadhyay (2006). Electronics (fundamentals And Applications). ISBN 978-81-224-1780-7. and Balkanski and Wallis (2000-09-01). Semiconductor Physics and Applications. ISBN 978-0-19-851740-5.

- Wysokiński, Carol; et al. (2003). "Spin triplet superconductivity in Sr_2RuO_4" (PDF). Physica Status Solidi. 236 (2). arXiv:cond-mat/0211199 . Bibcode:2003PSSBR.236..325W. doi:10.1002/pssb.200301672. Retrieved 8 April 2012.

Permissions

All chapters in this book are published with permission under the Creative Commons Attribution Share Alike License or equivalent. Every chapter published in this book has been scrutinized by our experts. Their significance has been extensively debated. The topics covered herein carry significant information for a comprehensive understanding. They may even be implemented as practical applications or may be referred to as a beginning point for further studies.

We would like to thank the editorial team for lending their expertise to make the book truly unique. They have played a crucial role in the development of this book. Without their invaluable contributions this book wouldn't have been possible. They have made vital efforts to compile up to date information on the varied aspects of this subject to make this book a valuable addition to the collection of many professionals and students.

This book was conceptualized with the vision of imparting up-to-date and integrated information in this field. To ensure the same, a matchless editorial board was set up. Every individual on the board went through rigorous rounds of assessment to prove their worth. After which they invested a large part of their time researching and compiling the most relevant data for our readers.

The editorial board has been involved in producing this book since its inception. They have spent rigorous hours researching and exploring the diverse topics which have resulted in the successful publishing of this book. They have passed on their knowledge of decades through this book. To expedite this challenging task, the publisher supported the team at every step. A small team of assistant editors was also appointed to further simplify the editing procedure and attain best results for the readers.

Apart from the editorial board, the designing team has also invested a significant amount of their time in understanding the subject and creating the most relevant covers. They scrutinized every image to scout for the most suitable representation of the subject and create an appropriate cover for the book.

The publishing team has been an ardent support to the editorial, designing and production team. Their endless efforts to recruit the best for this project, has resulted in the accomplishment of this book. They are a veteran in the field of academics and their pool of knowledge is as vast as their experience in printing. Their expertise and guidance has proved useful at every step. Their uncompromising quality standards have made this book an exceptional effort. Their encouragement from time to time has been an inspiration for everyone.

The publisher and the editorial board hope that this book will prove to be a valuable piece of knowledge for students, practitioners and scholars across the globe.

Index